ACPL ITEM
DISCARDED

PRENTICE-HALL VOCATIONAL AGRICULTURE SERIES

BEEF PRODUCTION. Diggins and Bundy

CROP PRODUCTION. Delorit and Ahlgren

DAIRY PRODUCTION. Diggins and Bundy

EXPLORING AGRICULTURE. Evans and Donahue

FRUIT GROWING. Schneider and Scarborough

JUDGING LIVESTOCK, DAIRY CATTLE, POULTRY, AND CROPS. Youtz and Carlson

LEADERSHIP TRAINING AND PARLIAMENTARY PROCEDURE FOR FFA. Gray and Jackson

LIVESTOCK AND POULTRY PRODUCTION. Bundy and Diggins

MODERN FARM BUILDINGS. Ashby, Dodge, and Shedd

MODERN FARM POWER. Promersberger, Bishop, and Priebe

POULTRY PRODUCTION. Bundy and Diggins

PROFITABLE FARM MANAGEMENT. Hamilton and Bryant

PROFITABLE FARM MARKETING. Snowden and Donahoo

PROFITABLE SOIL MANAGEMENT. Knuti, Korpi, and Hide

RECORDS FOR FARM MANAGEMENT. Hopkins and Turner

SHEEP PRODUCTION. Diggins and Bundy

SOIL—USE AND IMPROVEMENT. Stallings

SWINE PRODUCTION. Bundy and Diggins

USING ELECTRICITY. Hamilton

YOUR FUTURE IN POULTRY FARMING. Goodman and Tudor

SECOND EDITION

USING ELECTRICITY

J. ROLAND HAMILTON

Professor
Department of Agricultural Education
Mississippi State University
Member American Society of Agricultural Engineers

PRENTICE-HALL, INC., ENGLEWOOD CLIFFS, NEW JERSEY

Second Edition
USING ELECTRICITY, J. Roland Hamilton
(Formerly USING ELECTRICITY ON THE FARM)

© 1959, 1971 by Prentice-Hall, Inc., Englewood Cliffs, N. J. All rights reserved. No part of this book may be reproduced in any form or by any means without permission in writing from the publisher. Printed in the United States of America. ISBN 0-13-939637-3 Cover photograph by Grant Heilman

10 9 8 7 6 5 4 3 2 1

PRENTICE-HALL INTERNATIONAL, INC., London PRENTICE-HALL OF AUSTRALIA, PTY. LTD., Sydney PRENTICE-HALL OF CANADA, LTD., Toronto PRENTICE-HALL OF INDIA PRIVATE LTD., New Delhi PRENTICE-HALL OF JAPAN, INC., Tokyo

ABOUT THIS BOOK

Using Electricity, Second Edition, like its predecessor, *Using Electricity on the Farm,* is a simplified reference and how-to-do-it guide for vocational students, club workers, homeowners, and people engaged in all kinds of agri-businesses. This book has many practical suggestions for taking advantage of electric energy to improve agri-business and the home. The abundance of illustrations and simplified examples makes it easy to read and to understand.

There are six problem-units concerning the following phases of electricity and electrification: (1) opportunities for using electric power; (2) basic principles of electricity in simple language for all consumers of electric energy; (3) farmstead and home wiring; (4) electric motors for farm, ranch, and home; (5) electric water systems, including irrigation, and lighting; and (6) selected areas of electric equipment for increasing production and lowering labor costs on the farm and ranch.

1. In assessing *opportunities* for using electric energy to improve agriculture and the home, a number of actual cases of all-electric homes and electrified farming are discussed in detail. At the same time, however, questions concerning practicability and economy are kept before the reader in every case. Examples show how to estimate the cost of electric equipment and its operation.

2. In presenting *basic electricity,* the book includes only those principles that are necessary for students and users of electricity to know in order to obtain maximum returns safely. For example, such aspects of electricity as *voltage, amperage, wattage,* and *resistance* are presented in a manner that even persons with no formal education in electricity can understand. Proper selection of wire size and electric equipment for safety, economy, convenience, and quality of electrical service are tied in with these basic principles of electricity. Each basic principle is clearly illustrated and explained.

3. *Farmstead and home wiring* is given strong emphasis because this aspect of electricity appears to be the first and most fundamental problem in getting quality electric service. This section has been completely rewritten since the first edition. The problem-unit begins with planning the exterior wiring system and shows by illustration how to select and install adequate exterior wiring. Interior wiring, including actual skills, is dealt with by the liberal use of illustrations and explanations.

4. The section on *electric motors* presents illustrations, including cutaway photos of all major types of motors, and discusses factors to consider in selecting motors for specific uses. The context explains and illustrates how to "gear up" motors properly to obtain optimum results in different situations. A trouble-shooting chart is included near the end of the unit.

5. The problem-unit on *water systems and lighting* is built largely around examples and illustrations of how to provide adequate light and water for the home, farm, and ranch.

6. The last section of the text focuses upon selecting, operating, and caring for electric equipment for agri-business use, including farm and ranch production. Major farming "areas" covered are dairy, poultry, livestock, crops, and shop work.

Practical questions and suggested projects are included at the end of each chapter and/or problem-unit.

Over 200 persons and organizations contributed valuable information and illustrations for use in this book. Credits for illustrations are listed on page 350. I wish to express my gratitude to all who helped to make this book possible.

Special thanks are extended to the following persons who assisted directly in the preparation of the manuscript: my wife Frances Hamilton; Mrs. Charlotte Wrighton, my secretary; Dr. O. L. Snowden, Head of the Department of Agricultural Education at Mississippi State University, and I. O. Templeton, University electrician.

Also to those who were kind enough to review the manuscript of the first edition, I express my appreciation: Stanley S. Richardson, Professor, Agricultural Education, Utah State University; L. B. Swaney, Department of Vocational Agriculture, Clinton (Missouri) High School; Bob E. Taylor, State Supervisor Agricultural Education, Phoenix Arizona; Curtis R. Weston, Instructor, Agricultural Education and Agricultural Engineering, University of Missouri; Walter E. Jeske, Vo-Ag Instructor, Waverly (Iowa) Community Schools; Wilbur R. Bryant, Canton (South Dakota) High School.

J. Roland Hamilton

CONTENTS

PROBLEM-UNIT I HOW TO USE ELECTRICITY TO IMPROVE AGRICULTURE AND THE HOME   1

1 **DISCOVERING AND USING OPPORTUNITIES IN FARM ELECTRIFICATION   2**

How Adequately Are Agri-Businessmen Using Their Opportunities in Electrification? How Can You Use Electricity to Improve Your Home and Agri-Business? What Farming Operations Should Be Electrified? Suggested Projects for Problem-Unit One.

PROBLEM-UNIT II HOW TO APPLY BASIC PRINCIPLES OF ELECTRICITY   24

2 **THE NATURE OF ELECTRICITY—WHAT IT IS AND HOW TO MEASURE IT   25**

How Can the Movement of Particles in the Atom Produce an Electric Current? How Does an Electric Circuit Resemble a Water System? How Is the Cost of Electricity Determined?

3 **HOW ELECTRICITY IS PRODUCED AND TRANSMITTED TO THE HOME AND FARM   43**

How Can Static Electricity Cause Lightning? How Can a Storage Battery Produce Electricity? How Can a Generator Produce Electricity? How Can Electric Energy Be Converted into Heat, Light, and Motive Power? How Can Electricity Be Transmitted from the Power Plant to Your Farm or Home? How Can You Ground and Polarize Your Electrical System? What Should You Know About Transmission Materials and Equipment? How Does a Transformer Work?

4 **HOW TO COMPLY WITH SAFETY RULES   61**

What Are the Common Causes of Electrical Accidents? How Much Electricity Is Necessary to Kill a Person? What Practices Should You Follow to Avoid Electrical Hazards? What Measures Should You Take to Protect Your Property from Electrical Hazards? What Should You Know About National Protection Agencies? Whan Can You Do for a Person Who Has Received Electric Shock? Suggested Projects for Problem-Unit Two. Glossary for Problem-Unit Two.

PROBLEM-UNIT III HOW TO PLAN AND DO "ORDINARY" WIRING 78

5 HOW TO SELECT AND ARRANGE THE EXTERIOR WIRING SYSTEM 79

What Is Good "Electrical Insurance"? How Serious Can Fuse-Blowing Trouble Be? What Does the Exterior Wiring System Include? What Equipment and Materials Are Used in Exterior Wiring? How Can Proper Location of Meter Be Determined?

6 BRANCH CIRCUITS AND OUTLETS FOR THE RESIDENCE AND FARM SERVICE BUILDINGS 113

What Recommendations Should You Follow in Planning All Branch Circuits? What Types of Branch Circuits Will You Need? How Many Circuits of Different Types Will Be Needed for the Residence? What Outlets and Accessories Will Be Needed for the Home? What Branch Circuits and Outlets Will Be Needed for Each Farm Service Building?

7 HOW TO GET READY FOR A WIRING JOB 145

What Wiring Regulations Should You Observe? What Kind of Wiring Materials Should You Choose? How Should You Choose the Proper Style of Wire? What Kind of Exterior Wiring Should You Choose? What Wire Should You Choose for Special Purposes? What Kind of Interior Wiring Should You Choose? What Tools and Equipment Are Needed for a Wiring Job? How Can You Work Up a Wiring Diagram and a Bill of Materials? What Materials Are Needed for Wiring a Twelve-Circuit Home?

8 **HOW TO DO FARMSTEAD WIRING   160**

Should You Undertake to Do a Wiring Job Yourself? How Should Wire Be Cut, Spliced, and Connected to Terminals? How Should the Exterior Distribution Equipment Be Installed? How Should Interior Wiring Operations Be Done? Suggested Projects for Problem-Unit Three. Glossary for Problem-Unit Three.

PROBLEM-UNIT IV **HOW TO SELECT AND CARE FOR ELECTRIC MOTORS   204**

9 **HOW TO SELECT ELECTRIC MOTORS AND MOTOR DRIVES   206**

What Points Should Be Considered in Selecting Electric Motors? What Type of Motor Should You Choose? How Will the Type of Electric Service Influence Motor Selection? What Motor Accessories Should You Select? What Type and Size of Motor Drive Should You Have?

10 **HOW TO CARE FOR AN ELECTRIC MOTOR   228**

How Can You Protect Your Motors from Overloading? How Should an Electric Motor Be Installed? How Should You Care for Your Pulleys and Belts? What Routine Maintenance Jobs Should You Do? How Should You Lubricate Your Electric Motors? Suggested Projects for Problem-Unit Four. Glossary for Problem-Unit Four.

PROBLEM-UNIT V **FARMSTEAD LIGHTING AND WATER SYSTEMS   248**

11 **HOW TO SELECT AND CARE FOR ELECTRIC WATER PUMPS AND RELATED EQUIPMENT   249**

What Should You Consider in Choosing the Location and Type of Well for the Farmstead? What Size Pump Should You Have? What Kind of Pump Should You Choose? What Size Water Pipe Will You Need? What Size and Type of Motor Will You Need? What Kind of Pump House Should

You Have? How Should the Pump Motor Be Wired? What Care Should Be Given to the Farmstead Water System? What Electric Equipment Is Needed for Irrigation?

12 HOW TO PROVIDE GOOD LIGHT FOR HOMES, GROUNDS, AND FARM SERVICE BUILDINGS 277

What Is Good Light? How Can You Provide Good Light in Your Home? How Can You Provide Good Light for Your Home and Farm Grounds? How Can You Provide Good Light for Your Farm Service Buildings? Suggested Projects for Problem-Unit Five.

PROBLEM-UNIT VI HOW TO SELECT AND CARE FOR OTHER ELECTRIC FARMING AND AGRI-BUSINESS EQUIPMENT 300

13 HOW TO SELECT AND CARE FOR ELECTRIC EQUIPMENT FOR DAIRY, POULTRY, AND LIVESTOCK AGRI-BUSINESS 302

What Major Electric Appliances Should You Have for your Dairy? What Equipment Is Needed for Feeding Livestock or Dairy Cattle? What Apliances Should You Have for Your Poultry Farm?

14 HOW TO SELECT AND CARE FOR ELECTRIC EQUIPMENT FOR CROP DRYING AND FOR THE HOME AND FARM SHOP 330

What Electric Equipment Is Needed for Drying Farm Crops? What Power Tools Are Needed for the Home and Farm Shop? Suggested Projects for Problem-Unit Six.

ILLUSTRATION CREDITS 350

INDEX 351

FIGURE 1. During the 150 years from pioneer days in A to all-electric, automated (Illinois) farm in B, more progress in agriculture was made than during the preceding 4,000 years.

PROBLEM-UNIT I

HOW TO USE ELECTRICITY TO IMPROVE AGRICULTURE AND THE HOME

The world recognizes the American agri-businessman as tops in his field. Undoubtedly he deserves this honor. Within the span of one hundred years, agricultural progress in the United States has surpassed all such achievements of the previous four thousand years.

The American farmer's ability to out-produce other farmers of the world was a major factor in the winning of the two world wars. This same ability, along with the famous ingenuity of American agri-businessmen, has brought to this country the highest standard of living ever enjoyed by any nation.

In view of the past centuries of famine and drudgery in the world, how has the farmer accomplished so much in such a short period? The answer can be summarized as follows: American agri-business has *applied scientific methods and mechanical power to the farm.* The use of electricity in agriculture, while still in its infancy, is an important part of the American farm-power story. This book is intended to help you discover and take advantage of the many benefits that you can enjoy from wise use of the power lines that serve your home or your business.

1 DISCOVERING AND USING OPPORTUNITIES IN FARM ELECTRIFICATION

In a recent poll,[*] leading farmers throughout the United States gave first rank to mechanical power as a contributor to farm progress. Table 1 shows how 1,235 farmers voted on the question "What development has made the greatest contribution to fifty years of agricultural progress?" Notice that both items at the top of the list—Improved Farm Machinery and Farm Electrification—belong in the farm-power area. These two power items together received almost two-thirds of the total votes cast for first place. There were less than two percentage points of difference between first and second rank.

Although the importance of the other five items on the list is not questioned, mechanical power has played the leading part in reducing human labor in farming. This point is verified by the following facts: In 1830, when nearly all farm work was done by man and beast, almost 80 percent of the total labor force in this country worked on the farm. At the present time, less than 10 percent of the nation's workers are engaged in farm work; mechanical power does the rest. What has happened to the "extra" workers who used to do farm work? They are manning the factories and doing other jobs that have helped to bring to this country its high standard of living.

HOW ADEQUATELY ARE AGRI-BUSINESSMEN USING THEIR OPPORTUNITIES IN ELECTRIFICATION?

When a farm is connected to receive electricity from a central-station

[*] Lloyd E. Partain, *Personal Report on National Farm Study.* Philadelphia, The Curtis Publishing Company (no date).

TABLE 1 THE MOST VALUABLE CONTRIBUTION TO 50 YEARS OF AGRICULTURAL PROGRESS, RANKED ACCORDING TO PERCENT OF VOTES CAST *

Placing	Development	Percent Giving First Rank
1	Improved Farm Machinery	32.4
2	Farm Electrification	30.8
3	Improved Soil and Water Conservation	20.7
4	Better Livestock Breeding	6.5
5	Better Varieties of Crops	5.1
6	New and Improved Fertilizers	2.5
7	New Pest Control Products	2.0

* Based on 1,235 farmers' votes.

power line, it is classified as being "electrified." This does not imply that it is adequately electrified. Indeed, the present use of electricity in agri-business is lagging far behind the opportunities that are available. This fact is obvious when you drive around the countryside and observe that many agri-businesses and homes are using electricity only for lighting. Others have home lighting plus a water system and nothing else. Are these rural people taking full advantage of their opportunities for a better way of life? Obviously not!

FIGURE 2. In less than a century the handling of grain on the farm has progressed from the hand-powered grain thresher in A to the operator setting dials for automatic feed grinding and mixing in B. In C, cured grain pours into storage.

FIGURE 3. Handling silage and manure by electric power in A and B releases man labor for other work.

FIGURE 4. Thomas Edison, in B, invented the first practical light bulb, thus starting the widespread use of good light for all types of inside work. Two employees of a Mississippi food processing plant, in A, need daylight quality of light for grading mustard and spinach for human consumption. "Greens" in this scene will be frozen and packaged for retail stores.

Electricity Is Available to a Majority of Farms in the United States

Table 2 shows that electricity has come to almost all American farms. The amazing thing about this movement is that it has taken place with such unbelievable speed. A study of Table 2 will show this.

There is no point in showing the number on percent of electrified farms before 1920, since in that year only 1.6 percent were connected. But the real roots of rural electrification in the United States can be traced to Edison's first workable incandescent light bulb (1879), followed by the opening of his Pearl Street power plant in 1882. That plant is shown in Figure 5.

Some electricity was used in farming prior to 1900, but this was confined mostly to irrigation in the western part of the country. Gradually, however, a few farms in scattered sections of the United States began to tie onto community power lines near city stations; then from 1920 to 1925 the farm electrification movement gained momentum. By 1935 about one farm in ten in this country was electrified, and this figure jumped to 34 percent by 1940. Since that time, the electrification of rural America has been rapid and widespread. At the present time almost all of the accessible farms in the United States are connected to receive electricity. The 2 percent not electrified either are voluntarily so or cannot conveniently be reached by a power line.

HOW CAN YOU USE ELECTRICITY TO IMPROVE YOUR HOME AND AGRI-BUSINESS?

The increase in mechanical power on the farm is shown in Table 3. Electric power is included in these figures. You will notice that the horsepower per farm worker was 12.7 in 1930, but has since increased to more than 80. What does this mean in terms of manpower?

TABLE 2 TOTAL FARMS IN THE UNITED STATES, AND NUMBER AND PERCENT ELECTRIFIED, BY 5-YEAR INTERVALS *

Year	Total Farms	Electrified Farms	Percent Electrified
1920	6,448,343	103,000	1.6
1925	6,371,640	204,800	3.2
1930	6,288,648	596,000	9.5
1935	6,821,350	743,952	10.9
1940	6,096,799	1,853,249	34.0
1945	5,859,169	2,679,184	45.8
1950	5,382,134	4,154,359	77.2
1955	4,792,393	4,468,043	93.4
1969 **	3,059,000	2,997,820	98.0

* U. S. Census and USDA Outlook Charts.
** USDA Statistical Reporting Service, Projection.

TABLE 3 TOTAL FARM EMPLOYMENT AND HORSEPOWER PER FARM WORKER ON U.S. FARMS, BY 10-YEAR INTERVALS *

Year	Total Farm Employment (Millions) **	Horsepower Per Worker
1870	8.0	1.6
1880	10.1	1.8
1890	11.7	2.2
1900	12.8	2.2
1910	13.6	2.2
1920	13.4	5.3
1930	12.5	12.7
1940	11.0	20.0
1950	9.3	37.0
1955	8.2	45.0
1970 ***	6.0	80.0 +

* Data covering the period 1870 through 1930 were taken from USDA Miscellaneous Publication 157. Data for the period 1940 through 1955 are estimates of the USDA Agricultural Research Service.
** USDA 1968 Agricultural Outlook Charts.
*** USDA projection.

For practical purposes one man is considered equal to a ¼-hp motor. Using this figure, it is easy to estimate the power available to the average farm worker today, as follows: 1 horsepower = 4 men; therefore, 80 horsepower = 80 × 4 men = 320 men. This means that the average farm worker has at his disposal the working power of 320 men.

No doubt this vast amount of power on the farm will force you, if you farm, to do most of your work with mechanical power. Many cost studies in doing farm work with electric power have shown that a grown man can earn only 3 to 5 cents per hour in competition with an electric motor and machine. For example, while pumping water by hand pump you can earn only 5 cents per 1,000 gallons in competition with an electric pump!

How to Find and Use Your Opportunities in Electrification

In order to use your opportunities in electrification, you must find out *what they are* and then decide *what to do*. The first step in finding opportunities is to make a survey of the conditions of the electrical system on your home farm. Table 4 is a suggested form for making this study. The form is simple and can be copied onto a sheet of note paper in a few minutes' time. Study the examples listed in Table 4 before attempting to complete your own form. Then take the form home and fill in the two left-hand columns. Do not fill in the two right-hand columns until you have had time to give thorough study to your needs. Some of the items that should be listed in these columns on the right are covered in detail further on in the book. Moreover, you may need to obtain advice and instruction on some of these items before completing your form.

Study Your Needs for the Use of Electricity in Farming. One interesting way to find out whether you need to increase your use of electricity is to keep records of man labor spent in various farm jobs for one month. Then use these data to estimate the annual man labor required in the various chores. Convert these figures to dollars and cents. Next, obtain estimates of the cost of electric equipment, installation, and cost of electricity to do some of these farm jobs. Figure about 12 years of useful life for an electric machine, and refer to Chapter 2 for instructions on how to estimate cost of electricity. You should then be in a position to say whether or not it would pay you to electrify a given job or perhaps to increase the size of some of your equipment.

FIGURE 5. Edison's Pearl Street dynamo A used about 10 pounds of coal in producing 1 kilowatthour of electricity. A modern steam turbine B produces 1 kilowatthour with less than 1 pound of coal.

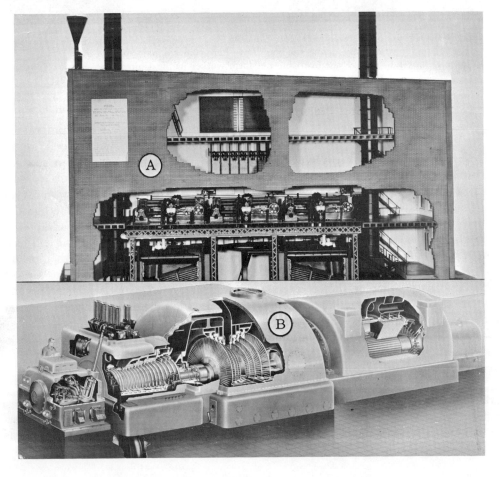

8   USING ELECTRICITY

FIGURE 6. From power plants such as the one shown here comes a major part of the more than 80 horsepower (average available per farm worker in the United States).

Decide What Electrical Projects You Should Undertake

Throughout this book you will find instructions on how to do many things in farm electrification. These range from the simple repairing of an electric cord to the wring of a house. As you turn through the book you may decide to wire a farm building or to construct and install a yard light. You might construct a portable electric motor project or install an automatic feed-handling system. In fact, your opportunities for using electricity as a means of improving farming and other agriculture are almost unlimited.

Investigate National Awards Programs. Before you complete your plan for electrical work it may pay you to investigate the national awards programs. The National FFA Foundation sponsors a national awards program in farm electrification, and the Westinghouse Educational Foundation sponsors a similar program for 4-H Club workers. By entering one of these programs you may win state or national recognition for your electrical work in addition to considerable prize money. You will have your electrical projects whether or not you win any prizes. Moreover, as a result of your project work you may become interested in the electrical field and make a career of it.

WHAT FARMING OPERATIONS SHOULD BE ELECTRIFIED?

From a modest beginning as a source of light for the farm home, electricity has now been used in more than

TABLE 4 SURVEY OF PRESENT USES AND FUTURE NEEDS FOR ELECTRICITY IN AGRICULTURE

	Present *		Future	
Major Farm Use		Description (Number, size, etc.)	Major Farm Use	Description (Number, size, etc.)
EXAMPLES				
1. Service entrance		1 main service, 60 amp	1. Service equipment	1 main, 200-amp, and 4 individual service entrances, 100-amp, 100-amp, 60-amp, 60-amp
2. Farmstead lighting		1 yardpole lamp, 200w	2. Farmstead lighting	2 yardpole lamps, 200w and 300w each
3. Farmstead water system		¼-hp motor, 120-gph pump, 2 faucets	3. Farmstead water system	¾-hp motor, 500-gph pump, 5 faucets
4. Feed handling (grain)		none	4. Feed handling (grain)	5-hp motor, crimper-mixer with feed dispenser and portable elevator
5. Feed handling (silage)		none	5. Feed handling (silage)	3-hp unloader and 1½-hp dispenser
6. Dairy		conventional ½-hp milker, pail type	6. Dairy	1½-hp pipeline milker
7. Other				
8.				
9.				

* Make your own form and complete the left-hand columns at home; then complete the right-hand columns as a school project.

five hundred ways on the farm. Although more than 50 percent of these uses have been in the farm home, the use of electricity in farm production, processing, and storage is increasing widely and rapidly.

Authorities in the farm-power field are predicting an era of push-button, electric, automatic farming in the years ahead. Whether or not you can have such a farm will depend upon many factors, but the size of your operations will have the greatest influence on whether you can electrify. For example, it is uneconomical to own and operate a $3,000 combine milker for a herd of 20 cows, or to buy a $3,500 automatic feeding system for a 25-steer feeder project. The conclusion is clear: if you expect to have an electrified farm, your operations must be large enough to justify the cost of the equipment and the cost of its operation.

On the other hand, rural people throughout the United States have improved their farms—both large and small—with the help of electrical projects. One good example of this is shown in Figure 7. The homemade elevator in this scene was built at a total cost of about $120, including $25 for the ¼-hp motor. On the basis of 12 years' useful life, depreciation is charged at $10 per year. Assuming 20 days of use each year, the daily charge for depreciation is 50 cents. Since repairs are negligible and are made at home, no charge is included for them.

The cost of electricity for operating the motor for 8 hours is figured as follows: A ¼-hp motor draws about 600 watts; therefore:

$$600w \times 8 \text{ hr} = 4,800 \text{ watthours}$$

$$\frac{4,800 \text{ watthours (wh)}}{1,000} = 4.8 \text{ kilowatthours (kwh)}$$

(See Chapter 2 for instructions on how to compute the cost of electricity.) At 3 cents per kwh the cost of 4.8 kwh is 14.4 cents ($3 \times 4.8 = 14.4$).

The total daily cost of owning and operating the elevator in Figure 7 is $50 + 14.4 = 64.4$ cents.

FIGURE 7. This homemade elevator, powered by a ¼-hp motor, is constructed of scrap metal (old bridge parts) so as to last for many years, with only minor repairs. Total cost about $120. Elevator replaces one hired man for about 65 cents per day.

The point of the example is this: In many feed-handling jobs, this elevator has been used to replace a hired man valued at $12 to $15 per day. Don't you think this kind of project is a paying proposition, even for a small farm?

Amount of Electrified Farming on United States Farms

Some knowledge of what others are doing in electrification should help you to decide whether or not to electrify your own agri-business. Records of the use of electricity on several hundred farms in ten major type-of-farming areas are available for this purpose. The USDA has done much research on the amount and cost of electrification on randomly selected farms throughout the United States—some large and some small.

The average annual electric bills ranged from $36.17 for the small farms to $157.52 for the large farms. The price of electricity averaged from 1.54 cents to 4 cents per kwh, depending on the amount used. (NOTE: These "average" farms are not representative of the heaviest users of electricity; for example, on some irrigated farms the monthly electric bill runs to $150 or more.)

The heaviest users of electricity in the study were the dairy and poultry producers, followed by grain-livestock farms. The lightest users were the small farms, where there was little or no opportunity to put electricity to work outside the farm home.

A study of the electric bills revealed that the use of electricity has been increasing by 15 percent per year on certain types of farms. Much of this increase is going into farm production, processing, and storage.

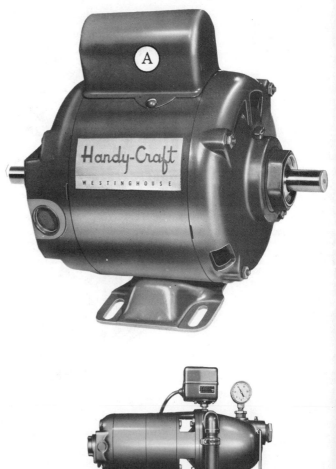

FIGURE 8. The ¼-hp motor in *A* can do more work than one man when properly geared to the shallow-well pump in *B*.

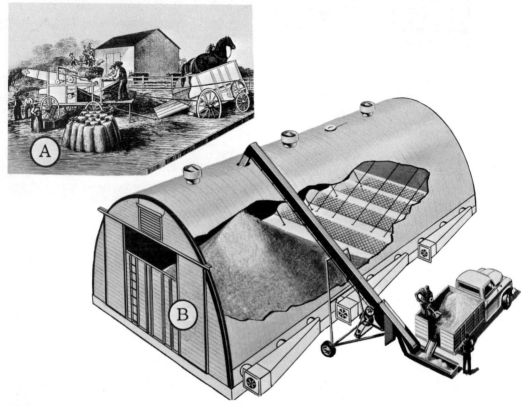

FIGURE 9. (A) The old method of handling grain depended on horse and man labor. (B) The modern method takes advantage of low-cost electricity in handling, drying, and storing.

The investment in electric equipment on these farms varied widely. Dairy, poultry, and livestock farms topped the list with an investment figure of $9,900. For example, on a certain well-equipped dairy farm in Wisconsin, $3,200 went for household equipment; $5,700 covered the cost of a combine milker, a gutter cleaner, a silo unloader, a bulk-milk tank, a water heater, barn ventilators, shop equipment, and miscellaneous items. An additional $1,500 to $2,500 was spent for farmstead wiring. This layout is considered adequate for a modern electrified 50-cow dairy operated by one man. (Total $11,000.)

The study also brought out the fact that farmers generally do not acquire a modern electrified farm in one or two years. They grow into this status by buying their electric equipment a piece or two at a time. They even wire their farmstead by stages.

The amount of electrified farming carried out by farmers, according to this study, was controlled largely by farm economics. You, too, must consider these factors in deciding whether or not to electrify your farm. Some of the most important ones follow.

How the Money Question Relates to Electrification

When you are making a decision on whether to buy a major piece of electric equipment, the money question will have three important aspects: (1) cost and returns based largely on increased production; (2) market demands and better prices; and (3) insurance against risk.

Cost and Returns. This is the most important consideration for most farmers. Unless you are financially able to indulge in luxury, a piece of farm equipment must be justified on the basis of money returns. Of course, returns may take several different forms, as you shall presently see. The following examples will illustrate this point:

One Alabama farmer operates a laying-hen unit of 36,000 birds without hired help. He is able to do this with the help of electric equipment that cost about $5,000. This includes electric feeders, waterers, and egg-handling machines. His investment in this modern equipment has made it possible for him to triple the size of his flock in comparison with that possible with hand methods.

Another farmer, in Indiana, handles 36,000 broilers per turn (four turns per year) without outside help. His automatic feed grinding and dispensing machinery and other electric equipment have helped to reduce labor requirements to two-fifths of a minute per bird. The cost of the mill, bins, and installation was about $900, and the cost of operation of the mill was approximately $42 per turn of birds. He saved $1,353 on feed costs during the first turn of broilers. This operation is based in large measure upon electric equipment to save labor.

Experiments have shown that a dairyman can handle about 50 percent more cows with the same effort after installing a pipeline (combine) milker. The cost of this machine, including the in-place cleaning equipment, is about $2,500 to $3,000 for a two-unit outfit. Further labor reductions, which would make possible further increases in herd size, can be effected by the addition of a silo unloader, a barn cleaner, a cow trainer, an automatic waterer, and other miscellaneous dairy equipment.

There are, of course, hundreds of additional opportunities for putting electric equipment to work in farm production. The hard economic fact here is that

FIGURE 10. Electric automatic feeders, waterers, night lighting, and ventilation equipment are required in modern egg production on this Ohio farm.

14 USING ELECTRICITY

FIGURE 11. *Top:* Electric automatic equipment has greatly reduced labor and risk in broiler production: *(A)* hover brooding; *(B)* heatlamp brooding.

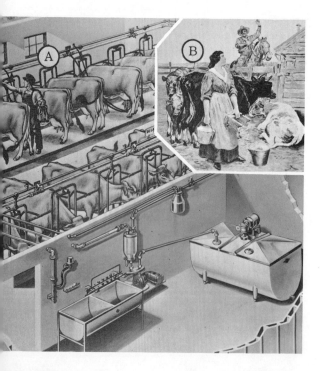

FIGURE 12. *Left:* The early settlers in *B* never thought of milking cows with electric energy as the dairyman in *A* is doing. Electricity also takes care of feeding, watering, washing milk lines, storing and cooling milk, and lighting in this modern dairy.

a certain volume of production is necessary in order to justify the cost of the equipment.

Market Considerations. Market demands may force you to comply with certain trends or else give up that type of farming. For example, in most areas there is little or no market for milk in cans; it can be sold only in bulk form. This trend could force you to purchase a bulk-milk tank whether or not it is

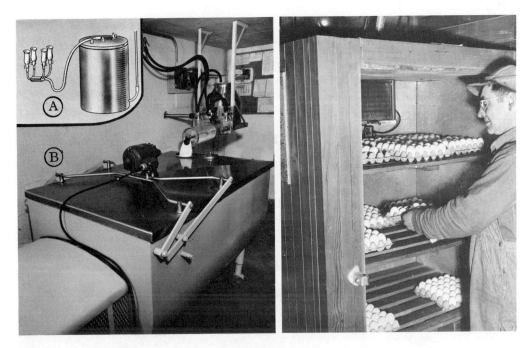

FIGURE 13. *Left:* The combine milker releasing milk directly into a bulk-tank cooler in *B* is a great improvement over the 1878 model milker in *A*. The modern system reduces labor, improves sanitation, and eliminates milk cans.

FIGURE 14. *Right:* An egg cooler normally will pay for itself in 1 to 3 years, depending upon volume to be handled, from increased prices for better quality eggs.

favorable to do so. A bulk tank large enough to take care of a 50-cow dairy costs from $2,500 to $3,000. Better prices for bulk milk, along with other savings, should pay for this tank in three to five years.

Farmers are finding that the better-quality farm products resulting from the proper processing and storing of them bring better prices sufficient to justify the cost of the equipment. For example, the increase in market value of eggs, as reported by numerous poultrymen, pays for the equipment used in cleaning, grading, and storing in one to three years; that is, if the volume is adequate.

Again, it is clear that larger operators have the advantage in electrifying their farms.

Insurance Against Risk. For an investment of $1,500 to $2,000 in drying equipment, you can take the worry and risk out of storing grain, hay, and seed. The cost of electricity would be about $1 to $1.50 per ton for hay; and total cost, including depreciation, repairs, and so on would be about $4.00 per ton. Other gains would include better market prices, better selection of harvesting period, and less loss of crops in the field.

Other electric equipment for insurance against risk includes brooders for

FIGURE 15. *Top:* An electric fan *(A)* blowing air through a wire duct *(B)* can take the risk out of storing green hay. *(C)* Forage crops were plowed by oxen and harvested by hand methods 150 years ago.

FIGURE 16. *Left:* The heat lamp and wiring for this pig brooder cost less than $10 and save an average of 1 to 3 more pigs per litter.

pigs, lambs, and other young animals; heater cable and heaters for the water system; burglar alarms; egg coolers; various storages for other farm products; and similar items. A well-lighted farmstead may easily pay for itself by preventing accidents.

Whether or not you should buy hay or grain dryers or other expensive electrical equipment will depend upon the size of your farming operations. You cannot afford to invest $2,000 in hay- or grain-drying equipment for just a few tons of hay or a few bushels of grain.

How the Labor Problem Relates to Farm Electrification

In this day of expensive and often undependable labor, farmers who use hired workers are gearing up every possible operation to electric power. This makes sense when you realize how little it costs to operate an electric motor. By referring to Table 8 you will see that one Michigan farmer milked 20 cows an average of twice a day for a whole year at a cost of $8.02 for electricity. Another farmer in New York recently invested less than $2,500 in a barn cleaner and feed-handling equipment that allowed him to release his $250-a-month hired hand.

The trend toward part-time farming in this country has been made possible, in large measure, by the use of electric and sometimes automatic equipment. A farmer can handle the milking chore for 50 cows in 1 to 1½ hours if his barn is fully electrified. A large poultry operation can be carried on, along with a full-time off-the-farm job, if the houses have proper electric appliances.

Electric, automatic feed-handling equipment can be purchased now that will enable one man to handle the feeding chore for 900 to 1,500 head of beef cattle; or 4,000 to 5,000 head of sheep; or up to 36,000 laying hens. A part-time operation of fewer head could be handled with smaller equipment.

FIGURE 17. *Top:* This barn cleaner takes the backache out of manure handling and reduces labor.

FIGURE 18. *Bottom:* Mississippi dairyman milks 36 cows per hour with the aid of his "pipe-line" milk-house installation. He is discussing with the county agricultural extension agent the in-place electric cleaning system that reduces human labor and increases sanitation.

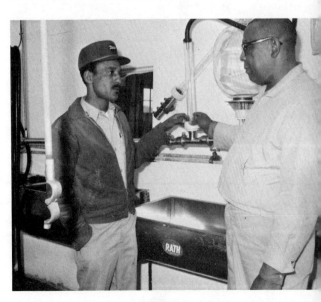

Statistics show that the trend in farming is toward *larger operations* and *less human labor*. Electric equipment for stationary farm jobs holds out real opportunities for you in the coming age of automation.

How Personal Matters Relate to Electrification

Prevailing standards in your community will influence your decision on whether or not to buy certain equipment. For example, if "everybody" has a television set, you are likely to get one even if you must buy it on the installment plan. This community pressure, no doubt, influences many farmers to purchase things that could hardly be justified on a monetary basis.

Farmers, like everyone else, surround themselves with as many comforts and conveniences as they can afford. Thus, appliance sales have now reached a total of more than a billion dollars a year in the United States. Included in this are food freezers, refrigerators, ranges, air-conditioners, radios, vacuum cleaners, television sets, electric blankets, toasters, and many other appliances.

FIGURE 19. *Left:* An Indiana farmer setting dials for proportioning, mixing, and grinding 1,200 pounds of feed per hour for his 32,000 broiler operation. The machine cuts itself off at the proper time.
FIGURE 20. *Right:* A combination feed mill, mixer, and auger controlled by a time clock greatly reduces the labor required to feed 2,500 head of sheep.

Another factor that influences farmers to purchase electric equipment, or indeed any kind of equipment, is that of personal prestige and respect in the community. Studies have shown that farmers place a high value on the esteem of their neighbors.

Some appliances are bought and justified on the basis of the family's health. A home-size humidifier is an example of this.

Personal considerations, as you have seen, certainly enter into all decisions on buying major equipment for the farm or farm home. Your financial backing will be the limiting factor here.

Conclusion on Whether or Not to Purchase Electric Appliances

The preceding discussion makes it clear that the decision on whether or not to electrify your farm or any part of it is not an easy one. It will be necessary to take account of the total cost of a machine, but it will be equally important to consider the use of this machine in expanding farm operations, improving market value, reducing risk, and improving the labor situation on the farm. And finally, your personal tastes and ambitions will certainly influence your decisions. It has been shown that the addition of one key piece of equipment often sets off a "chain reaction" that ends only with re-equipping a whole farm operation.

Considering the almost unlimited opportunities in farm electrification, no doubt you will decide to live and farm electrically in the years ahead. The limitations in electrification rest mostly with you. The possibilities for a better way of

FIGURE 21. Housewife adjusts one of the 16 million color television sets owned by people in the United States. There are about 59 million black and white sets in the United States.

FIGURE 22. A central heat pump not only heats and cools residence but aids family health by filtering the air.

farm life through electrification have not been fully determined at this time. Indeed, many of the known benefits have not been tried out by a majority of American farmers and agri-business men.

SUMMARY

The electrification of rural America stands as one of the milestones in fifty years of agricultural progress. Approximately 98 percent of the farms in the U.S. are now connected to receive electricity. Electricity, together with other sources of power, now provides the average farm worker with the equivalent working power of 300 to 400 men. This is about five to six times the total power available per worker in 1930.

The first step in using your opportunities in electrification is to make a study of the need to expand the use of electricity on the farm. This process requires the making of a survey of all parts of the present electrical system, including all pieces of electrical equipment. Also, jobs now done by hand should be studied to determine whether or not it would be economical to electrify them. Several national awards programs in farm electrification are available to farm boys and girls in every state.

When properly applied, a ¼-hp motor will do as much work as a grown man. This motor, geared up to a homemade elevator at a combined cost of about $120, can be used to replace a hired man in handling feed-stuff or grain. The cost of electricity, at 3 cents per kwh, is about 15 cents for 8 hours of operation. Depreciation on the motor and elevator is about 50 cents per day of operation.

Electricity has been used in more than five hundred ways in agri-business. Many of these uses are economical for small farms, but the greatest advantages in using electricity to do farm work lie with larger farms. This fact is especially true in farm production, processing, and storage.

Several USDA studies show that investment in electric equipment for a modern 50-cow dairy, a farm shop, household appliances, a water system, and the like runs as high as $9,900. Equipment for choring and other farming operations accounted for $6,500 of this amount, while household appliances cost $3,400. Farmstead wiring cost between $2,000 and $3,000. Getting a farm of this kind electrified usually requires a period of years.

Opportunities in electrification are influenced or controlled by the cost balanced against possible returns. The purchase of an electric appliance is often justified on the basis of increased production from larger herds, flocks, and other operations; better market prices; and insurance against risk.

The labor situation on a farm may demand the addition of electric appliances. The replacement of hired help or the release of a farmer's time to do other work may justify the cost of new electric equipment or a whole new set-up.

Personal values often influence farmers to buy new appliances: comfort, convenience, health, and personal prestige are taken into account along with economics in deciding whether to buy a new appliance, or to electrify the home or agri-business.

There is no practical limit to your opportunities in electrification except in your own mind.

PROJECTS FOR PROBLEM-UNIT ONE

QUESTIONS

1. Do you think that farmers were correct in placing rural electrification and improved farm machinery at the top of the list of greatest contributions to fifty years of agricultural progress? Why?
2. In what ways has mechanical power affected the farm labor situation? The total economy of the country?
3. How can you use a farm survey in planning for future farm electrification?
4. What advantages does a large farm have over small farms in using electricity for farm production?
5. In what ways can an electric appliance contribute to farm income? To personal prestige in the community?

ADDITIONAL READINGS

Burgess, Constance, *Household Equipment Series on: Automatic Washer, Household Range, Refrigerator, Automatic Clothes Dryer, Automatic Dishwasher.* Extension Home Management Specialist, University of California, Berkeley, Calif. (no date).

Davis, Joe F., *Use of Electricity on Farms.* Agricultural Information Bulletin No. 161, U.S. Department of Agriculture, Washington, D.C., 1956.

Edison Electric Institute, *Questions and Answers About The Electric Utility Industry.* New York, N.Y., 1968.

Leonard, Harry and Paul Johnson, *Aids to Using Electricity on Indiana Farms.* Agricultural Engineering Department, Purdue University, Lafayette, Indiana, 1962.

Noll, Edward M., *Practical Science Projects In Electricity/Electronics.* Theodore Audel and Company, Indianapolis, Indiana, 1966.

U.S. Department of Agriculture, Rural Electrification Administration, *How Electric Farming Can Help Your Business.* REA Bulletin No. 140-1. Washington, D.C.

SUGGESTED PROJECTS FOR PROBLEM-UNIT ONE

Do you want to take advantage of low-cost electricity to reduce labor, increase production, and make other farm improvements? If so, study the National Electrical Awards Program to see whether you might qualify for awards while carrying out some electrical projects. There are two awards programs in electrification that are national in scope, yet operate at the state and local levels:

1. The National FFA Foundation sponsors an electrical awards program for FFA boys. This program provides several $200 Regional Awards and a $250 National Award. You may win one of these prizes. Moreover, electrical organizations give additional awards at the state and local levels. Thus your chances of winning prize money are much greater if you live in an area where one of the contests is in operation.

22 USING ELECTRICITY

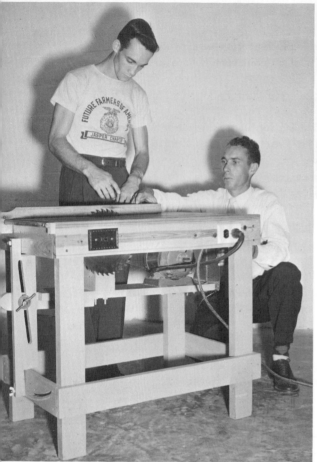

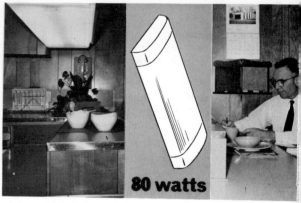

FIGURE 23. *Left:* An Alabama FFA boy and his agriculture teacher adjusting the guide of a homemade table saw which is part of a collection of electrical projects that won a National FFA award.

FIGURE 24. *Top:* Deluxe lighting project. Starting with 80w fluorescent center fixture, author constructed dinette-bar lamp (arrow). Good light for eating, studying, and general light. Cost, $25.00.

2. The other national program is the 4-H Electric Awards Program, sponsored by the Westinghouse Educational Foundation. The Rural Electrification Administration and numerous power suppliers throughout the United States assist in this 4-H work. Each year ten college scholarships are awarded to national winners; also, state winners receive a trip to the 4-H Club Congress in Chicago each year. Numerous awards are given to district and local winners.

Winning Projects. In Figure 23 you will see one of the national winners in a FFA Foundation Awards Program. This Alabama Future Farmer is shown with his vocational teacher adjusting a homemade saw—a project he used in winning a national award. He used the saw to improve his farm and home.

Other Projects. Figure 24 shows a lighting project constructed from a 80w center-type fluorescent fixture adapted to fit space underneath kitchen cabinets and provide dinette light. Total cost, about $25.

PROBLEM-UNIT II

HOW TO APPLY BASIC PRINCIPLES OF ELECTRICITY

FIGURE 26. Improper wiring and lack of lightning protection are two major causes of farm fires. This costly fire could have been prevented.

You do not have to be a scientist or an electrician to use electricity and to enjoy its benefits. However, you can lose your life or property, or you can waste money on high electric bills, if you violate certain basic principles of electricity. The purpose of this problem-unit is to help you to understand some of these principles so that you can apply them in your all-electric home and farm.

Is it true that no one knows what electricity is? Yes, because no one has ever seen it. However, scientists have established beyond doubt one fundamental principle of electricity—namely, that it is associated with the *structure and stability of all matter*. Through the application of scientific theories, man can produce electric current; he can control it; he can use it to do all kinds of useful work. So theory has become practice.

Knowing that you will likely be living in an all-electric home, or operating an electrified farm or agri-business, you should learn all that you can about the principles of electricity in order to use it safely, economically, and productively. In so doing, you will be able to provide a better way of life for yourself and your family.

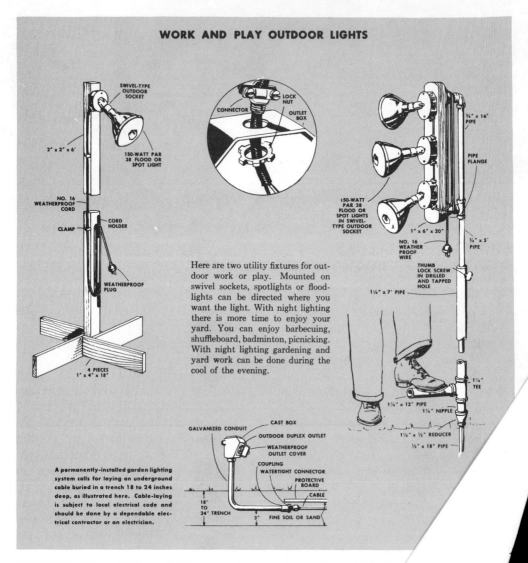

FIGURE 25. The portable yard lights shown here can be constructed in a few [hours?] at very little cost.

Home and Farmstead Improvement. The first a[rea] to start with in electrification would be to improve [...] Any or all stationary jobs should be considered. B[...] you may eventually electrify a whole farming sy[stem includ]ing of hogs or beef cattle. Refer to Table 4 fo[r...] of needs.

2 THE NATURE OF ELECTRICITY—WHAT IT IS AND HOW TO MEASURE IT

Scientists have been able to prove that a movement called an *electric current* can be made to *flow* through certain substances referred to as conductors (copper, aluminum, et cetera). But such current will not flow unless the conductor (electric wire) is in the form of a continuous circuit; and even then some kind of *force* must be brought to bear on the conductor to *cause the current to flow.* Once you get an electric current flowing, it has most of the characteristics of water flowing through a pressurized water pipe. In fact, a point by point comparison of an electric circuit with a water system is presented later in this chapter as a means of explaining certain basic principles of electricity. However, the comparison explains an electric current, not electricity itself.

Probably the first thing that comes to mind in discussing electricity is energy—namely, that energy is produced by an electric current—*heat, light,* and *motive power.* To say that electricity is one form of energy is an easy way out, but this definition is only a partial explanation. When you attempt to explain electricity, you are dealing with a much broader term than an "electric current," for electricity is involved in the innermost composition of all matter—the *atom.* Science has demonstrated, by various means, that all matter consists of atoms (usually billions of them in one inch of No. 12 copper wire) and that *electrons are one part of the atom. Protons* form another essential part of the atom, especially important so far as electricity is concerned. (See Figure 28.) Later, you will study how these electrons *actually flow* and why the movement is in fact a kind of current. So, it is only natural that discussions of electricity usu-

FIGURE 27. Owner and electric-utility experts examine stand-by generator to operate 800-cow dairy in case of power failure. Inset at lower right shows hand-handling of silage less than half a century ago.

ally focus mostly on the electric current rather than on electricity itself. To understand electricity you would have to *look into nature itself!*

Considering the complex nature of electricity, you can readily see why prehistoric man was able to harness and use other forms of energy long before he could put electricity to work for humankind. For example, it was not necessary to understand the principles of mechanics in order to use a lever to lift heavy objects, or to use horses to operate machinery that was too heavy to operate by hand.

Electricity was different because of its complexity. Man was not able to harness it for practical use until he acquired some understanding of its nature. The study of electricity, even the electric current, was hampered by its lack of the usual measurable characteristics such as weight, color, odor, length, volume, et cetera.

During the past 150 years some of our greatest scientists have devoted their lives to the study of electricity and scientific experiments—necessary before we could have the all-electric home and the automated agri-business which characterize America today.

HOW CAN THE MOVEMENT OF PARTICLES IN THE ATOM PRODUCE AN ELECTRIC CURRENT?

As previously mentioned, the most plausible theory of electricity grows out of the composition of matter, involving its innermost parts. All matter is composed of very minute particles called *molecules*. A molecule is the smallest identifiable particle of a substance, too small even to be seen with a high-power microscope. Yet, scientists have proved that a piece of copper wire, for instance, is not solid, but is composed of separate molecules.

But the molecule is composed of even smaller particles called *atoms*, and atoms are known to be the *building blocks* of all matter. These particles of matter are so small that billions of atoms of copper, or other substances, could be placed on the head of a pin. Electricity, as such, can only be explained by the nature of atoms and the manner in which they are constructed and in which they function when subjected to various forces and conditions.

Protons and Electrons, in the Atom

Figure 28 is a rough portrayal of the physicist's concept of the structure of an atom. The center consists of a sphere-like arrangement containing *protons*—the *same number* as the number of *electrons*. Electrons are arrayed in circles, or shells, throughout the atom. No one knows why they assume this arrangement, also the same in every atom of a given substance.

The protons at the center carry *positive* charges, while the electrons carry *negative* charges. The electrons are in constant movement around the nucleus, which contains the protons, and are held in orbit by the *attraction between the positive and negative charges* of the electrons and protons. (See Figure 43 and explanation of "magnetic attraction.") All matter would disintegrate without this natural attraction between positive and negative charges.

What Is a "Free" Electron?

Referring again to Figure 28, notice that the outer ring of the atom contains only *one* electron. Certain substances

FIGURE 28. A diagram of the theoretical arrangement of electrons in an atom. The lone electron in the outer ring can be displaced by a generator or a battery.

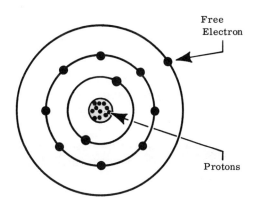

have this type of atomic structure, the lone electron being referred to as "free," or loosely held. This means that such an electron can be forced out of its normal position by various means, leaving an imbalance in the atom. That is, if an electron is forced out of its normal position, the protons would then out-number the electrons. In short, the atom would then have greater *positive* attraction than *negative*.

Electron Movement

During the natural constant orbiting of electrons around the nucleus, billions of them in a small piece of copper wire, free electrons from adjacent atoms tend to move into vacant spaces left by those forced out of normal position by a generator or a battery. When free electrons move to different atoms, a sort of chain reaction is thus created. Furthermore, if great numbers of electrons are moved out of normal position, by chemical or mechanical action, it is possible to create a "flowing" situation: that is, electrons move en masse along a conductor, or circuit, and an electric current has been started.

The flow of electrons cannot be maintained unless there is a continuous path such as a wire connected in a continuous *circuit*. That is, the ends of the wire must be joined, usually to an appliance. (See simple circuit in Figure 29.) Electricity is usually defined as *energy*. It is easier to define it as a *movement of electrons* than to define it as a substance.

Not all substances have a free electron in their atomic structure. Such substances will *not* conduct an electric current—for example, rubber, glass, plastics, and similar materials. These *nonconductors* are used as a cover for metallic conductors and are referred to as *insulators* or insulation material.

What Factors Affect Rate of Flow of Electrons?

The rate of flow of electrons through a conductor is determined by the rate of displacement of electrons at some point in the circuit, the size of the wire carrying the current, the length of the wire, and the resistance in the wiring or appliance. Power suppliers produce the great supply of electric current, used in modern living, by employing huge generators to force electrons out of their normal position at one point in a circuit of large electric wire. (For detailed discussion of generators, see page 46.)

The resulting imbalance of electrons produced by a generator causes a difference of pressure between two (or more) electric wires, and this pressure (voltage) causes a flowing of electrons. They will move from the high-pressure ("hot") wire toward the low-pressure ("neutral") wire. In short, the electrical pressure on the "high" lines serving your home or farm is greater on one wire than on the other. This difference in pressure is referred to as electromotive force or *voltage*. (See further discussion of voltage on page 30.)

Electrons can be set in motion also by batteries in which chemical solutions act on unlike metallic plates connected to a circuit of wire. (Refer to Figure 75, "Simple Circuit".) Batteries also will generate electrical pressure but not enough to supply the high lines in a community; generators are needed for this.

An understanding of the (over-simplified) electron theory of electricity discussed in the preceding topic should enable you to make safer and more economical applications of the principles of electricity when you plan your future electrical system.

HOW DOES AN ELECTRIC CIRCUIT RESEMBLE A WATER SYSTEM?

In Figure 29, a point by point comparison of the corresponding principles of electric circuits and water systems is shown.

Notice that the circuit in Figure 29 consists of one black (hot) wire and one white (neutral or ground) wire. When the circuit is closed, electricity flows through the black wire, from the source (generator) at 115-volts pressure, to the iron. After being "used" by the iron, the current flows, at very low voltage, through the neutral (white) wire back to the source. A switch installed in the black wire (not in the neutral) provides a means of breaking the circuit and thus stopping the current. So that electricity may be available at all times, the generator must operate continuously; this kind of electricity (a-c) cannot be stored.

In the water system shown in Figure 29, water is supplied at 40-pounds pressure per square inch. It is this pressure (PSI) that causes the water to flow.

FIGURE 29. The "flowing" of electricity closely resembles a water system.

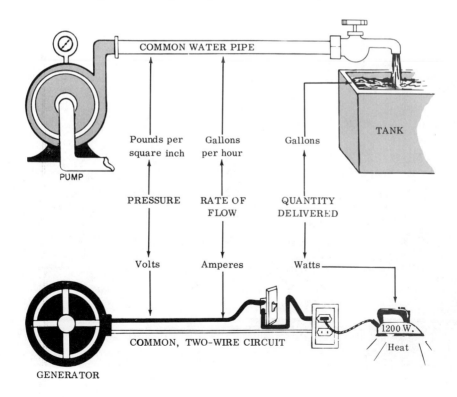

The faucet corresponds to the switch in the electric circuit. A pump and pressure tank maintain the proper pressure while the water is flowing.

How Voltage Corresponds to Pressure in a Water System

Just as the pressure in a water system causes the water to flow, so voltage causes an electric current to flow; more simply stated, *voltage is electrical pressure.* The electrical pressure needed to operate most household appliances is 115 volts; a range, a water heater, or a large motor, however, needs 230 volts. Some motors operate on 440 volts. Voltages (electrical pressures) at these levels are produced by generators.

All conductor wires offer some *resistance* to the flow of electricity. Consequently, transmission lines that span hundreds of miles carry a high voltage, which may be as much as 13,000 volts

FIGURE 30. *(A)* The "high" line coming into a farm carries 6,900 volts to the transformer at the top of the yardpole. The transformer reduces voltage to 230 volts for a three-wire circuit; 115 volts for a two-wire circuit. *(B)* Diagram shows how voltage is reduced by using a greater number of turns of wire in the primary winding (at top).

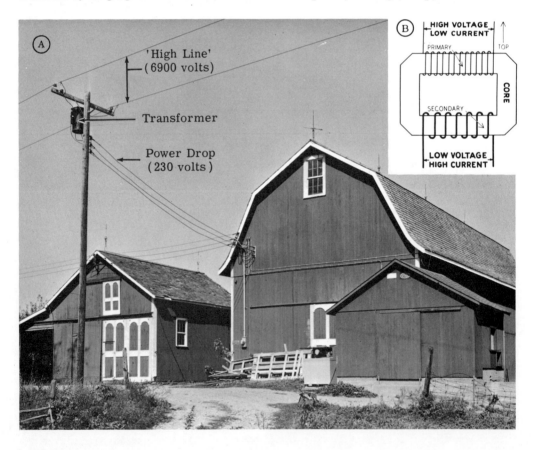

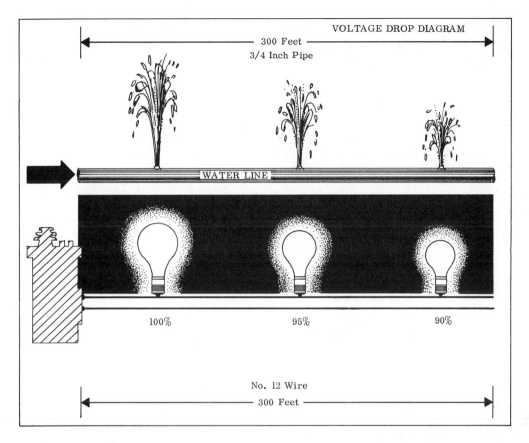

FIGURE 31. An electric current, like water flowing in a pipe, loses some of its pressure as it flows along the circuit. Diagram illustrates this.

on rural lines. High voltage must then be reduced to a usable and safe voltage at your home, probably to 115 or 230 volts. This is accomplished by the use of a transformer, which is placed near the "service drop" (wires) that connects to your meter. Figure 30 shows a cutaway section of an ordinary home-type transformer that changes high voltage-low amperage to relatively low voltage-high amperage (as needed).

What Is Voltage Drop? When an electric current flows through a conductor, it meets with some *natural resistance*. Figure 31 illustrates the result —namely, the voltage, or electrical pressure, drops slightly as the distance increases because of this resistance. Notice the similarity to the loss of water pressure as the length of pipe increases. The loss of pressure in the water pipe is due to friction, comparable in some respects to electrical resistance.

The results are similar: less electric current is delivered at the end of the circuit; less water is delivered at the end of the water pipe, as distance increases. (Note: Most of the voltage drop in a

115-volt current occurs at the *point of use,* as at the electric motor, lamp, et cetera, but the voltage drop in a run of wire is also important. Otherwise, wire-size would not be important in wiring.)

Practical Result. To ignore voltage drop in selecting electric wire is to invite trouble. Assume that you run a feeder circuit 300 feet to operate a 1,750w pig brooder (about 7.5 amperes). In order to "save" a few dollars, you select a three-wire 230v service and choose No. 12 wire. The result is that your voltage drops to 207 and amperage to 6.7. So, your wattage is $207 \times 6.7 = 1,387$, not 1,750. The result could be loss of pigs and loss of profit. (Reference to wire-size tables indicates No. 10 copper conductors for 7.5 amperes, 300 feet.) Also, you pay for electric energy lost in the heating of the undersized wires. This would be false economy.

How Amperage Corresponds to Gallons Per Minute

The 250-gallon tank in Figure 29 will be filled in 50 minutes, provided that the rate of flow continues at 5 gallons per minute. An electric current is measured in a way similar to this. That is, *the rate at which electricity flows through a circuit is measured in amperes.* This corresponds to the measure of gallons per minute in the water system. In both cases, *time* is necessary to utilize these measures.

In both illustrations the pressure is one of the main factors that affects the rate of flow. However, there are other things that tend to restrict or reduce the rate of flow in a water pipe as well as in an electric circuit. These are (1) long runs of pipe or wire; (2) poor condition of pipe or wire; (3) small pipe or wire; and (4) resistance.

To illustrate this important point, assume that the rate of flow of water is 5 gallons per minute for a 50-foot length of ¾-inch pipe. What happens if a ½-inch pipe is substituted, while the other factors remain the same? The rate of flow will be reduced, of course. If the pipe (or the wire) is lengthened, a further reduction in the rate of flow results. If the inside walls of a pipe become rusted, or if a circuit wire is partially

FIGURE 32. This combination meter is used to measure the exact voltage and amperage flowing in a circuit or an electric motor.

cut, there will be a further reduction of pressure in the current or rate of water flow.

When you begin to plan your own electric circuits, remember the important principle that amperage is affected not only by voltage but also by *wire size, length of circuit, condition of wire,* and overall *resistance*.

The 1,200w iron in Figure 29 draws at the rate of approximately 10.4 amperes. This is found as follows:

$$\frac{1,200 \text{ watts}}{115 \text{ volts}} = 10.4 \text{ amperes}$$

If the wiring for the iron is too small, however, the voltage may drop to 100 volts or less. At 100 volts the current flows at approximately 8 amperes, and the wattage output is as follows: 100 volts × 8 amperes = 800 watts, a loss of 400 watts. This lost wattage is paid for, yet does no useful work. At the same time, the undersized wire may be damaged by over-heating. (Wattage can also be found on the name plate of an appliance.)

Amperage, then, is referred to as the *rate at which a current flows* and, when multiplied by voltage, equals the "expenditure" of electric energy—namely, *watts*—usually expressed as kilowatts (1,000 watts). However, a 1,200w iron has no relevance to measurement for electric energy expended until *time* is included. In water, you could measure total quantity by GPM (gallons per minute) × No. minutes; but you could also measure it by the gallon, which you cannot do in measuring electric energy expended, since it has no measurable characteristics except *computed* characteristics.

How Electric Power Is Measured (in Watts)

The *rate of doing work* is called *power*. The unit of measure of electric power is the *watt*. One watt = 1 volt × 1 ampere. Thus, if the amperage rating of an appliance is known, you can find its wattage (electric power) by *multiplying the amperage rating by the voltage*. To illustrate, if the nameplate on an electric motor specifies "10 amperes at 115 volts," the wattage rating would be 10v × 115 amps = 1,150 watts. (This is the motor size in *watts*, although motors are rated in *horsepower*, with voltage and amperage specified.)

Most electrical appliances and heating equipment have the wattage rating specified on the label. Your water heater may specify "4,500 watts." But this figure only tells you the watt capacity, similar to water-pump capacity in gallons per hour, so wattage has no value in computing electric energy used until *time* is injected. The product of time (hours) × watts is stated in *watthours* (whr). To illustrate, electric energy consumed by the water heater for three hours would be 4,500w × 3 hr = 13,500 whr. Since the watt is such a small unit, electric energy is measured in terms of *kilowatts*, or 1,000-watt units. To convert 13,500 whr to *kilowatthours* (kwhr or kwh), divide by 1,000. The "amount of electricity" (technically not a good term) to be paid for here is 13.5 kwhr. The electric meter makes the kwhr computation automatically. The iron shown in Figure 29 would expend:

$$1,200\text{w} \times 3 \text{ hrs} = 3,600 \text{ whr}$$

$$\frac{3,600 \text{ whr}}{1,000} = 3.6 \text{ kwhr or kwh}$$

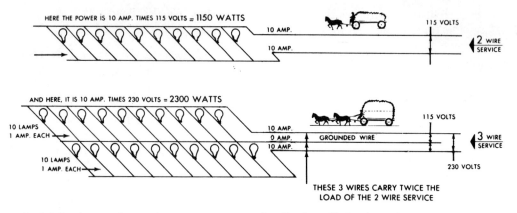

FIGURE 33. A two-wire service at 10 amperes and 115 volts will develop 1 hp. By adding 1 "hot" wire the resulting three-wire service will develop twice as much power as the two-wire circuit.

@ 3¢ per kwhr the cost would be 3.6 (kwhr) × 3 = 10.8¢, total cost for 3 hours operation.

Special Note on Electric Motors. In terms of electric energy, 746 watts is equal to one horsepower. However, energy is lost in operation in friction inside the motor, wind resistance in the motor, and in heat. Hence, the expenditure of electric energy by a 1-hp motor is about 1,800 watts. One 5-hp motor is more efficient than five 1-hp motors—that is, the wattage draft of one 5-hp motor is 6,440 watts, but five 1-hp motors would expend about 9,000 watts. (See Table 12 for amp-data on motors.)

Formulas and Other Relationships. For solving most ordinary electric-energy problems the following relationships should be kept in mind:

1. Watts = volts × amperes
2. Volts = $\dfrac{\text{watts}}{\text{amperes}}$
3. Amperes = $\dfrac{\text{watts}}{\text{volts}}$

Resistance is involved in all use of electricity, but it is not used much in ordinary consumer computations. *One ohm* is the resistance that is in effect when one ampere is flowing under one volt (of pressure). Resistance varies with wire size.

Figure 33 depicts the difference in amperage when 115v service is changed to 230v; only one additional wire is necessary to double amperage.

HOW IS THE COST OF ELECTRICITY DETERMINED?

The average selling price of electricity now is less than one-fifth what it was in 1882. Great improvements in the manufacture and distribution of electricity have made this reduction possible. For less than 15 cents today, you can purchase enough electric energy to operate a ¼-hp motor for 8 hours; and, in many farming operations, you can thus replace a grown man.

Electricity is measured by the kilowatthour, but the price usually varies in accordance with the amount used during one month.

How to Figure Kilowatthours of Use

You have already seen that the watt is the measure of electric power. One watt, however, is too small to use in calculations, so the *kilowatt* (1,000 watts) is used in figuring electric bills. That is, ten 100w lamps = 10 × 100 watts, or 1,000 watts. If you used these 10 lamps continuously for one hour, you would use *1 kwh of electricity* (1,000 watts used for one hour = 1 kwh).

To estimate the amount of electric energy used by an appliance in one month, do the following:
1. Determine the wattage rating of the appliance. (See label.)
2. Estimate the total number of hours used during the month.
3. Multiply wattage rating by the total number of hours.
4. Divide by 1,000.

EXAMPLE ONE:
1. A range oven is rated at 8,000 watts.
2. By checking the time on several different days it is found that the oven is used one-half hour per day, or a total of 15 hours during one month.
3. Then, 8,000 watts × 15 hours = 120,000 watthours.
4. 120,000 ÷ 1,000 = 120 kwh.

Assuming an average price of 2 cents per kwh, the cost of operating the oven for one month would be 120 × 2 cents = $2.40.

EXAMPLE TWO: In order to estimate the cost of using an electric motor, it is necessary to convert horsepower to watts before you can apply the four-step formula in the preceding example. The data in Table 5 can be used as a quick means of making this conversion.

Assume that you operate a ½-hp grain elevator as follows: first day, 6 hours; second day, 5 hours; third day, 8 hours; fourth day, 4 hours. How much electric energy will be expended?

TABLE 5 ELECTRIC MOTOR SIZE IN HORSEPOWER AND WATTAGE EQUIVALENTS

Horsepower	Watts *	@ 3¢ per kwh
1	1,800	5.4
¾	1,600	4.8
½	1,100	3.3
⅓	800	2.4
¼	600	1.8
⅙	500	1.5
⅛	300	.9

* Wattage data shown are not exact. Use only for estimating loads. See motor nameplate to find exact rating. See also National Electric Code.

36 USING ELECTRICITY

FIGURE 34. *Top:* The ¼-hp motor shown here will do the work of one hired man, while expending only about 600 watts of electric energy.

FIGURE 35. *Bottom:* The typical watthour meter measures the amount of electricity used.

ANSWER: Apply the four steps as outlined in Example One.
1. 1,100 watts (½-hp motor)
2. 23 hours used (6 + 5 + 8 + 4)
3. 23 × 1,100 = 25,300 watthours
4. 25,300 ÷ 1,000 = 25.3 kwh

Assuming 2 cents per kwh, the cost of operation would be 50.6 cents for electricity; 2¢ × 25.3 = 50.6 cents for electric energy.

In computing electric bills, power companies use actual meter readings. The preceding examples were for estimating purposes only.

How Electric Rates Vary

Power companies usually charge more for the first 25 to 50 kwh. The next block of 50 or so, and each succeeding block, may cost you less per kwh. This is what the power supplier calls a "sliding scale." The company's expenses on power lines, poles, and transformers continue whether you use any electricity or not. Therefore, it is only fair that every customer pays a minimum charge per month; you use the electrical equipment whether or not you use any electricity. This minimum charge usually covers the first 50 to 100 kwh.

Table 6 shows the rate charged by one electric cooperative in a southern state. The total cost by brackets is also figured for the first 1,400 kwh. (NOTE: The price of the first 50 kwh is 4 cents each, or a total of $2. The last bracket of 1,000 kwh, at .4 cents, costs only $4. This gives a spread of ten to one between the price in the first bracket and that in the last bracket.)

You should study your power rates and decide whether you are using a sufficient amount of electricity to get into

THE NATURE OF ELECTRICITY

TABLE 6 GENERAL RURAL RATE IN A SOUTHERN STATE

Kwh Used			Price	Cost of This Bracket	Total Cost
First bracket	first	50	4 cents	$2.00	$ 2.00
Second bracket	next	50	3	1.50	3.50
Third bracket	next	100	2	2.00	5.50
Fourth bracket	next	200	1	2.00	7.50
Fifth bracket	next	1,000	.4	4.00	11.50 *

* The 1,400 kwh covered by the five brackets would cost $11.50, as shown by the figure in the last column. This would average .8 cents per kwh for the first 1,400 used. (This is a more favorable rate than the average throughout the country.)

the more favorable price brackets. In view of the low cost of electric energy, it may be to your advantage to put more electricity to work in your agricultural work and in your home.

The Way to Read a Meter

Figure 36 shows a meter face as of March 1 and April 1. The first reading —2,822—is the total of all the electricity

FIGURE 36. Meter readings taken one month apart.

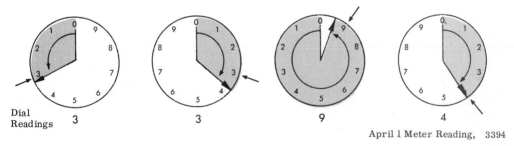

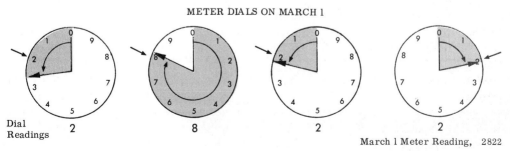

```
April Reading    3394
March    "       2822
Kilowatt Hours Used   572
```

38 USING ELECTRICITY

TABLE 7 TYPICAL ELECTRIC BILL

Meter reading April 1		3,394
Meter reading March 1		2,822
Kwh used		572

		Amt. to Pay
First bracket	first 50 kwh at 4¢	$2.00
Second bracket	next 50 kwh at 3¢	1.50
Third bracket	next 100 kwh at 2¢	2.00
Fourth bracket	next 200 kwh at 1¢	2.00
Fifth bracket	next 172 * kwh at .4¢	.69
		$8.19 **

* This is the amount in excess of the first 400 kwh and is paid for in the fifth bracket at .4 cents each.
** The average cost of 572 kwh would be $8.19 ÷ 572 = 1.4 cents.

that has been measured by that meter up to March 1. In other words, the meter simply adds the March consumption to this total and the difference is the number of kilowatthours used during that month.

Therefore, it is necessary to take a reading of the meter on the first day of each month (if the electric bill is paid monthly) in order to determine the amount used. Then it is a simple matter of subtracting the first reading from the last.

Referring to the meter faces in Figure 36, locate the four dials. Beginning on the left and reading to the right, each dial contributes one digit to the meter reading. For April 1, the meter reading is 3-3-9-4, and for March 1, the reading is 2-8-2-2.

Notice that the pointers on two of the dials (1 and 3) turn to the left while the other two (2 and 4) turn to the right. A number is counted only when the dial is at or fully past that number. Check this by looking at the dial next on your right.

How to Figure the Monthly Bill

After you have learned to read the meter, you may want to figure your monthly bill. Using the readings on the first day of each month, subtract the smaller amount from the larger and the result will be the total number of kwh used that month.

You can now use the bracket system shown in Table 7 to figure your bill. On the basis of the readings previously mentioned, the amount is 572 kwh. This covers the first four brackets (400 kwh) and leaves 172 kwh to fall into the fifth bracket. The last column on the right shows the cost of each bracket. The present bill includes 172 kwh in the fifth bracket at .4 cents each, or 69 cents.

TABLE 8 AMOUNT AND COST OF ELECTRICITY FOR VARIOUS USES DURING ONE YEAR AT THE RAY LOTT FARM *

Farm or Farm Home Use	Kwh Used During 1 Year	Annual Cost @ 2 Cents	Work Done
Barn & Milkhouse Lights, Radio, Clippers	1,341	$26.82	Radio & lights for barn, silo, & milkhouse
Gutter Cleaner	120	2.40	Manure removed for 38 head of livestock
Milking Machine	401	8.02	Used twice a day—average 20 cows
Ventilators in Barn	741	14.82	Improves condition of stable
Milk Cooler	2,727	54.54	Cooled approximately 24,000 lb of milk
Milk Cooler, Pump	184	3.68	Circulates spray in milk cooler
Uskon Heating Panels	771	15.42	Maintains milkhouse temperature above 38°
Implement Shed and Bull Barn Stock Tank De-icer	572	11.44	Lights
Farm Shop	94	1.88	Lights, welder, power tools
Chicken House	2,357	47.14	5 hr of light each day— poultry water warmers
Ultraviolet Lighting in Chicken House	261	5.22	Improves condition of air; dimmer for night lighting
Range	1,179	23.58	Meals and baking—4 people
Refrigerator	429	8.58	10 cubic foot size
Deep Freezer	812	16.24	18 cubic foot size
Oil Burner & Circulating Water Pump	388	7.76	Heating system in house
Water System Pump, House & Yard Lights, Radio, Washer, Iron & Small Appliances, Water Heater (off peak)	1,296	25.92	Flat rate $4.69 for 90-gallon tank
Total Amount and Cost of Electricity for One Year	13,673	$273.46	

* Cooperative demonstration project conducted by Ingham County Agricultural Agent, Agricultural Engineering Department, Cooperative Extension Service, Michigan State University, and the Detroit Edison Company, 1952. (The Lott Farm is located at Mason, Mich.) Rechecked 1970; herd tripled.

Therefore, the total bill for the month of March is $8.19, which averages 1.4 cents per kwh.

Remember, though, that the per-kwh cost would be 4 cents if less than 50 kwh were used, and you would, of course, pay $2 whether you use any electricity or not.

The power company may allow a discount for prompt payment. In a year's time the discount may come to a considerable amount. If your service is discontinued, it is customary for the company to charge a re-connection fee. If you are beginning a new electric service you are likely to be charged a deposit on your meter.

FIGURE 37. A demand charge may be required by the power supplier to serve a large irrigation pump motor like the one in this installation.

Special Rates and Demand Charges

The rural service rate used in the preceding example may not apply for irrigation equipment or other heavy installations. When power companies have to install extra-heavy service equipment, they may require a *demand charge*, regardless of the amount of electricity used. This charge is similar to a rental charge to take care of the extra expense of improvement on property. You should find out about demand charges before planning electric irrigation or other heavy service.

Special rates, sometimes referred to as *off-peak* rates, may be available in your area for use in heating water. Check with your power supplier about this also. If such a rate is available, you can save a considerable amount on your electric bill in a year's time.

SUMMARY

There are a few everyday principles of electricity which you must understand if you are to enjoy its full benefits. For example, you should know about wire size and its effect on voltage drop; two-wire versus three-wire service and its effect on economy and quality of electrical service. As you electrify your agricultural operations and home in the years ahead, an understanding and use of the common principles of electricity will reward you with a safer situation, with many financial gains, and with greater comfort for you and your family.

An electric current is the movement of "free" electrons within the atomic structure of metals and other conductors.

These free electrons are loosely held and can therefore be set in motion by chemical action. There are other ways to set electrons in motion. A *generator* operates on the principle of creating electrical pressure.

In both chemical and magnetic generation, electrical pressure is built up when free electrons are forced out of place at one point in a circuit of wire. This is similar to creating pressure at one point in a pipe, causing air or water to move.

Electrical pressure is called *voltage* and may be compared to pressure in a water system. The rate at which a current flows is referred to as *amperage* and is often compared to the measure, gallons per minute, in water.

The measure of electric power is the *watt* and is found by multiplying volts by amperes; for example, a 1-ampere light bulb operating on 115 volts would use electricity at the rate of 115 watt-hours (1 ampere $\times$ 115 volts $=$ 115 watts). It would be entirely correct to say that this light bulb is expending *115 watts of electric power*. To convert watts to horsepower, figure 746 watts $=$ 1 hp. Of course, a 1-hp motor actually uses more than 746 watts, because of power losses caused by friction, vibration, and wind resistance. Normally about 1,800 watts are drawn from the line by a 1-hp motor, yet it delivers only 746 watts of power.

Electricity is paid for by the *kilowatthour*. One kilowatthour is 1,000 watts times one hour; ten 100w light bulbs burning for one hour is 1,000 watt-hours or 1 kwh. Divide watthours by 1,000 to obtain kilowatthours. A watt-hour meter automatically makes these calculations as electricity is used.

While the average price of electricity for the country is about 3 cents per kwh, rates vary from one part of the country to another. Moreover, power suppliers usually have a "sliding scale" on electric rates; that is, the first 1 to 50 kwh cost more than the next 50, and so on. Because of this, large consumers pay less per kwh than do those who use small amounts. Some power suppliers allow off-peak use of electricity at reduced rates, usually for heating water. Another rate is called a demand charge. This applies where the power supplier is required to install extra-heavy transformers and wiring.

QUESTIONS

1 What is meant by a "loosely" held electron? In what way does it relate to the flowing of an electric current? Can a current flow through your body?
2 How can you show that voltage corresponds to pressure in a water system?
3 In what way does the size of wire affect voltage? Amperage? Wattage?
4 How can it be shown that amperage corresponds to gallons-per-minute in a water system?
5 What is meant by a demand charge? An off-peak rate? A sliding scale?

ADDITIONAL READINGS

Alabama Power Company, *Experimenting with Electricity*. Rural Sales Section, Alabama Power Company, 1964.

American Association for Agricultural Engineering and Vocational Agriculture, *Electrical Terms, Their Meaning and Use*. Agricultural Engineering Center, Athens, Ga., 1962.

Delco-Remy Division of General Motors Corporation, *Fundamentals of Electricity and Magnetism*. Anderson, Ind., 1956.

Howard W. Sams and Company, *The ABC's of Electricity*. Indianapolis, Ind., 1963.

Richter, H.P., *Practical Electrical Wiring*, 7th ed. McGraw-Hill Book Co. New York, N.Y., 1967.

Texas Education Agency, *Farm Electrification*. Austin, Texas, 1960.

3 HOW ELECTRICITY IS PRODUCED AND TRANSMITTED TO THE HOME AND FARM

Even though electrification is recognized as a real milestone of progress in the United States, it has created some serious farm problems throughout the nation. Electrification of farms and farm homes has developed so rapidly that many farmers, although living in all-electric homes and doing most of their home chores electrically, do not yet understand the principles of transmission. This has resulted in much dangerous wiring on the nation's farms, and has caused loss of life and property in every section of the country each year since rural electrification began on a large scale.

In addition, dangerous wiring is causing damage to electric equipment throughout the country. This is holding back farm progress. The many opportunities in farm and farm home electrification depend upon your understanding of the principles of transmission of electrical energy. In order to understand these principles, you must have a reasonable knowledge of how electricity is produced.

HOW CAN STATIC ELECTRICITY CAUSE LIGHTNING?

Although static electricity is similar in many ways to ordinary electricity, it does not flow in the manner of alternating or direct current. The static charge is considered an elementary form of electricity, however. While static electricity has little or no practical value, its presence in all substances has provided the scientist with a ready means of studying the nature of electricity.

One of the first things the scientist noticed about electricity was that there are two kinds—positive and negative. By

FIGURE 38. A flash of lightning is a giant spark jumping across an air gap between a cloud and some object on the earth.

FIGURE 39. Lightning set this house afire. Proper lightning rods could have prevented the fire.

rubbing a glass rod with a piece of silk cloth, you can cause the glass to pick up one kind of charge while the silk will take on the opposite charge. If you continue to stroke the rod with the piece of cloth for a few minutes, the two will cling together. This demonstrates a common law of electricity: *Unlike charges attract each other; like charges repel each other.* This principle is at work when your hair sticks to the comb on a cool, dry morning, or when you get a "shock" in your auto after sliding across a plastic seat cover while wearing woolen clothes. Aside from its value as a scientific area of study, static electricity will concern you most in the form of lightning.

Benjamin Franklin proved that lightning is the result of static charges. A flash of lightning is a giant electric spark that results from an extreme static buildup. In some way, a cloud gains or loses electrons, thus building up an excessive charge of positive or negative electricity. At the same time, another cloud (or some other object) becomes charged with the opposite kind of electricity. As the charging process continues, a point is reached at which the difference in electrical balance between the two clouds causes a giant electric spark to jump across the gap separating them. Sometimes opposite charges occur between a cloud and some object on the earth—a building, perhaps. When this happens, the spark may actually strike the building and set it afire. The use of lightning rods on the building "grounds" it, thereby conducting the excess charge into the air where it is neutralized. The lightning neutralizes the difference in electrical pressure for the time being, but the same forces may begin to build up electrical charges again so that the process will be repeated.

HOW CAN A STORAGE BATTERY PRODUCE ELECTRICITY?

Less than one hundred years ago the main supply of the world's electricity came from storage batteries. This was before the discovery of the induced current. Electricity produced by a battery is called *direct current,* because the electricity flows in one direction—in a straight line.

An electric current can be started by the displacement of electrons in the inner structure of copper or other conductor metals if there is a complete circuit of wire. Figure 41 shows an electric cell that can do this. This simple cell consists of a glass beaker containing a solution of water and sulfuric acid, a copper plate, a zinc plate, a piece of insulated wire, and a small lamp. The acid solution attacks the metals and displaces electrons in the atomic structure. One of the metals, however, is broken down more rapidly than the other. This process produces an unbalanced condition of electrons and therefore causes them to move through the wire circuit. Electrons move from the negative (zinc) plate, through the wire and lamp, back to the positive (copper) plate, and then through the acid solution back to the zinc plate. This movement is an electric current.

Electricity continues to flow until the acid has broken down the metal plates. In a tractor or automobile battery, a generator takes over and recharges the plates as the acid breaks them down. The electric current from

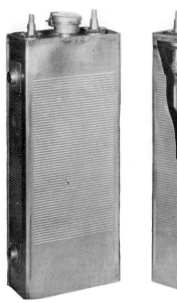

FIGURE 40. In 1916, this type of battery was widely used as a source of electricity. Power plants and community high lines have largely replaced it.

FIGURE 41. The principle involved in this simple electric cell is the same principle that makes a modern storage battery operate.

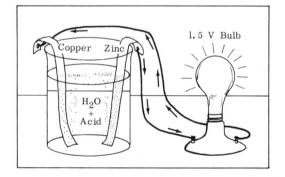

the generator has the power to stimulate chemical action in the battery and thus rebuild these plates. This is why a battery may last for several months or even years.

46 USING ELECTRICITY

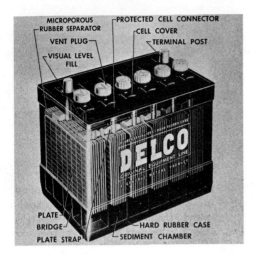

FIGURE 42. Cutaway of a storage battery for a modern auto or tractor.

FIGURE 43. If a bar magnet is suspended freely in the air, the N-pole of another bar magnet will repel (push) the N-pole of the suspended magnet and will attract the S-pole of the suspended bar. Electric motors and generators make use of this principle in their operation.

Although batteries have many uses on the farm, they are not satisfactory for operating a modern electrical system. The limited voltage and direct current from a storage battery can be transmitted only two or three miles. The alternating current (a-c) which serves your home comes from a generating plant which may be located one hundred or more miles away. Figure 42 shows a cutaway section of a storage battery that is suitable for use with a tractor.

HOW CAN A GENERATOR PRODUCE ELECTRICITY?

About two thousand years ago a Greek tribe, the Magnesites, found some pieces of ore which had the power to attract other particles of metal. This power of attraction was later named "magnetism" in honor of the Greeks who discovered the original ore.

Little value was placed on magnetism for centuries after its discovery, although early scientists believed there was a relationship between electricity

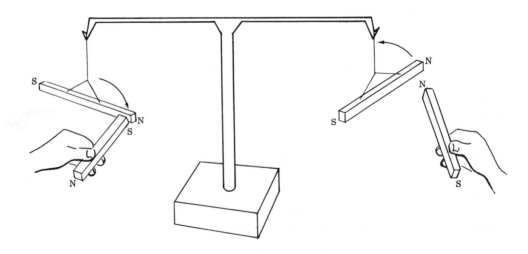

and magnetism. This was due to the similarity between the two; *both* electricity and magnetism have opposite poles that will attract or repel north and south poles. These poles are referred to as *positive* and *negative*.

In 1819, Hans Christian Oersted, a Danish scientist, proved that there is a basic relation between magnetism and electricity. His discovery paved the way for Michael Faraday to produce an electric current from magnetism. This happened in 1831. Faraday's experiment consisted of a magnet with north and south poles opposite each other, and a means by which he could thrust a coil of wire back and forth between the two poles. In the experiment, each time the coil passed between the two magnetic poles a small charge of electricity was generated. This process is referred to as *inducing a current,* and the electrical flow so produced is called *induced current.* Figure 44 shows how this may be done. It is possible to reverse this process and produce magnetism from an electric current. This latter principle is employed in electric motors, transformers, and electric arc welders.

How a Generator Produces A-C Electricity

Commercial generators, usually called *alternators,* produce most of the world's supply of electricity. A modern power plant gives the appearance of being extremely complicated; however, it produces electricity by the simple principle of induction, first demonstrated by Faraday.

By studying the series of diagrams in Figure 45, you will see how electrical pressure (voltage) is induced in a loop of wire. This loop represents the coils of wire in the rotating part of an alternator or generator.

The magnet, with its north and south poles, represents the stator, or stationary part of the alternator. Notice the magnetic field of force (flux) which is shown by the dotted lines. This field is produced by the attraction between the north (positive) and south (negative) poles of the magnet.

Remember that Faraday's experiment showed that a loop of wire will have an electrical charge induced in it when it cuts through the lines of force in a magnetic field, such as you see represented by the dotted lines in Figure 45.

After locating point "D" on the wire loop in Figure 45, observe what happens while the loop is making one complete revolution through the magnetic field. At the starting position the voltage is zero, but as the loop begins to turn, it cuts through some of the magnetic lines of force. This induces a voltage in the wire (*positive*), and it builds up to a peak, which is reached at the ¼-cycle position. When the loop has reached the ½-cycle position, the voltage has come

FIGURE 44. An electric current is induced (generated) when a coil of wire is thrust through magnetic lines of force between the N- and S-poles of a magnet.

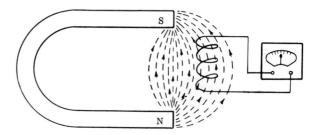

48 USING ELECTRICITY

back to zero. Why is this so? At this point the loop is not cutting across any magnetic lines. Study the position of point "D" again and you will see that the movement at the ½-cycle position is not *across* the magnetic lines but is *with* the direction of these lines.

As the loop starts into the third ¼-cycle a rather strange thing happens. Notice that point "D" is again cutting through some of the magnetic lines but in the opposite direction from that in the first ¼-cycle. The voltage rises again, similar to its action in the first part of

FIGURE 45. Electricity from a power plant comes out in a wavy line called alternating current. These diagrams show why.

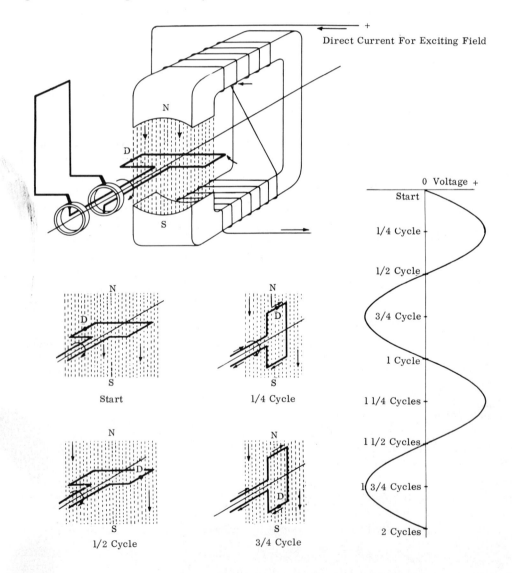

the cycle, but this time it is *negative*. This is shown by the up-and-down voltage line crossing first to the positive, then to the negative side of the zero line. Study the voltage line at the right in Figure 45 to see how this works. The peak of the negative charge is reached at the ¾-cycle position; that is, when the wire is cutting directly across the magnetic lines again. This is exactly like the positive charge as it occurred at the ¼-cycle position except that at this point the voltage is negative.

Finally, the voltage begins to fall as point "D" enters the last ¼-cycle; by the time the loop reaches the starting position again, the voltage has dropped to zero.

Thus one complete revolution of the loop (point "D") constitutes one cycle. By studying the wavy current line in Figure 45, you will see that each cycle contains one positive peak and one negative peak, starting at zero voltage and finally coming back to zero.

A modern alternator has many more loops (coils) and more magnetic fields than are shown in the diagram in Figure 45. As a matter of fact, most commercial alternators are constructed so as to produce 60-cycle current. This means that the current reaches 60 positive peaks and 60 negative peaks per *second*. Also, the voltage in 60-cycle current is zero 120 times per second, but the change is so rapid that you cannot detect the zero voltage in a light bulb.

Electricity produced by a generator (alternator) is referred to as *alternating current* because it is negative half the time and positive the other half. Figure 46 shows a modern electric power plant

FIGURE 46. The a-c electricity which operates the portable grinder at right is coming from the generating station at left which is almost 500 miles away. To cause current to flow across this great distance, the voltage must be several thousand volts when leaving the station.

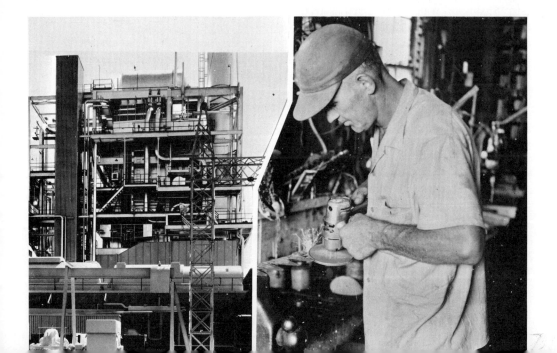

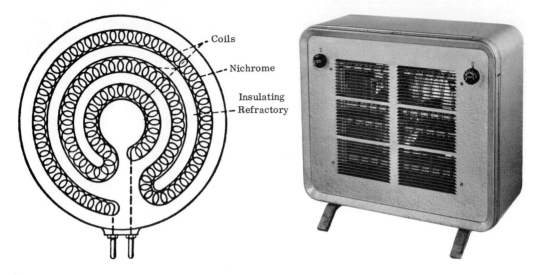

FIGURE 47. The heater at right contains coils similar to the one at left. Current causes the coils to become red hot.

where a-c electricity is produced and sent out under several thousand volts pressure. The rotor in one of these huge alternators is over 25 feet in diameter. It contains heavy masses of inductance coils, which require great power to rotate. Steam or water power is used to turn most of the alternators that make commercial electricity in the United States.

How Alternating Current Differs from Direct Current

In the preceding explanation, you observed that alternating current flows in a wavy line; that is, first to a positive peak in voltage, then to a negative peak. Direct current does not act this way at all. It flows in a straight line. At a voltage equal to a given a-c voltage, direct current flows in a straight line at the peak of the a-c line, either positive or negative. That is, the voltage in d-c current does not rise and fall as it does in a-c current.

HOW CAN ELECTRIC ENERGY BE CONVERTED INTO HEAT, LIGHT, AND MOTIVE POWER?

Practically all of the electricity used on the farm and in the residence is first converted into heat, light, or motive power.

How Electricity May Be Converted Into Heat Energy

When an electric current flows through a conductor it meets with some resistance in that conductor. This resistance (opposition to the flowing current) produces heat in the conductor. For a given size of conductor, a greater flow of current produces more heat; for a given flow of current, a smaller conductor produces more heat.

In getting heat from electricity, these principles are applied by using a coil of small wire through which a current is made to flow. The use of a coil of wire provides more surface for heating. The wattage rating of a heating element is determined by the size and length of the wire that forms the coil and by the voltage-amperage rating of its current. The size and kind of wire in the element must be such that it will not burn up in normal usage.

How Electricity May Be Converted Into Light

Observe the diagram of the incandescent light bulb in Figure 48. Incandescent light is produced in much the same manner as is heat, with one exception. A filament in the bulb, corresponding to a heater coil, carries an electric current in a space devoid of air. The space inside the bulb contains other gases, however, and in the presence of these gases the filament, which is made of a special metal, glows with a white heat and gives off light, yet the filament does not burn up.

How Electricity May Be Converted Into Motive Power

By examining the cutaway section of a common motor, Figure 49, you will notice that it is similar to a generator. In fact, the two are very nearly identical in construction and operation. Both have two basic parts: (1) a stationary winding called a stator; and (2) a rotating part called a rotor or armature.

Generators and motors both operate on the principle of magnetic induction. You have already seen how this principle works with a generator. You will now see how magnetic induction can be used to make an electric motor run.

The Rotor and Stator Constitute a Spinning Magnet. You are acquainted with the principle that an electric current flowing through a coil of wire creates a magnetic field about the coil. When a conductor (loop or coil of wire) cuts through a magnetic field, an electric current is induced in the conductor. This principle is involved in the turning of an electric motor. The stationary part of a motor (stator) contains several coils of insulated wire, and, when an electric current flows through them, a magnetic field is created about the coils. The rotor, which also contains metal conductors, turns inside this magnetic field. What

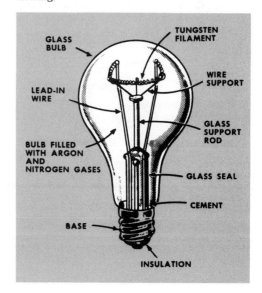

FIGURE 48. The tungsten filament inside this light bulb becomes white hot when a current is flowing.

makes it turn? The rotor is actually an electromagnet, getting its magnetism from the electric current that flows through it. Since the rotor has magnetism, it has *north* and *south* poles. The stator, too, has *north* and *south* poles. (Positive and negative are used interchangeably with north and south.)

You have previously studied the principle of like poles repelling each other and unlike poles attracting each other. Here is what happens when the current starts to flow through a motor.

Assume that a particular point on the rotor, marked "D," is a north pole. This pole will be "pushed" by another north pole in the stator if the relative position of the two is just right. This pushing force causes the rotor to turn, but, except for a special arrangement, it would go no farther than the next south pole in the stator. Just at the right instant, the north pole "D" is reversed, by a special mechanism in the motor, and becomes a south pole. The result is that the south pole in the stator gives the newly created south pole "D" another "push" and the rotor goes on its way.

The rotor in an electric motor spins around inside its case at approximately 1,725 or 3,450 rpm, the speed being determined by the number of pairs of poles. Its poles are constantly changing at just the right instant to receive the maximum "push and pull" forces. Thus, a motor can be constructed to run at a given speed and to develop a given horsepower. The number of pairs of poles and cycles determine speed and cycles; size of windings determines power.

HOW CAN ELECTRICITY BE TRANSMITTED FROM THE POWER PLANT TO YOUR FARM OR HOME?

You are already familiar with a simple two-wire electric circuit. As illus-

FIGURE 49. The two basic parts of an electric motor are (1) the stator and (2) the rotor. Magnetic force operates in both parts, causing the rotor to turn.

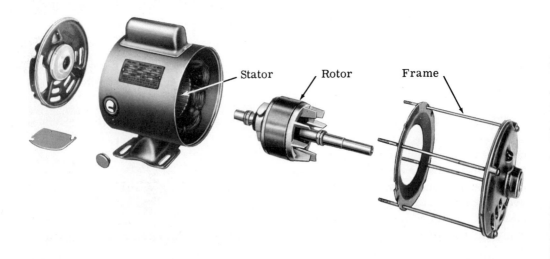

trated in Figure 30, the two-wire "high" line, which may carry up to 13,000 volts, is similar to a simple two-wire circuit; that is, it has one "hot" wire and one neutral. The high voltage is on the hot line, and the current flows through this line to an appliance and then back through the neutral wire to the generating plant. The neutral is sometimes called the "ground" because it is connected to a ground rod at the meter pole or at the point of entry of each building. Some "grounds" are not the same as a neutral.

The two-wire "high" line in Figure 30 will supply single-phase current (usually 60 cycles) at 115 or 230 volts. If you need 230 volts, you must have two hot wires and one neutral wire leading in from the transformer. One hot wire and one neutral wire from the transformer will furnish only 115 volts.

In some sections of the country, three-phase current is supplied by high lines having three hot wires and one neutral wire. Increasing single-phase to three-phase current is similar to increasing the number of cylinders of an automobile from four to twelve. In single-phase, 60-cycle current, you get 120 "pushes" per second, but three-phase current produces 360 "pushes" per second. This provides smoother running and quicker starting of electric motors. Also the *first* cost of a three-phase motor is less than the same size in a single-phase type. Unfortunately, three-phase current is not available in rural areas because of the expense involved in line construction and cost of additional equipment.

Since high lines carry up to 13,000 volts, they are dangerous, therefore, it is necessary for the wires to have sufficient strength and quality to make them safe

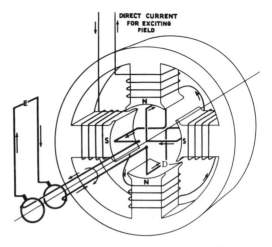

FIGURE 50. When a current is flowing through the windings, push-pull forces between the stator and rotor poles go to work, and the rotor becomes a spinning magnet.

FIGURE 51. The high-voltage electricity produced in the power plant at upper left passes through several reduction stations before reaching the farm.

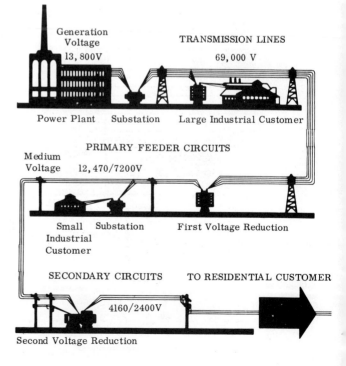

54 USING ELECTRICITY

in every respect. In addition, high lines must be large enough to provide for an adequate voltage and amperage to flow. High-quality insulators are required on the power poles to provide for further safety.

HOW CAN YOU GROUND AND POLARIZE YOUR ELECTRICAL SYSTEM?

Although the voltage at your meter will usually test 115 volts for a two-wire service or 230 volts for a three-wire service, this may build up to several thousand volts for a few seconds, because of lightning or other unusual conditions. *Grounding*, the method used to protect life and property against this hazard, refers to the neutral wire being connected to permanently moist earth. When excess electrical pressure is present in the wiring, it will cause high amperage in the conductor to flow into the earth through the ground rod.

Two methods of grounding are practiced: (1) in the city, the neutral wire is connected to the city water pipes by means of a copper ground wire and special clamps made for this purpose; (2) on the farm, it is customary to establish a permanent ground by driving a ¾-inch or larger rod (galvanized type) into the earth to a sufficient depth to assure a moist connection. (See Figure 52.) (For detailed instructions on installing a ground wire, see page 175.)

In addition to grounding the service entrance at each building, all electric motors larger than ½-hp should have an individual and separate ground. This can be done by attaching a ground wire from the frame of the motor to the exterior ground rod or to city water pipes. Additional grounding assures protection against "shorted" motors. Make certain that the connections are made properly. An exterior attachment should be made with a standard ground clamp.

How to Polarize Your Wiring

Since one wire in each two-wire circuit in your buildings will be hot, it is important that you be able to tell which is hot and which is neutral. This is done by using black insulation for the hot

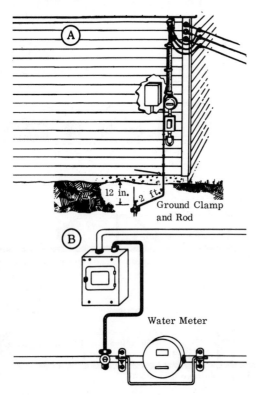

FIGURE 52. In *A*, the electrical system is grounded by means of a ground wire connecting a neutral wire to a ground rod; the wire is fastened to the rod by means of a special ground clamp. In *B*, the system is grounded to the city water pipes. Note the special clamp for fastening the ground wire to the water pipe and also the jumper wire around the water meter.

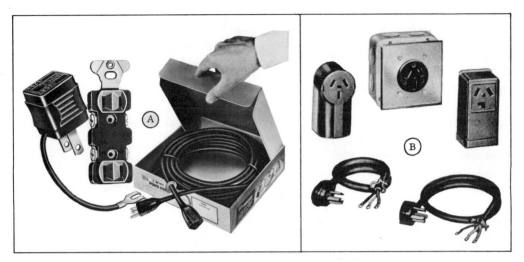

FIGURE 53. The equipment in *A* provides three-wire service for 115-volt portable machines. The adapter at left plugs into the receptacle at center and the clip connects to the cover-plate screw. The three-prong plug on the cord (right) then plugs into the adapter, while the other end receives the three-prong plug-in of the machine. The plug-ins and receptacles in *B* are for 230-volt appliances. The plug-in automatically polarizes the circuit since it will fit in only one position.

wire and white for the neutral. When a circuit is wired into the main switch, the white wire should always be connected to a silver-colored terminal, which is usually located in the center section of the switch. This word of caution is in order: in actual practice, colors sometimes get crossed. Therefore, the only *safe policy* is to *disconnect the main switch* and not depend on code colors when working on electrical wires.

If two hot wires are required for a circuit, one is generally black and the other red. Green is used to indicate a ground wire.

It is dangerous to get these colors mixed, as this could result in installing a switch or fuse in the neutral wire. If this happens, an open switch or blown fuse in the neutral wire would have the effect of destroying the ground. The white (or gray) wire must run an *un-interrupted course* from the main switch to every outlet in the circuit. If a white wire must be cut, as is necessary at a junction box, it must be spliced (rejoined) inside the box in accordance with the National Electrical Code.

Switches must always be installed in the hot line. (NOTE: In a few types of connections to three- and four-way switches, the white wire is allowed to function as a hot wire; therefore, it will *not* appear as black at the switch box.)

In damp locations, the operation of portable electric tools can be a dangerous practice. This will be true if the ground wire inside the motor becomes shorted. Electrocution can result if the current runs through your body and into the damp floor. A special adapter plug-in can be used as a means of providing a third wire in conduit and metallic cable. (See Figure 53.) This extra wire is a safety measure. Use grounded-type receptacles if non-metallic cable is used. A three-prong plug-in is called *polarizing* a current. Plug-ins for 230-volt appliances are equipped with blades which permit them to be inserted in only one position; that is, the two hot leads from the line are connected to the two hot leads in the appliance. (See Figure 53-B.)

WHAT SHOULD YOU KNOW ABOUT TRANSMISSION MATERIALS AND EQUIPMENT?

A conductor wire must have proper conducting quality and must be properly insulated. For certain uses, wiring must have special protection in the form of a metal casement or a waterproof covering. Figure 54 shows details of construction in a good grade of electric wire.

The most widely used conductor is copper; however, aluminum or aluminum-alloy wire has become popular in recent years. Aluminum will conduct electricity at about 85 percent the rate of copper for the same size wire, so aluminum wire must be larger than copper wire for carrying equal loads. (See wire-size tables.)

It is essential that conductors be rust-proof and corrosion resistant and have sufficient strength to withstand both wind and ice storms. In recent years, manufacturers have produced a better quality of high-line wire, thus reducing the number of poles required per mile. This development has helped to reduce the cost of construction of rural high lines.

Insulation Materials

Electricity will not pass through rubbber, plastic, and similar material. Therefore, such covering is used on electric wires to insulate them and make them safe. It is used to prevent fires and to reduce voltage loss through "shorts."

Some plastic-coated wire is weatherproof and rot resistant and therefore may be buried directly in the ground. Although this type of wire is more expensive than other types, it is widely used for certain purposes. Rubber is commonly used on wiring inside a building where people are likely to come in close contact with it. If the wire is to be handled frequently, as is necessary with extension cords, rubber covering is recommended because of its durability.

Non-metallic sheathed cable for dry locations inside buildings has a weather-resistant covering used as insulation. This type of wiring has a coal-tar-coated, fibrous covering that is tough and will

FIGURE 54. Service entrance cable (SE) is constructed so as to provide for the safety and long life of the wire.

Sheath   Tape   Neutral   Braid   Rubber   Copper

last many years if not otherwise damaged. The insulation may rot, however, in extremely wet and corrosive conditions. (For further discussion on wiring materials, see Chapter 7.)

HOW DOES A TRANSFORMER WORK?

The purpose of the high-line transformer was mentioned in Chapter 2, although its operation was not discussed. It is important to know a few basic things about this part of your electrical system since it controls the entire service.

The "iron pot" located on the power pole near your home is rather simple in

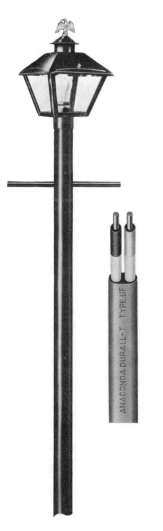

FIGURE 55. The plastic-covered wire at right (Type UF) is ideal for underground wiring. It can be buried directly in the ground for wiring a post lantern.

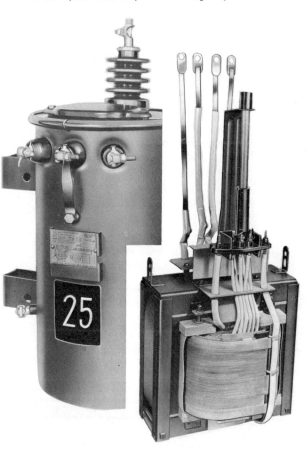

FIGURE 56. Iron pot at left houses transformer windings at right, stepping voltage down to safe level for home and farmstead use. Size is 25 KVA, same as 25,000-watts capacity.

construction. Actually, it consists of two windings called (1) a primary and (2) a secondary. The *primary* receives the current from the line and passes it on to the *secondary* through the mysterious process referred to as magnetic induction. That is, the current passes from the primary to the secondary without benefit of a metal connection of any kind. Here is how the transformer works:

Earlier in the book, the relationship between magnetism and electricity was discussed. One thing mentioned was that a magnetic field is formed around a wire carrying an electric current. Therefore a much greater magnetic field is formed about a *coil* of wire carrying a current, since each wire adds its magnetic force to the total field.

Now if the primary winding (coil) is energized and a secondary coil is placed within the magnetic field around the primary, a current will flow in the secondary, even though there is no metal connection between the two windings. Therefore, the current flowing from the transformer to your meter is an "induced" current; that is, it jumps across the space between the primary and the secondary through the medium of magnetic force. The iron core in the transformer is used to increase the magnetic force, not to conduct a current from one coil to the other.

If the size of wire and number of turns in both windings are equal, the induced current will be approximately the same. If, however, the ratio is varied, the induced current will vary also. For example, if the number of turns in the secondary is one-tenth the number in the primary, the voltage induced in the secondary will be one-tenth that coming into the primary. The size of the wire in the secondary in comparison to that in the primary determines amperage of the induced current and affects voltage.

Assuming that the voltage on the high line at your farm is 6,900, the ratio of the windings in your transformer is probably 30 to 1, reducing the voltage to 230. This is a matter of simple arithmetic; that is, $6{,}900 \div 30 = 230$. The 6,900 volts pressure going into the transformer is stepped down to 230. Ordinarily a three-wire power drop—the wires from the power pole into the farm —splits the 230 volts into two 115-volt circuits. One hot and one neutral wire make a 115-volt circuit. Only one neutral is needed for the two hot wires, and this hookup is usually referred to as 230-volt service.

Three-phase service is different. Generally, three hot wires and one neu-

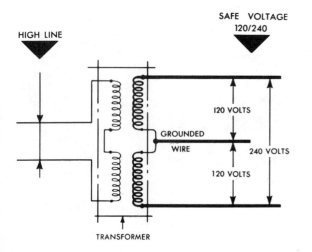

FIGURE 57. The ratio and the wire size of the coil at left, in comparison with the coil at right, determine the voltage and amperage of the current in the wires leading to the meter (three heavy lines at right).

tral are provided in a three-phase service and either two or three transformers are used. The advantage of three-phase service is that you can use much larger motors than is allowable on single phase lines, and they cost less.

The amount of electric energy needed determines the size of the transformer. A 25kv size is most often used for a farm. A large electrified farm may need a 37.5kv size.

Another farm use of a transformer is the farm arc welder. This machine takes 37.5 amperes at 230 volts from the high line and converts this to 180 amperes at 25 to 40 volts. This transformer produces an arc hot enough to weld steel.

In the Problem-Unit on farmstead wiring, you can find additional information on other transmission equipment needed in the farm distribution system.

FIGURE 58. A farm arc welder uses a step-down transformer to convert 230-volt current to low voltage-high amperage.

SUMMARY

Static electricity exists in all substances but is not noticeable except when unlike substances are rubbed together. The unlike charges generated cause the two to cling together, whereas like charges repel each other. Lightning is the result of a static charge jumping across an air gap between a cloud and a building or an object on earth. (It has no practicable value.)

A storage battery generates direct-current electricity; that is, a current that flows in a straight line. The current from a battery is the result of chemical (acid) action on unlike electrodes. The acid action creates an imbalance between the electrodes, thus causing electrons to flow from one electrode to the other.

It is necessary for the electrodes to be restored while in use; otherwise they soon lose their power and the battery is then "dead" or useless. A generator is used for the restoring process.

A generator is rather simple in construction and operation. It has a rotor (armature) that turns inside a magnetic field created by stator coils. As the rotor turns, an electric current is induced in its coils. This current is then sent out to the battery, where it causes a rather complicated chemical action that restores the electrodes.

Commercial electricity is produced by giant generators. These generators (sometimes called alternators) produce a current at several thousand volts. The current, however, is not direct but flows, rather, in a wavy line, first positive, then negative. This type of electric service is called alternating current.

The high voltage on a rural electric high line must be reduced to a safe level for farm use. Transformers are used for reducing the current, usually to 115 or 230 volts. This reduction is accomplished by the use of two windings called the primary and secondary. A high-voltage current flowing through the primary induces a lower voltage in the secondary winding. The farm arc welder operates on a similar principle, but produces low voltage, high amperage.

Inside the farm home or other farm building, it is necessary to have a system whereby a hot wire can be distinguished from a neutral, or ground, wire. This is done by using black and red for the hot side of a circuit and white for the neutral, called *polarizing*.

For heavy electrical loads, three wires carrying 230 volts are sometimes used. Good insulation and good quality wire are necessary in making the electrical service safe and economical.

QUESTIONS

1. Why is a generator necessary in maintaining a storage battery?
2. How could you show by simple demonstration that a magnetic field can produce electricity?
3. What principle is involved in causing the rotor in an electric motor to turn? What determines its top speed?
4. Why must a neutral wire never be fused? Never interrupted by a switch?
5. Why is a transformer sometimes called a "step down" or "step up" transformer?

ADDITIONAL READINGS

Fairbanks Morse and Company, *Electrical Machinery Catechism.* Freeport, Ill., (no date).

Graham, Kennard C., *Fundamentals of Electricity,* Fifth Edition. American Technical Society, Chicago, Ill.

Richter, Herbert P., *Practical Electricity and House Wiring.* Frederick J. Drake and Company, Wilmette, Ill., 1952.

Southern Association of Agricultural Engineering and Vocational Agriculture, *Electrical Terms: Their Meaning and Use.* University of Georgia, Athens, Ga., 1967.

Westinghouse Electric & Manufacturing Company and Science Service, Inc. *Fundamentals of Electricity,* (Ed. Morton Mott-Smith, Ph.D.), 306 Fourth Ave., Pittsburgh, Pa., 1943.

Wright, Forrest B., *Electricity in the Home and on the Farm,* Third Edition. John Wiley and Sons, New York, N.Y., 1950.

4 HOW TO COMPLY WITH SAFETY RULES

Reports given out by the National Safety Council indicate that electricity has a killing power about forty times greater than other farm hazards. Whereas only one in two hundred ordinary farm accidents is fatal, one in each five injuries from electric shock ends in death. The total number killed by electricity each year shows that people apparently do not realize how dangerous it is. In one recent year 1,056 persons in the United States lost their lives in electrical accidents, and this rate is approximately the average for any given year. From 3 to 4 percent of all accidental deaths in the United States can be traced to electricity.

In addition to personal injuries and deaths, electricity starts fires that destroy over $80 million worth of property in this country each year. An additional $17 million worth of appliances are destroyed or damaged beyond use every year by faulty wiring and other electrical hazards. Electric machinery also causes thousands of non-fatal accidents, such as loss of eyesight, permanent crippling, loss of hands or fingers, and many others.

WHAT ARE THE COMMON CAUSES OF ELECTRICAL ACCIDENTS?

It is possible to avoid most of the accidents, both to persons and to property, resulting from electricity. In fact, studies show that the two most common causes are inexcusable. These are (1) *carelessness* and (2) *misuse*. The proof of this can be found in hazardous wiring, overloaded circuits, improper fuses, abuse of appliances, and many other unsafe practices in the handling of electricity on farms and in farm homes. These abuses can and do result in deaths,

FIGURE 59. Engineer checks current on transformer-type electric fence. Principle of operation involves using high voltage-low amperage current, in "pulsation" flow. Too much amperage could be fatal.

FIGURE 60. Portable electric tools may be fatal, especially when the operator is standing on damp concrete or on the ground. Use a three-wire cord and adapter to reduce the chance of shock from a shorted circuit inside the machine.

injuries, fires, and unsatisfactory electrical service.

As farm electrification increases, it will be more important than ever to use safe practices. Even in doing simple wiring jobs, it is essential that you understand the common principles of electricity and use reasonable skill in doing the job. Remember that a faulty circuit can result in a death or a costly fire. Even the simple job of putting in a fuse can be dangerous unless it is done properly. When you operate a portable power tool, you may be taking a chance unless the tool is protected properly by the right kind of wiring and plug-in arrangement.

HOW MUCH ELECTRICITY IS NECESSARY TO KILL A PERSON?

Only a very small amount of electricity is necessary to kill a person. The current or amount of electricity is the thing that kills, not the voltage, although voltage is involved. Less than ¼ ampere will stop the heart of the average man. This is the amount of current flowing in a 25w lamp at 115 volts.

The amperage in a current flowing in a 100w lamp will be fatal if your body is in good contact with the ground or with any conductor that is in contact with the ground. Because of this hazard, it is extremely dangerous to stand on wet ground or pavement and touch an electric circuit.

It is the purpose of the safety suggestions in this chapter to help reduce the common hazards of electric energy and to develop skill and judgment in using electricity.

If some of the rules discussed here seem trivial, remember that violations of simple precautions have caused the greatest number of deaths and the most property damage. You are likely to be more alert to the obvious dangers than to the simpler ones. For example, you would be more apt to use a penny to replace a blown fuse than to take hold of a broken power line. Yet the penny represents just as great a danger, since this might result in a fire that could take the lives of your entire family.

Your all-electric farm and farm home will make it necessary for you to handle electricity and electric equipment almost constantly. You should learn now to avoid the hazards that could result in tragedy for you and your family.

WHAT PRACTICES SHOULD YOU FOLLOW TO AVOID ELECTRICAL HAZARDS?

To prevent injury caused by ignorant or careless use of electricity, keep the following safety hints in mind:

1. The first rule of safety with electricity is to become *safety conscious*. Only one hot wire is necessary for a current to flow through your body. Remember that electricity is present in certain hot wires of electric circuits all the time unless the main switch is off or the main fuses are out. Your body can complete a circuit between a hot wire and the earth.

2. Always respect the danger of electricity in any form and never take chances with it.

3. Avoid working on hot wires or appliances that are connected to a source of electricity; disconnect the appliance; open the main switch; pull out main fuses; and take all other measures to make your work safe.

4. Give your heart a break! It has been found that some people cannot stand even slight electric shock; it stops their hearts.

5. Before operating a large permanently mounted electric motor, be certain that it has an individual ground.

6. Before operating a portable electric machine, such as a portable saw,

FIGURE 61. *Top:* A boy holding a metal socket and a 25w bulb could be killed if the socket becomes shorted. The current flowing through a 25w bulb (115 volts) is sufficient to kill a grown man.

FIGURE 62. *Bottom:* The ordinary 15-ampere plug fuse is intended to protect life and property. Do not destroy this protection by using a penny underneath.

be certain that the circuit is properly grounded to the frame of the machine. Polarized plug-ins are available to make it easy to connect an extra ground wire to portable electric machines. The cost is reasonable and could save your life.

7. Avoid probing into a convenience outlet with a screwdriver or similar object.

8. Avoid touching electric wires, switches, drop cords, or appliances while standing in water. Switches and other electrical devices should not be installed within reach of the bath tub or shower. Wet shoes or wet concrete floors may prove equally hazardous. If it is necessary to use a portable electric tool while standing on damp concrete, first place a dry board on the floor. Good rubber gloves will make the use of portable electric machines safer where the working area is damp.

FIGURE 63. *Top:* The repairman working on the convenience outlet (B) would be inviting suicide without first cutting off the main service switch (A). The arrow indicates the proper direction for this.

FIGURE 64. *Bottom:* The large fan motor in this hay-drying installation has an individual ground in addition to a magnetic switch for overload protection.

9. Before cutting an electric wire, make certain that it is disconnected from the power source.

10. Assume that all electric wires are hot until proven otherwise.

11. Avoid touching or holding an electric fence. People as well as farm animals have been killed in this way.

12. Demonstrate your good judgment by refusing to take part in any horseplay with electricity. Never cause another person to be shocked; practical jokes with electricity have ended in serious injury to others.

13. Although a farm arc welder operates on low voltage, faulty wiring in the machine could cause you to receive the full load from the line—230 volts at 37.5 amperes. This amount of electricity would almost certainly be fatal if you were standing on damp ground. Check your welder periodically to make cer-

tain that connections are tight and not corroded. Blow out dust with a low-pressure air hose. This will help to prevent short circuits in the welder.

14. Avoid touching broken wires that are dangling from a power line, since this would be virtual suicide.

WHAT MEASURES SHOULD YOU TAKE TO PROTECT YOUR PROPERTY FROM ELECTRICAL HAZARDS?

Except for losses of farm animals and farm buildings due to lightning, the bulk of property loss from electricity is caused by fires. Of the major causes of electrical fires, faulty wiring is the main culprit. This may be due to mechanical damage, overheating, or using wire that is too small for a given electrical load. Enclosed wires often become shorted by improper installation or by corrosion resulting from poor splicing. "Cheap,"

FIGURE 65. Although the maximum voltage of the farm welder in *B* is 65, it is possible for the machine to become short circuited and cause the death of the person in *A*.

FIGURE 66. *Left:* Dangling, frayed lamp cord used for house wiring is a hazard to property and life.

FIGURE 67. *Right:* The child got out but the house burned. Faulty wiring was blamed.

poor-grade wire in the original job nearly always causes trouble in the end.

In addition, faulty electric appliances take their roll in fire damage. The wiring in an electric appliance may simply wear out and the insulation becomes hard and cracked from age and use, especially if the wire is exposed to heat and moisture. Switches may be dangerous when used after they begin to show signs of failure.

Electric services on the farm have expanded so much in recent years that the wiring system is often overloaded. Figure 68 shows an overloaded main switch box. This switch became so hot it started a fire. Check your wiring system to determine whether too many appliances are connected to one circuit.

The only answer to overloaded electric wires is to add new circuits at the main switch, but this may require a new entrance switch. The old one may be too small with all circuit positions already taken. The lesson to be learned from an overloaded electrical system is that good planning should be done before it is installed. A well-planned system is less expensive and is less of a fire hazard in the long run.

Other property losses can be accounted for in the loss of efficiency, through low voltage and poor contacts, of the electric current to do farm work. Electric appliances wear out too soon when the voltage is below normal. In short, if your electrical system is not up to standard, you pay for electricity that does no useful work. Also, your electric appliances may wear out before they should.

Some of the common rules that can make farm property safer are:

1. In all phases of the farm and home wiring, follow the National Electrical Code.

2. Check the present electrical load on all circuits before adding other devices to it. Do not overload. For example, the load on No. 12 wire should not exceed 2,300 watts. If more than this amount is being used, the fuse should blow: (1) toaster, 700w + iron, 1,200w + lamp, 100w = 2,000 watts. This is a safe load. But, (2) toaster, 700w + iron, 1,200w + percolator, 600w = 2,500 watts. This circuit is over-loaded for No. 12 wire. The fuse should blow if it is of proper (20-ampere) size.

3. Keep the proper size fuses in your fuse panel. Never use a penny as a substitue. This practice causes fires.

4. Check the cause of a circuit breaker tripping or a fuse blowing. The

FIGURE 68. *Top:* Overloaded circuits, heavily fused, built up enough heat to start a fire.
FIGURE 69. *Bottom:* A Texas farmer examines a 200-ampere mainswitch. Adequate wiring is the first principle of safety.

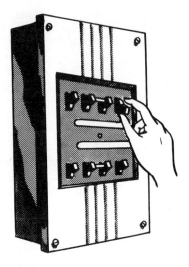

FIGURE 70. *Top:* Before resetting a breaker be sure to correct the cause of tripping.

FIGURE 71. *Bottom:* (A) A Fustat is said to be tamper-proof because it cannot be exchanged for a larger size. (B) A magnetic switch protects large motors by tripping when overloads occur. This manual-type switch requires resetting by hand after tripping. (C) A Fusetron allows overload for starting electric motors yet protects the running load.

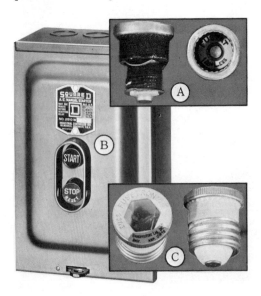

purpose of a fuse is to provide protection from overloads. A penny or oversize fuse destroys the safety feature provided by the fuse system.

5. Use time-delay fuses or other appropriate devices to protect electric motors and their circuits. Regular household fuses are not satisfactory for electric motors.

6. Check the cause of blackened, overheated wires. A smoking wire or electric appliance is cause for alarm. Electric motors may become overheated for numerous reasons and may set buildings afire unless the circuit is properly protected. After continued operation, an electric motor should feel "hot" though bearable to the hand.

7. Study the electrical system at least once a year to find damaged wire, faulty switches, and other potential trouble spots. Repair worn or damaged cords as soon as they are discovered. (See Figure 77 in project section.) In installing new electric circuits, be certain that the wire size is adequate for the load that it will carry. (See Chapter 5 for instructions on how to determine wire size.)

8. In purchasing electric appliances and materials, be certain the quality is up to standard. "Cheap" fixtures will be costly if they cause a fire. The Underwriters' Laboratories (UL) label on an appliance is good insurance even though it is no guarantee that you will not have a fire.

9. Farmers and farm boys should be capable of installing a new switch or adding new circuits to the present electrical system. For a major wiring job, however, it is wise to employ a certified electrician, unless you have had some wiring experience.

10. Install correct-size wire for all electric motors and other electric equipment.

11. Store gasoline, paint, and other inflammable materials away from the vicinity of a farm arc welder.

12. All entrance switches should be enclosed and may require locking.

WHAT SHOULD YOU KNOW ABOUT NATIONAL PROTECTION AGENCIES?

Everybody suffers when accidents occur and when property is destroyed by fires. The national hazard from electricity is so great that the insurance companies and other professional groups have banded together to help make electric service safer for everybody. The quality of many electrical devices and materials on the market today is better because of the work of these organizations.

The Underwriters' Laboratories

If you have ever seen a label on an electrical device bearing the letters "UL," you may have wondered what this meant. These letters signify that the item has been tested by the Underwriters' Laboratories. The presence of the "UL" label means that the item measures up to safety standards for that item.

The Underwriters' Laboratories organization was established by the National Board of Fire Underwriters as a national testing service. Their purpose is to eliminate or reduce fires that are started by faulty wiring and inferior electrical products. Manufacturers are not required to have their products tested by the Underwriters' Laboratories; however, increasing numbers of buyers are insisting on products that have been so tested.

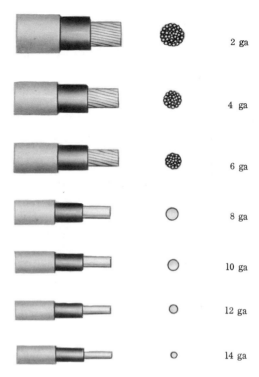

FIGURE 72. Common sizes of wire used on the farm include No. 12 to No. 2 and larger. No. 14 is seldom used except for small motors and extension cords.

FIGURE 73. The UL label on an appliance indicates that the item has been tested and approved by Underwriters' Laboratories.

The result is that a majority of manufacturers are now using this service.

Products that are sent to these laboratories undergo tests of much greater severity than actual use. As an example of this, rubber-covered wire bearing the label "Approved by Underwriters' Laboratories" has been tested by soaking in water for 12 hours, then subjected to 1,500 volts of electric pressure. Wire that stands up under such conditions would rarely break down in normal use.

In fairness to some reputable manufacturers, it should be stated that some of them do not have their products processed through UL, claiming that the expense is too great for value received. However, it is only fair for the consumer to use this protection if he so desires.

To be approved by Underwriters' Laboratories means that a particular item meets the minimum standards for that type of article. The "UL" label does not claim that one article is better than another; for example, two electric motors having "UL" labels may be of different quality. The label simply signifies that safety standards have been met.

National Electrical Code

Tragic fires like the one shown in Figure 67 may be prevented by adequate wiring. The National Electrical Code is designed to prevent or reduce such disasters. This code is a set of rules and regulations that have been drawn up by engineers, fire prevention specialists, insurance authorities, and others. The object of the code is to bring about safe and efficient electrical service. This code is revised every two years or so.

FIGURE 74. An engineer tests a television set for safety. The high voltage required for the picture tube makes TV dangerous unless safety standards are met.

The Code sets forth minimum standards of safety for wiring and materials, but it does not guarantee adequate wiring plans.

Electricians are expected to be familiar with the rules and regulations of the National Electrical Code and should see that requirements are met with respect to materials and installation. Many cities and other localities have their own electrical codes, which are often stricter than the National Code.

It is not practical to include in a book of this nature all regulations of the National Code. If you plan to do some electrical work around your home, however, it would be wise to obtain a copy and study the regulations covering the type of jobs you plan to do. Some cities require that new electrical wiring be inspected for conformance with local and national codes before it can be used.

The safest policy in wiring is to follow the rules and regulations of the Code. In fact the violation of the Code is an invitation to start fires and perhaps leave yourself open to electrical injury. For example, it may seem safe to cut a wire, splice in another circuit, tape the completed joint, and leave it at that. The Code will not permit this practice unless the wire is properly spliced or joined, taped, and enclosed in a junction box. Otherwise the splice may eventually work loose or corrode and cause a fire.

WHAT CAN YOU DO FOR A PERSON WHO HAS RECEIVED ELECTRIC SHOCK?

The impulse you may have to help someone who is receiving electric shock could cause you to be electrocuted. On the other hand, you may be able to save some person's life without undue hazard to yourself if you follow a few rules.

The muscles of a person receiving electric shock become paralyzed, causing him to "freeze" to the hot wire. Sometimes the victim is not able to free himself from the conductor. Getting him loose may not always be easy but may be accomplished in the following manner:

1. Look for the "live" wire or source of electric current which is causing the shock, but do not take hold of the person's body with your bare hands under any circumstances.

2. Decide immediately whether it would be easier to move the person or to move the conductor. If the conductor is to be moved, use a dry stick to push it loose; if the person's body is to be moved, use twelve to fifteen thickness of dry newspaper or cloth as an insulator.

3. With the cloth or paper in the hands, grasp the person's arm or leg and quickly pull him free of the conductor.

First Aid and Treatment

Have someone call a doctor or an ambulance or both. The effect of electric shock is to damage the heart and stop its action as well as to stop the breathing. Burns on various parts of the body often occur as side effects, and these may be serious in themselves. Severe burns may cause the skin to come off the entire surface of the feet, face, or both.

The only effective treatment for an electrocuted person is to use artificial respiration or a Pulmotor as a means of reviving his breathing. If it is to be effective, artificial respiration must begin almost immediately and continue for several hours or until the person is breathing normally. Refer to a first aid manual for further instructions on how to apply artificial respiration.

SUMMARY

Each year over 1,000 persons in the United States are killed by electricity. Either carelessness or misuse is involved in most of these deaths.

The death rate of electrical accidents, in comparson with ordinary farm accidents, is about forty times greater. This is not strange when it is realized that the amount of electricity flowing through a 25w bulb is sufficient to kill a grown man.

Approximately $100 million worth of property is destroyed annually by electrically started fires and other electrical damage. Overloaded wiring often gets hot enough to start a fire. Again, carelessness and misuse are the main causes.

The best safety precaution is to be safety conscious and show a healthy respect for electricity at all times. Circuits should never be overloaded or overfused. Damaged lamp cords and other troubles should be taken care of as soon as they are found.

There are two national organizations in this country that promote safety in electrical wiring and electrical appliances. These are the Underwriters' Laboratories and a group of specialists who sponsor a National Electrical Code.

Items that have been tested by the Underwriters' Laboratories will be stamped with the letters "UL." This label indicates that the item has passed safety specifications as to materials and construction.

The National Electrical Code sets forth minimum standards for wiring materials and procedures. The object here is to reduce hazards to both life and property. This code should never be violated, since fire or death may result.

The best treatment for a person who has received electric shock is to give artificial respiration if breathing has stopped. Then the person should be covered with a blanket and kept as quiet as possible until a doctor arrives.

QUESTIONS

1. Why is electricity so much more likely to be fatal than other farm hazards?
2. Why is electricity more dangerous around water?
3. How is it sometimes possible for a wire to become hot enough to start a fire?
4. What happens in the process of reviving an electrocuted person by artificial respiration?

ADDITIONAL READINGS

Bussmann Manufacturing Division of McGraw-Edison Company, *Protection Handbook*. University at Jefferson, St. Louis, Mo., 1964.

Edison Electric Institute, *Reduce Crime and Highway Accidents with Street Lights*. Street and Highway Lighting Committee, New York, N.Y., 1968.

Hamilton, C.L., *Current Follies.* National Safety Council, Chicago, Ill., 1955.
National Safety Council, Farm Division, *Farm Safety Review.* Chicago, Ill. (All issues).
Texas Education Agency, *Farm Electrification Lesson Plans, Wiring and Safety.* Austin, Texas, 1967.

SUGGESTED PROJECTS FOR PROBLEM-UNIT TWO

Electrical Demonstrations Suitable for Class Work and Programs. Your class or club can put on interesting demonstrations to illustrate electric circuits and electrical principles at very little expense by using the following ideas. The demonstrations can be used for club programs, contests, or for shop projects.

1. *Construct a Simple Circuit.* Materials needed for this project include a dry-cell battery, a laboratory-type lamp socket and a small bulb, 6 feet of No. 18-gauge single-conductor cord, and a simple knife switch. You might be able to borrow these materials from the physics laboratory at your school, or you can purchase them for about $2 at any good hardware store.

Cut the wire into three equal lengths and prepare the ends of each piece for connections, as directed on page 161. Refer to Figure 75 and connect three lengths of wire as shown in the diagram; that is, the first length from battery to switch; the second length from switch to lamp socket; and the third length from lamp socket to battery.

If the battery, bulb, switch, and wiring are in proper order, the lamp should light up when the switch is closed (in down position). Point out that the flow of current is from the battery to the switch to the bulb and back through the battery again.

FIGURE 75. A two-wire circuit operated by a dry-cell battery. A knife switch opens the circuit as shown in the diagram. The bulb will light up when the switch is closed.

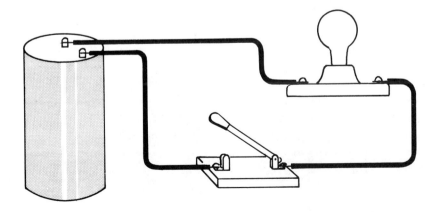

Explain that a lighting circuit in the farm home is essentially the same as this simple hookup. The difference is in the higher voltage-amperage of the household current and in the larger wire required in the household circuit–preferably No. 12 gauge.

2. *Demonstrate Voltage Drop.* Voltage drop is a constant problem on most farms; the following demonstration should be valuable as well as interesting.

Have on hand a household fan, five or six strips of 1-inch tissue paper about 12 inches in length, and a 100-foot extension cord of No. 18 gauge wire. Tie the strips of paper to the center of the fan guard and perform the test in two steps as follows:

Step 1. Plug the fan cord into a convenience outlet and turn on the fan. Observe that the strips of paper blow straight out. Remove the fan plug-in.

Step 2. In this test, first plug the 100-foot extension cord into a convenience outlet, then plug the fan cord into the extension receptacle. Now the current must flow through the extension cord before reaching the fan motor. Turn on the fan and observe the angle at which the strips are blowing outward. Notice that they do not blow straight out in this test because of voltage drop. Repeat Steps 1 and 2 to make certain that everyone in the audience sees the difference between the tests.

Explain that voltage lost in the 100-foot cord caused the fan motor to run slower in the second test, as indicated by less action of the paper strips. Explain further that the small wire and fan motor overheat and that both will be damaged unless the circuit is properly fused. Point out that the obvious cure is wire of proper size.

Safety Projects.

1. *Repair Damaged Extension Cord.* Refer to Figure 77 and follow the self-explanatory steps in repairing a cord that is damaged in the middle. This simple job may prevent a costly fire.

2. *Repair Cord Ends and Replace Plugs.* Refer to Figure 78 and follow the self-explanatory steps in repairing cords that are damaged at the ends; also replace the plug in accordance with instructions given in the illustration. This job too may prevent a fire.

FIGURE 76. The voltage drop through 100 feet of No. 18 wire results in the slower speed of the fan.

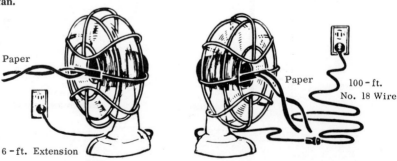

PROJECTS FOR PROBLEM-UNIT TWO 75

... if cord is frayed in the middle

(1) Cut frayed edges clean.

(2) With electrician's rubber tape or plastic electrical tape wrap each wire *separately* from end to end.

(3) With friction tape (tire tape) bind the two wires together. If plastic tape is used, tape the re-insulated wires together.

FIGURE 77. *Left:* Avoid fires by following the three easy steps shown here.

FIGURE 78. *Bottom:* Follow these instructions to do a good job of repairing the frayed ends of an extension cord or replacing a plug.

... if cord is frayed at the plug or plug has to be replaced

(1) Release cord by loosening screw posts inside plug.

(2) Cut off frayed end of cord.

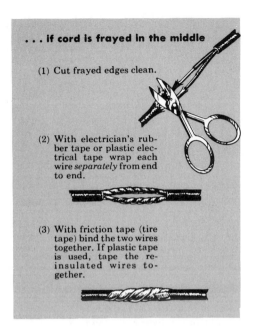

(3) Strip two inches of outer fabric covering from cord. Moulded rubber cord can be split back two inches from the end. Be sure that when the cords are split no copper wires are exposed through the protective insulation.

2" COVERED CORD

MOLDED RUBBER CORD SPLIT APART

(4) Strip insulation from ends of cords to expose approximately three-fourths inch of bare wire. Twist strands together.

¾"
TWIST STRANDS OF WIRE

(5) Slip cord through plug and tie cord ends into an underwriter's knot. Then if cord is jerked (it shouldn't be!) there will be less chance of pulling the copper wires away from under the screw posts. Pull knot down inside plug.

(6) Wrap wires clockwise around prongs to form an S. Loop bare copper ends clockwise around the screw posts. Hold in place and tighten screws firmly.

... if plug doesn't fit tightly

If the prongs of the plug slip out of the outlet or receptacle, just separate prongs a little with your fingers to get a tighter fit. Don't use too much pressure.

3. *Install a Yard Light.* Remove farmstead darkness hazards by installing one or more yard lights. The project shown in Figure 79 calls for a 20-foot length of 3- or 4-inch used iron pipe. Weld a metal bracket to the top of the pole and anchor an outside-type lamp fixture to this. Select a fixture carrying a sufficient number of PAR-38 lamps to provide light for the entire farmstead. The project shown in Figure 79 can be improved by welding metal climbing cleats to the side of the pole. Wiring is threaded through the inside of the pipe before the light is installed. Make certain that your wire is of the proper size.

4. *Install Tractor Lights.* If you drive your tractor on the highway, you should have it equipped with night lights. These lighting sets, including fixtures and wiring, can be purchased from your implement dealer. Instructions for installing lighting equipment are furnished with the set. A workman is shown installing a night light in Figure 80. His tractor will be safer when he is driving on the highway after dark.

5. *Change a Blown Fuse.* Changing a blown fuse can be dangerous, especially in a damp location. Follow the practice of standing on a dry board when

FIGURE 79. *Left:* One multi-unit yardlight may pay for itself many times over by preventing after-dark accidents. Materials needed include (1) a 20-foot length of 3-inch used pipe, (2) a multi-socket outside-type fixture, (3) par-38 light bulbs, and (4) a quantity of wire. The light pole at upper right could be improved by welding climbing cleats to the side.

FIGURE 80. *Right:* This workman has little trouble installing a night light on his tractor.

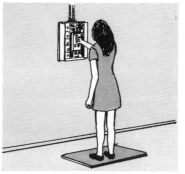

FIGURE 81. If the mainswitch is located in a basement or other damp location, stand on a dry board when removing a blown fuse. Avoid touching anything but the fuse (right).

changing a fuse or setting a circuit breaker. Avoid touching metal parts of the switch or breaker box. It is also a good safety measure to disconnect the main switch before changing fuses or setting breakers. (See Figure 81.)

GLOSSARY FOR PROBLEM-UNIT TWO

Ampere   The measure of the rate at which electricity flows. Often abbreviated *amp*.
Volt   The unit of measure of electrical pressure which causes a current to flow.
Watt   A unit of electric power, equal to 1/746 hp. 1 volt $\times$ 1 ampere = 1 watt.
Kilowatt   1,000 watts, or 1.34 hp.
Killowatthour   1,000 watts being used through one hour; often designated *kwh*. 1,000 watts $\times$ 1 hour = 1 kilowatthour.
Conductors   Copper, aluminum, silver, and other metals through which electrons can be easily moved.
Insulators   Glass, rubber, plastic, porcelain, and other substances through which electrons will not move. Also known as non-conductors.
Circuit   A complete path providing for a continuous flow of electric current. Usually consists of two wires: one hot wire to bring the current from the source of supply, and one neutral for conducting the current back to the source.
Phase   Refers to the timing of an alternating current. When two alternating currents reach their zero, maximum, and intermediate values at exactly the same instant, they are said to be in phase. A two-phase current consists of two separate alternating currents 90 degrees apart. A three-phase current consists of three separate alternating currents 120 degrees apart.
Cycle   A complete series of changes that take place in the flow of an electric current and recur at regular intervals. Thus in 60-cycle alternating current, the current builds up to a positive peak and then to a negative peak 60 times each second.
   NOTE: This glossary was adapted in part from a publication issued by the Tennessee Valley Authority. (Now out of print.)

PROBLEM-UNIT III

HOW TO PLAN AND DO "ORDINARY" WIRING

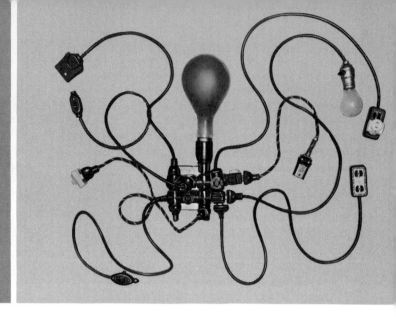

FIGURE 82. Common species of "electric octopus."

In the first chapter of the book you saw how a Michigan farmer was able to release one hired man for the cost of two months' electrification (electric bill, depreciation on equipment, and interest). And in addition to greater farming efficiency, his family enjoyed greater convenience and comfort in their work and daily living.

You also studied how a ¼-hp motor is capable of doing the work of a grown person. So, a 5-hp feed-handling setup has the potential working capacity of twenty men! However, you are faced with the problem of getting the necessary electric service to the place where it is needed.

Getting the home and agriculture fully electrified, and partially automated, has been hampered from the beginning of electrification on a national scale by inadequate and otherwise faulty wiring. Far too often the home owner or agri-businessman has found out too late that his wiring system would not take on a new air-conditioning system, an additional 30,000 broiler unit, or another irrigation pump. Why? His electrical system was already loaded to capacity. The result frequently was to forego the additional equipment or appliance because the entire wiring system would have to be replaced or overhauled.

Inadequate electrical service is not always the fault of the power supplier. It is the responsibility of the owner to plan ahead, to build into the wiring system some spare capacity to take care of needed expansion. Needs for at least five years in advance should be considered in the original plan; ten years in advance would probably pay big dividends in most cases.

This entire problem-unit is devoted to "wiring" but is not intended as a training course for electricians.

5 HOW TO SELECT AND ARRANGE THE EXTERIOR WIRING SYSTEM

This chapter is devoted to the exterior wiring system and focuses upon three problems that have hindered the full use of low-cost energy—namely, *inferior or faulty* wiring materials, *inadequate or undersized* wiring materials, and *faulty wiring practices.*

Certainly the growth of electrification in the home and farmstead has been great. Figure 88 depicts this by showing four "services." in the 1930's and early '40's the average home and farmstead needed only the 30- to 60-ampere services shown on the left. Since then the average use has grown to the 100- and 200-ampere setups on the right. Now, large numbers of fully electrified homes and farmsteads are calling for 400-ampere service (not shown).

The objectives of this problem-unit are in accord with those stated in the *Farmstead Wiring Handbook*[1]—namely, (1) *safety,* by meeting Code Specifications; (2) *adequacy,* by providing enough electric energy where needed; (3) *expandability,* by wiring the system so that additions can be made without rewiring the whole system; and (4) *efficiency,* by so wiring the home and farmstead that additions can be made with reasonable ease and economy.

Other essentials to be kept in mind in planning the wiring system are nominal first cost, reasonable cost and ease of upkeep, and ease of making repairs.

WHAT IS GOOD "ELECTRICAL INSURANCE"?

Abiding by the National Code and using wiring materials that have the UL

[1] Excerpted from *Farmstead Wiring Handbook,* Edison Electric Institute, New York, N. Y., 1965.

79

FIGURE 83. A 20-ampere fuse for No. 12 wire = protection; a 30-ampere fuse for No. 12 wire = fire hazard.

label on them will amount to the best "insurance" you can get in the long run. (See Figure 73.) Also, using the *Farmstead Wiring Handbook* and *Residential Wiring Handbook* will increase your "free" insurance.

HOW SERIOUS CAN FUSE-BLOWING TROUBLE BE?

The combination of No. 12 copper conductor and 20-ampere fuse shown in Figure 83 makes for a safe circuit (2,300 watts), but if the 30-ampere fuse at right were substituted for the 20-ampere size, the current could then go up to 3,450 watts on No. 12 wire before blowing.

A Case In Point

One homeowner recently had "fuse-blowing" trouble, using a 20-ampere fuse on No. 12 wire. A power company service man came to check on the situation after the owner kept having the same trouble. After testing with a test lamp, he found that two new outlets had been wired to a kitchen appliance circuit— (1) one for a 1,650w heater and (2) one for a new 1,400w iron.

Seeing that the trouble was overloading, the service man had the owner list the loads on this one circuit:

1. Two ceiling lamps @100w [2]	200 watts
2. One lamp at sink @100w	100
3. One lamp at range @100w	100
4. One (new) heater @1,650w	1,650
5. One (new) iron @1,400w	1,400
Total	3,450 watts

He explained to the owner that No. 12 copper wire fused at 20 amperes would carry up to 2,300 watts (115v × 20 amps = 2,300w). But the addition of the heater and iron, both on at the same time, ran the load up to 3,450 watts, an excess load of 1,150w. If the 20-ampere fuse had failed to blow at 2,300w, the excess of the 1,150w may have started a fire. Fortunately the owner did not substitute the 30-ampere fuse, which would have made a fire possible.

The Solution. For temporary use the service man split the circuits, leaving the heater on one and reconnecting the iron outlet to a different circuit. He explained to the owner that the heater alone would

---

[2] Symbols and abbreviations such as "amps," "v," and "w" are standard usage for amperes, volts, watts, et cetera.

take up most of the 1,800 to 1,900 watts considered normal for No. 12 wire. So nothing else except one or two lights could be used while the heater was on. Similarly, a vacuum cleaner (1 hp) should not be used on the other circuit while the iron and several lights were in use.

A More Permanent Solution. The owner had no idea of the seriousness of his problem. All circuit spaces in his fuse panel were already taken. A new service switch would be needed. This would call for a larger capacity service entrance; then, a new service drop and a larger transformer would be needed. The service man suggested that the owners ask the power supplier to send out a planning expert. A complete rewiring job would be needed.

WHAT DOES THE EXTERIOR WIRING SYSTEM INCLUDE?

When the planner arrived, he brought with him some large, graph-like paper and a 100-foot tape. The owners, in the meantime, had listed their present electrical appliances and others to be added during the next five to ten years.

What Major Items Should the Plan Include?

The first thing the planner did was to refer the owners to Figure 84 and to identify the major parts of the exterior system. Any single part of the exterior system can frequently act as a bottleneck for the remainder of the electrical system. For example, the size of the service drop limits the total electrical energy available to the home or farmstead.

Parts of the Exterior System. Having in mind the needs of the owners, the planner referred them to Figures 84 and 91 and identified each major part of the exterior system, as follows:

1. The high-line alongside the highway carried about 13,000 volts at relatively low amperage, but a transformer on the high-line pole changed the current to a 230-volt service and regulated the amperage as to the capacity needed.

2. The lead-in wires from the transformer were identified as the *service drop*. These conductors, usually two hot and one neutral, must have sufficient capacity to get adequate current [3] from the transformer to the point of attachment at the meter pole or the building. The arrangement of the service drop may be (1) rack-style as in Figure 84, (2) a vertical arrangement of conductors on the poles, or (3) in a twisted or rope-like style as in Figure 90 (inset). Still another arrangement is to have the service drop buried, thereby eliminating overhead wiring. (See Chapter 8.) It is customary for the power supplier to furnish the service drop, but they may not use the underground system since it is more expensive.

3. *Service Entrance.* The service entrance wires, usually called SE conductors; the conduit tubing and fittings; the meter; and main disconnect switch [4] are collectively referred to as the *service entrance*. In pole metering all of the service wires from the top of the pole, to meter, to disconnect, and back to the top of the pole are referred to as the meter loop. (Some power suppliers furnish this, but usually the owner pays for everything

---

[3] Amp-capacity is determined by method used in Tables 9 and 10.
[4] A disconnect switch is usually considered a part of the exterior system even though it is located inside a building.

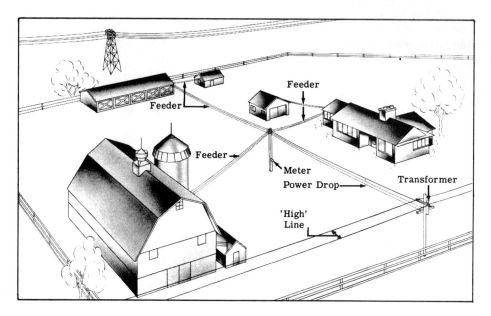

FIGURE 84. A central metering system provides the most economical wiring as well as better service.

beyond the point of attachment of the service drop except the meter socket and meter.)

4. *Disconnecting Means.* Several illustrations in the book show various types of *disconnecting means.* For a master service entrance in pole metering a disconnecting means, fused or fuseless, is usually installed so that all electrical service can be cut off at a single, central location on the farmstead. But the disconnect can also be a combination type, serving as a disconnecting means and fuse- or breaker-panel for wiring inside circuits. The more commonly used arrangement is to have one main disconnect with take-off lugs which supply other disconnects for the several buildings. Figure 97 shows such arrangement —one main and five disconnects for five other feeders.[5] Each of the five buildings would have a fused panel (or breaker) and "disconnect" inside. A similar arrangement can be installed on a meter pole, where pole metering is allowed. (In some areas pole metering is not allowed.) [6]

5. *Feeders.* The run of wiring from the master service entrance to each building is called *feeders;* each must be adequate in size to serve its particular building. In Figure 84, you can see several feeders leading from the meter pole to various buildings. These provide 3-wire, 230v service except two 2-wire, 115v sub-feeders.

---

[5] A disconnect switch is usually considered a part of the exterior system even though it is located inside a building.
[6] Feeders, main supply conductors from the main service entrance, are considered by the National Electrical Code to be the same as a service drop.

EXTERIOR WIRING SYSTEM 83

FIGURE 85. Two common types of wiring for yardpole metering. The service in *A* is wired with service entrance cable; in *B*, the yardpole is equipped with conduit. A small switch *(C)* controls the water pump and is independent of other switches.

84   USING ELECTRICITY

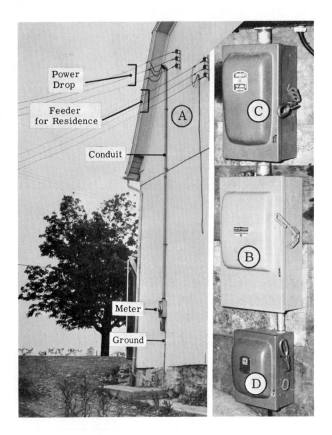

FIGURE 86. Metering at a farm building. Exterior parts of the installation are mounted on the barn wall (A). The three switches at right are mounted inside of the barn. The 200-ampere switch (B) supplies a 100-ampere service for the farm home through switch (C) and a 60-ampere service for the barn and feed mill through switch (D).

Feeders are treated by the N. E. Code as service drops, so must meet Code requirements, such as "at least 10 feet above ground level at point of attachment." (Also see Chapter 8 for information on feeders.) These runs may be of the same style and type as service drops.

6. *Individual Disconnects and/or Service Switches.* At each separate building there must be a standard service entrance and a service switch inside. These main switches are usually of the combination type (cutoff and circuit provisions). See Figure 95. This means that each building will need a disconnecting means at the master service entrance (except in pole metering) plus a service switch inside. Service switches come in standard sizes such as 60-, 100-, 150-ampere, up to 400-ampere and larger.[7]

7. *Special Pump Service.* If a farmstead has an individual water system, it is wise to install a feeder circuit ahead of the main disconnect so that the pump will operate with the main service cut off. Most farm owners prefer an underground feeder for the pump so as to avoid damage from falling objects. See Figure 85 for pump feeder on meter pole.

8. *Fuses and Fusing.* The example we used earlier was for a branch circuit but the exterior system also needs fused protection. In 230v service, two fuses are needed—one for each switch leg. The only way to determine the proper size of these fuses is to total the loads on 115v circuits and loads on 230v circuits, determine which loads are on which switch leg, then fuse each accordingly. Remember that these loads will change as new appliances and other electrical machinery are added. The same principle applies to the fuses for the service switch inside each building; fuses should be changed as loads change.

9. *Grounds and Grounding.* Details on installing grounds can be found in Chapter 8 and Figure 52. At this point

[7] See Fig. 98 for an assortment of fuses and disconnecting means.

it is imperative to emphasize the gravity of faulty grounds and grounding at the master service entrance and at the service entrance at each building. If city-system water pipes are available, the ground wire at each service entrance should be properly attached to these. See Figure 100 (inset at left) for details.

Where there is no suitable water system for grounding, it is necessary to use a driven ground. This should be either a copper-coated steel rod of proper size, or at least ¾-inch galvanized water pipe, both driven to permanent moisture level. (At least 8-feet deep.) See Figure 100 (inset at left) for a ground at a building without a suitable water system.

Some items such as large motors and dishwashers should have a separate ground *in addition* to the main ground at the service entrance.

10. *Poles, Insulators, and Other Auxiliary Equipment.* Serious accidents have been caused on farmsteads by power poles coming down during a storm. It is good practice to purchase poles that are manufactured and/or treated for use in electrical wiring. Naturally such poles should have sufficient height to meet Code requirements—proper clearance of feeders, or drop, to ground level (highest point along the line route). As mentioned before, underground feeders will eliminate poles, insulators, guy lines, et cetera, but an underground type of electrical wire would be needed for direct burial in earth.

FIGURE 87. Location of meter pole should be near the "Future" load center. Note dotted lines for location based on "Present" load.

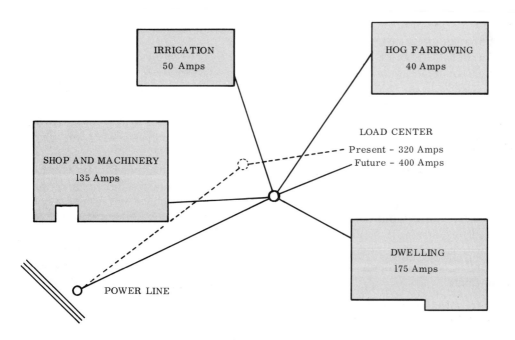

The power supplier or a competent lineman should be hired to assist with the installation of feeders. These experts will advise you as to type and size of poles, insulators, guy lines, et cetera. You will also need advice on installing underground feeders.

Should Pole Metering Be Used?

If you live in an area where pole metering is not allowed, you will not be concerned with the information about this type of installation. In that case you would select the best located building for the master service entrance. Again the power supplier should be consulted.

By studying Figures 84 and 87 you will see some of the advantages of pole metering. Most of all you would expect a less expensive system by using pole metering because some of the feeders can be smaller in size (meter pole can be located nearer load center). Also, you are likely to get more efficient service because of the shorter feeder runs to serve the additions that will come along unexpectedly. (Shorter runs are less likely to limit additions where voltage drop would do so.) Metering from a building usually means longer runs of wiring.

Disadvantages of Pole Metering. The major disadvantages in pole metering can be summed up as follows: The main service entrance is exposed to the weather, making repairs impossible at times. Also, the equipment on a pole is exposed to damage by vandals and is a potential hazard to children at play. Equipment on an outside pole has to be water-tight.

If the master service entrance is located inside a building, repairs can be made during bad weather and is not exposed to "pilfering."

The advantages of pole metering would appear to outweigh the disadvantages, but it may not be allowed by regulations in your area.

WHAT EQUIPMENT AND MATERIALS ARE USED IN EXTERIOR WIRING?

Probably the best way to learn how to plan the wiring for your home or farmstead is to work out actual cases. For your practice two situations are dealt with in detail—(1) a single building and (2) a farmstead that has four buildings to be served by one master service entrance. These cases are assumed to be pole metering but they could be converted to metering inside a building by making a few alterations.

What Assumptions Can You Use?

In both cases we have selected aluminum conductors, 230-volt, 3-wire service, for a 2-percent voltage drop. Thus we will allow a total of 4-percent drop in all the exterior wiring (service drop and feeders) and 1-percent for the interior wiring. We will, therefore, stay within the limits specified by the Code —namely, a total of 5-percent drop for the entire system.

We have specified certain types of wire for different uses in those examples, but your power supplier might suggest other types.[8] The selections and/or specifications are not recommendations.

---

[8] In the past decade many new types of wire have been manufactured for special uses and for different conditions of installation. (See discussion and Glossary, Chapter 7.)

EXTERIOR WIRING SYSTEM

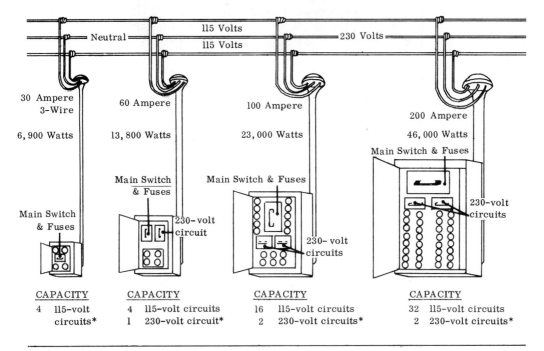

FIGURE 88. Comparison of 30-, 60-, 100-, and 200-ampere capacity service entrances.

How to Use Wire-Size Tables
(See Tables 9A, B, C, D)

The tables on pages 88–91 were taken from the *Farmstead Wiring Handbook;* however, some reference is made also to the National Electrical Code in selecting sizes of ground wire and conduit and in checking wire sizes. Data in Table 12 are used for computing electric motor loads.

Of the four tables included, 9A and 9B are for "Aluminum Conductors, 2-Percent Voltage Drop." Tables 9C and 9D are for Copper, same voltage drop. The Handbook also contains over twenty tables for other situations.

Briefly, the first thing to know when using Tables 9A, B, C, D is your *total ampere load* and locate this in column one of the table being used. Next, you will notice that the main body of each table is divided into two sections by *double vertical lines.* The data left of these lines are for *carrying capacity;* that is, the sizes indicated are safe for a given ampere-load up to 50 feet.

Conditions of Installation. Notice that data left of the double vertical lines are further divided into "In Cable, Conduit, Earth" and other columns refer to installation "Overhead in Air." Naturally, a given size of conductor will carry (safely) more amperage in open air than

TABLE 9A WIRE SIZES FOR CONDUCTORS, 2 PERCENT VOLTAGE DROP *

ALUMINUM CONDUCTORS, 115-120 VOLTS, SINGLE PHASE, 2 PERCENT VOLTAGE DROP

Load in Amperes	Minimum Allowable Size of Conductor			Length of Run in Feet — Compare size shown below with size shown to left of double line. Use the larger size.																			
	In Cable, Conduit, Earth — Types R, T, TW	Types RH, RHW, THW	Overhead in Air — Bare or Covered Conductors	30	40	50	60	75	100	125	150	175	200	225	250	275	300	350	400	450	500	550	600
5	12	12	10	12	12	12	12	12	10	10	8	8	8	8	6	6	6	6	6	4	4	4	4
7	12	12	10	12	12	12	12	10	10	8	8	6	6	6	6	4	4	4	4	4	2	2	1
10	12	12	10	12	12	10	10	8	8	6	6	4	4	4	4	2	2	2	1	1	0	0	0
15	12	12	10	12	10	10	8	8	6	4	4	4	2	2	2	1	0	0	00	00	000	000	00
20	10	12	10	10	8	8	6	6	4	4	2	2	1	0	0	0	00	000	000	000	000	4/0	4/0
25	10	10	10	10	8	8	6	4	4	2	2	1	0	0	00	00	000	4/0	4/0	4/0	4/0	250	250
30	8	8	10	8	6	6	6	4	4	2	1	0	0	00	00	000	000	4/0	4/0	250	250	300	300
35	6	8	10	8	6	6	4	4	2	2	0	00	00	000	000	4/0	4/0	250	250	300	300	350	350
40	6	8	10	6	6	4	4	2	2	1	0	00	000	000	4/0	4/0	250	300	300	300	350	400	400
45	4	6	10	6	4	4	4	2	1	0	00	000	000	4/0	4/0	250	250	300	350	350	400	400	500
50	4	6	8	6	4	4	2	2	0	0	00	000	4/0	4/0	250	250	300	350	350	400	400	500	500
60	2	4	6	4	4	2	2	1	0	00	000	4/0	4/0	250	250	300	300	350	400	500	500	500	600
70	1	2	6	4	4	2	2	0	0	000	4/0	4/0	250	250	300	300	350	400	500	500	600	600	700
80	0	2	6	4	2	2	1	0	00	4/0	4/0	250	300	300	350	400	400	500	600	600	700	700	800
90	0	2	4	4	2	1	0	0	000	4/0	250	300	300	350	400	400	500	600	600	700	750	800	900
100	—	1	4	2	1	1	0	00	000	250	250	300	350	400	400	500	600	600	700	750	800	900	1M
115	00	0	2	2	1	0	0	000	4/0	250	300	350	400	500	500	500	600	700	800	900	1M		
130	000	00	2	1	0	0	00	4/0	250	300	350	400	500	500	600	600	700	800	900	1M			
150	—	000	1	0	00	00	000	4/0	300	350	400	500	500	600	700	700	800	900	1M				
175	4/0	4/0	0	4/0	4/0	4/0	4/0	250	350	400	500	600	600	700	800	800	900	1M					
200	350	250	00	00	000	250	300	400	400	500	600	700	700	800	900	900	1M						

* Conductors in overhead spans must be at least No. 10 for spans up to 50 feet and No. 8 for longer spans, copper (one size larger for aluminum).
Note: Table 9A, B, C, D taken from *Farmstead Wiring Handbook*, Edison Electric Institute, New York, 1965.

TABLE 9B WIRE SIZES FOR CONDUCTORS, 2 PERCENT VOLTAGE DROP*

ALUMINUM CONDUCTORS, 230-240 VOLTS, SINGLE PHASE, 2 PERCENT VOLTAGE DROP

Load in Amperes	Types R, T, TW	Types RH, RHW, THW	Bare or Covered Conductors Single	Bare or Covered Conductors Triplex	50	75	100	125	150	175	200	225	250	275	300	350	400	450	500	550	600	650	700	750	800	900
												Compare size shown below with size shown to left of double line. Use the larger size.														
5	12	12	10		12	12	12	12	12	10	10	10	10	8	8	8	8	6	6	6	6	6	4	4	4	4
7	12	12	10		12	12	12	12	10	10	10	8	8	8	8	6	6	6	6	6	4	4	4	4	2	2
10	12	12	10		12	12	12	10	10	8	8	8	6	6	6	6	4	4	4	4	4	2	2	2	2	0
15	12	12	10		12	10	10	8	8	6	6	6	4	4	4	4	2	2	2	2	1	1	0	0	0	00
20	10	10	10		10	8	8	6	6	6	4	4	4	4	2	2	2	1	1	0	0	00	00	00	00	000
25	10	10	10		10	8	8	6	6	4	4	4	2	2	2	2	1	0	0	00	00	000	000	000	000	4/0
30	8	8	10		8	8	6	6	4	4	4	2	2	2	2	1	0	00	000	000	000	000	000	4/0	4/0	4/0
35	8	8	10		8	6	6	4	4	4	2	2	1	1	1	0	00	000	000	4/0	4/0	4/0	4/0	4/0	4/0	250
40	6	6	10		8	6	4	4	4	2	2	1	1	0	0	00	000	000	4/0	4/0	4/0	250	250	250	250	300
45	6	6	10		8	6	4	4	2	2	2	1	0	0	00	00	000	4/0	4/0	250	250	250	300	300	300	300
50	4	4	8		6	4	4	4	2	2	1	0	0	00	00	000	4/0	4/0	250	300	300	300	300	300	350	350
60	2	4	6	6	6	4	2	2	2	1	0	0	00	000	000	4/0	4/0	250	300	300	350	350	350	400	400	400
70	1	2(a)	6	4	4	4	2	2	1	0	00	00	000	000	4/0	4/0	250	300	350	350	400	400	500	500	500	500
80	0	2(a)	4	4	4	2	2	1	0	00	00	000	000	4/0	4/0	250	300	350	400	400	500	500	500	600	600	600
90	0	2(a)	4	2	4	2	1	0	0	00	000	000	4/0	4/0	250	300	350	400	500	500	500	500	600	600	600	600
100		1(a)	4	2	4	2	1	0	0	00	000	4/0	4/0	250	250	300	400	500	500	500	600	600	600	700	700	700
115	00	0(a)	2	1	2	1	0	00	000	000	4/0	250	300	300	350	400	500	500	600	600	700	700	700	800	800	
130	000	000(a)	2	0	2	0	00	000	000	4/0	250	250	300	350	400	500	600	600	700	700	800	800	800	900	900	1M
150		000(a)	1	0	1	0	00	000	4/0	250	250	300	350	400	500	500	600	700	800	800	900	900	1M	1M	1M	
175	4/0	4/0(a)	0	000	0	00	000	4/0	250	300	300	400	400	500	500	600	700	800	900	900	1M	1M				
200	300	250	00	4/0	00	000	4/0	250	300	350	400	500	500	500	600	700	800	900	1M	1M						
225	400	300			000	4/0	250	300	350	400	500	500	500	600	700	750	900	1M	1M							
250	500	350			000	4/0	300	350	400	500	500	500	600	700	750	900	1M									
275	600	500			4/0	300	350	400	500	500	500	600	700	750	800	900	1M									
300	700	500			250	300	400	500	500	600	600	700	800	800	900	1M										
325	800	600			300	300	400	500	500	600	700	700	800	900	1M											
350	900	700			300	350	500	500	600	700	700	800	900	900	1M											
375	1M	800			350	350	500	600	600	700	750	800	900	1M												
400		900			350	350	500	600	700	750	800	900	1M													

Length of Run in Feet

* Conductors in overhead spans must be at least No. 10 for spans up to 50 feet and No. 8 for spans up to 50 feet and No. 10 for longer spans, copper (one size larger for aluminum).

Note: For three-wire, single-phase service and sub-service circuits, the allowable current-carrying capacity of RH, RH-RW, RHH, RHW, and THW aluminum conductors shall be for sizes No. 2-100 amp, No. 1-110 amp, No. 0-125 amp, No. 00-150 amp, No. 000-170 amp and No. 4/0-200 amp.

TABLE 9C WIRE SIZES FOR CONDUCTORS, 2 PERCENT VOLTAGE DROP *

COPPER CONDUCTORS, 115-120 VOLTS, SINGLE PHASE, 2 PERCENT VOLTAGE DROP

Load in Amperes	Minimum Allowable Size of Conductor			Length of Run in Feet — Compare size shown below with size shown to left of double line. Use the larger size.																		
	In Cable, Conduit, Earth Types R, T, TW	Types RH, RHW, THW	Overhead in Air° Bare or Covered Conductors	30	40	50	75	100	125	150	175	200	225	250	275	300	350	400	450	500	550	600
5	14	14	10	14	14	14	14	12	12	10	10	10	8	8	8	8	8	6	6	6	6	4
7	14	14	10	14	14	14	12	12	10	10	8	8	8	8	6	6	6	6	4	4	4	4
10	14	14	10	14	14	12	12	10	10	8	8	6	6	6	6	4	4	4	3	3	2	2
15	14	14	10	14	12	12	10	8	8	6	6	6	4	4	4	4	3	2	2	1	1	1
20	12	12	10	12	10	10	8	6	6	6	4	4	4	3	3	3	2	1	1	0	0	0
25	10	10	10	10	8	8	8	6	6	4	4	4	3	3	2	2	1	1	0	0	00	00
30	10	10	10	10	8	8	6	6	4	4	3	3	2	2	1	1	0	0	00	00	000	000
35	8	10	10	10	8	8	6	6	4	4	3	2	2	1	1	0	0	00	000	000	000	4/0
40	8	8	10	8	8	6	6	4	4	3	2	1	1	0	0	00	00	000	000	4/0	4/0	4/0
45	6	8	10	8	6	6	6	4	3	2	1	1	0	0	00	00	000	000	4/0	4/0	250	250
50	6	6	10	8	6	6	4	4	2	2	1	0	0	00	00	000	000	4/0	4/0	250	250	300
60	4	6	8	8	6	4	4	2	1	1	0	00	00	000	000	4/0	4/0	250	300	300	300	350
70	4	4	6	6	6	4	3	1	1	0	00	000	000	4/0	4/0	250	250	300	350	350	400	400
80	2	4	6	6	4	4	2	1	0	00	000	000	4/0	4/0	250	300	300	350	400	400	500	500
90	2	3	6	6	4	4	2	0	0	000	000	4/0	4/0	250	300	300	350	400	400	500	500	500
100	1	3	6	6	4	3	2	0	00	000	4/0	4/0	250	250	300	350	400	400	500	500	500	600
115	0	2	4	4	4	2	1	0	000	4/0	4/0	250	250	300	350	350	400	500	500	600	600	700
130	00	1	4	4	2	1	0	00	4/0	4/0	250	250	300	350	400	400	500	600	600	700	700	700
150	00	0	2	2	2	1	0	000	4/0	250	300	300	350	400	500	500	600	700	700	750	750	—
175	4/0	00	2	2	1	0	00	4/0	250	300	350	350	400	500	600	600	700	—	—	—	—	—
200	250	000	1	1	0	00	000	250	300	350	400	500	500	600	700	700	—	—	—	—	—	—

* Conductors in overhead spans must be at least No. 10 for spans up to 50 feet and No. 8 for longer spans.

° Conductors in overhead spans.

TABLE 9D WIRE SIZES FOR CONDUCTORS, 2 PERCENT VOLTAGE DROP *

COPPER CONDUCTORS, 230–240 VOLTS, SINGLE PHASE, 2 PERCENT VOLTAGE DROP

Compare size shown below with size shown to left of double line. Use the larger size.

Load in Amperes	Minimum Allowable Size of Conductor: In Cable, Conduit, Earth — Types R, T, TW	Minimum Allowable Size of Conductor: Types RH, RHW, THW	Minimum Allowable Size of Conductor: Overhead in Air* Bare or Covered Conductors	Length of Run in Feet																				
				50	75	100	125	150	175	200	225	250	275	300	350	400	450	500	550	600	650	700	750	800	900
5	14	14	10	14	14	14	14	14	12	12	12	12	10	10	10	10	8	8	8	8	8	6	6	6	6
7	14	14	10	14	14	14	14	14	12	12	12	10	10	10	10	8	8	8	8	6	6	6	6	6	4
10	14	14	10	14	14	14	12	12	12	12	10	10	10	10	8	8	6	6	6	6	6	4	4	4	4
15	14	14	10	14	14	14	12	12	12	10	10	10	10	8	8	6	6	6	4	4	4	4	4	2	2
20	12	12	10	12	12	10	10	8	8	8	8	6	6	6	6	4	4	4	4	2	2	2	2	1	1
25	10	10	10	10	10	8	8	8	6	6	6	6	4	4	4	4	3	2	2	2	2	1	1	1	0
30	10	10	10	10	10	8	8	6	6	6	4	4	4	4	3	3	2	2	1	1	1	0	0	0	00
35	8	8	10	10	8	8	6	6	4	4	4	4	3	3	2	2	1	1	1	0	0	0	0	00	00
40	8	8	10	10	8	6	6	6	4	4	4	3	2	2	2	1	1	0	0	0	00	00	00	000	000
45	6	8	10	10	8	6	6	4	4	4	3	2	2	1	1	1	0	0	00	00	00	000	000	000	4/0
50	6	6	10	8	8	6	4	4	4	4	2	2	1	1	1	0	0	00	00	000	000	000	4/0	4/0	4/0
60	4	6	8	8	6	6	4	4	3	2	2	2	1	1	0	00	00	000	000	4/0	4/0	4/0	4/0	4/0	250
70	4	4	8	8	6	4	4	3	2	2	1	1	0	0	00	000	000	4/0	4/0	4/0	4/0	250	250	250	300
80	2	4	6	6	6	4	3	2	2	1	1	0	00	00	000	000	4/0	4/0	250	250	300	300	300	300	350
90	2	3	6	6	4	3	2	2	1	1	0	00	00	000	000	4/0	4/0	250	250	300	300	300	350	350	350
100	1	3	6	6	4	2	2	1	1	0	0	00	000	000	4/0	4/0	4/0	250	300	300	350	350	400	400	400
115	0	1	4	4	2	1	1	0	0	00	00	000	000	4/0	4/0	250	250	300	350	350	400	400	500	500	500
130	00	0	4	4	2	1	0	0	00	00	000	000	4/0	4/0	250	300	300	350	400	500	500	500	500	500	600
150	000	00	4	2	1	0	0	00	00	000	000	4/0	250	250	300	350	350	400	500	500	500	600	600	600	600
175	4/0	000	2	1	0	00	00	000	000	4/0	4/0	250	300	300	350	400	500	500	600	600	600	600	700	700	700
200	250	4/0	1	0	00	000	000	4/0	4/0	250	250	300	300	400	400	500	500	600	600	700	700	700	700	750	750
225	300	250	0	00	000	000	4/0	4/0	250	300	300	300	350	400	400	500	500	600	600	700	700	750	750		
250	350	300	00	000	000	4/0	4/0	250	300	300	350	350	400	400	500	500	600	600	700	700	750				
275	400	350	000	000	4/0	4/0	250	300	300	350	350	400	400	500	500	600	600	700	700	750					
300	500	400	000	000	4/0	250	250	300	350	400	400	500	500	500	600	600	700	750	750						
325	600	500	4/0	4/0	250	300	300	350	400	400	500	500	500	600	600	700	700								
350	600	500	4/0	4/0	250	300	350	350	400	500	500	500	600	600	700	700	750								
375	700	600	250	250	300	350	400	400	500	500	500	600	600	700	700	750									
400	750	600	250	250	300	350	400	500	500	500	600	600	600	700	700										

* Conductors in overhead spans must be at least No. 10 for spans up to 50 feet and No. 8 for longer spans.

° Conductors in overhead spans must be at least No. 10 for spans up to 50 feet and No. 8 for longer spans.

92 USING ELECTRICITY

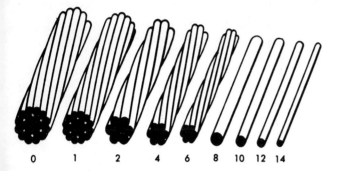

0   1   2   4   6   8   10 12 14

FIGURE 89. *Left:* Actual size of wire, No. 14 to No. 0, without insulation. Wire up to No. 8 is solid; above No. 8, it is stranded.

FIGURE 90. *Bottom:* A 200-ampere service entrance requires 3/0 (copper) SE cable for meter loop. Note "Triplex" style service drop in inset.

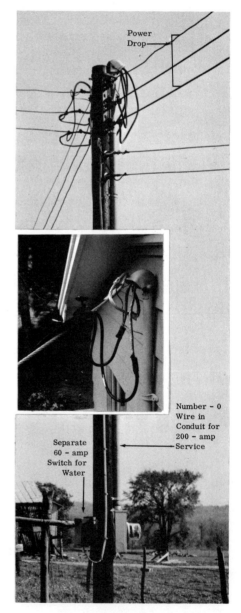

if buried in earth or enclosed in conduit, due to the accumulation of heat in the latter.

Notice also that several different *types* of wire are listed. These are not the only suitable types. In fact, there are dozens of newer types of wire on the market. Example, Type THWN (not listed in Tables 9A, B, C, D) has a relatively small outside diameter, so more wires of this type can be put into a given size conduit than some other types; or smaller size conduit can be used for a given wire size, using Type THWN. (See Glossary for Problem-Unit Two for details on wire types and symbols for each.)

Length of Run. The data to the right of the double vertical lines have been computed for varying distances, specified "Length of Run in Feet." Early in the book we studied the principle of voltage drop—that it *drops in direct relation to distance.* Therefore, if you know your ampere-load, locate this in Column 1, then read across the table to the "Length of Run." (A straight-edge is handy for use here.) For example, for 200-amperes, find 150 feet "Length of Run" (Table 9B). Size "250" is indicated. This refers to circular mils with three zeros omitted for convenience. So,

TABLE 10 QUICK REFERENCE FOR ESTIMATING *ONE* CONNECTED LOAD
(Example: Dairy and Feed Preparation Shed)

Appliance or Other Use	Load in Amperes at 115 volts	Load in Amperes at 230 volts
Lighting Outlets—20 at 1½ amperes	30.0	
Convenience Outlets—12 at 1½ amperes	18.0	
Haydryer, 5 hp (largest) 28 × 1.25 * (See Table 12)		35.0
Barn Cleaner, 3 hp		17.0
Silo Unloader, 5 hp *		28.0
Milk Cooler, 3 hp		17.0
Milker, 1½ hp		10.0
Water Pump, 1 hp		8.0
Water Heater, 40 Gal. (4,500w)		19.1
Milkhouse Heater (1½ kw)		6.9
Ventilating Fan, ⅓ hp	7.2	
	55.2	141.0
Convert all 115-ampere loads to 230 volts: 55.2 amperes at 115 volts to 230 volts = 27.6 amperes		27.6
TOTAL		168.6 amperes **

* Use 125% for largest motor only once in each building.
** Refer to Table 9B for proper size wire and service equipment.

the size 250 actually means 250 (000) cir mils.

Two measures are commonly used in wire sizes: (1) American Wire Gauge, specified AWG. Example, size 000 is written 3/0 AWG. No. 12 is larger than No. 14; No. 10 is larger than No. 12—that is, sizes below No. 1 are specified in *reverse order.* (2) Cir mil sizes are read from smaller to larger numerically—thus 300 (000) cir mils is larger than 250 (000) cir mils.

Example One, Table 10

By examining the data in Table 10, it is apparent that this is a *farm-service*

building, not a dwelling. Therefore, the *computed load* of *168.6 amperes* is taken as the *demand.* The N. E. Code does not provide a formula for computing a *maximum demand* for service buildings as is allowable for dwellings. (See Table 11 for details on maximum demand.)

Notice that 115 volt loads in Table 10 have been converted to 230 volt loads, then the total is stated in *amperes.* The result is 168.6, so our problem is to select wiring and equipment for this figure.

Service Entrance Conductors. We specify Type TW [9] (or THWN) for the SE conductors, conduit installation. Referring to Table 9B, for Aluminum Conductors (2-percent voltage drop) see

[9] Other types can be used, even SE Cable in some areas.

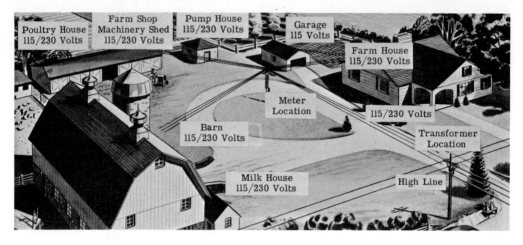

FIGURE 91. The probable maximum load of the entire farmstead takes account of the load that is likely to be on at one time in every building or service on the farmstead.

Column 1 and look for 168.6 amperes. The nearest figure is 175. So, for 175 amperes, Type TW in conduit, we find in Column 2, size 300. Remember that three zeros have been omitted in these tables, so the size is 300,000 cir mils; or 300 MCM as it is usually written.

Disconnecting Means. The nearest standard size disconnect or service switch is 200-ampere. In this case, where only one building is involved, the service switch would likely be a combination type disconnect and fuse panel, or breaker, suitable for serving the circuits indicated in Table 10, as well as for disconnecting the entire service.

Fuses and Fusing. A 200-ampere switch, 3-wire service, would have two "switch legs" to be fused. The only way to determine the exact size for each of the two main fuses is to use a diagram and total the ampere-load on each switch leg. Your power supplier or planning expert should be able to help you compute the sizes of fuses needed. If you had a perfect balance (not likely) you would simply divide the total by two and select the nearest size fuse for each switch leg.

Service Drop. As previously discussed, you may choose from several styles or arrangements of the service conductors as follows: (1) Rack style as shown in Figure 84; (2) twisted style such as Triplex, shown in Figure 90 (inset); (3) vertical arrangement as shown in Figure 84; or (4) underground installation. However, the supplier serving your home or farmstead may not use the underground method, so the latter may be ruled out. You have previously discussed the differences and advantages between the several methods or styles of installation of service drops and feeders (same as service drop). The increasing use of a twisted style (Triplex) service drop is due mostly to easier installation. Underground wiring, likewise, has several advantages already discussed —mostly the avoidance of overhead electrical wires, but this method is the most expensive of the four.

For convenience we specified Triplex and assumed 125 feet of run. See Table 9B and locate Triplex, in Column 5. Notice also the footnote at the bottom of this table. These data apply only to *carrying capacity* of aluminum conductors, 230 volt service, so will *not* affect the problem here because the length of run is more than 50 feet. Next, look for 168.6 amperes in Column 1. The nearest figure is 175 amperes. Reading horizontally across the table, locate the figure under 125 feet; size specified is 3/0. Check this against size in Column 5; it is 3/0. Since the *Handbook* states that the *larger* of the two sizes shall be used (also data at bottom in footnote shows 3/0), the 125-foot service drop should be size 3/0.

Ground. Reference to the N. E. Code indicates that this size service entrance should have No. 6 bare copper wire or No. 4 bare aluminum. See Figures 100 and 165 for details of installation of the ground.

Considerations About Installation. Not only should the service drop and ground be installed by a competent lineman but the feeders also. However, you should be knowledgeable concerning your electrical system so as to be able to work with the lineman and know when a particular job is up to standard. In some states the owner is permitted to do his own wiring including feeder circuits, and some persons do. For those who plan to do wiring, the major wiring skills are covered in Chapter 8. But first, you should study the examples in this chapter.

FIGURE 92. A 200-ampere disconnect switch (or service entrance switch) is needed for the load in the example.

Poles, Insulators, Conduit, and Miscellaneous. The selection and installation of poles and auxiliary equipment are another part of the exterior wiring that is usually handled by the power supplier or lineman, even though the owner usually pays for the conduit and other parts of the service entrance. Make certain that your feeder poles are suitable for life-long service, or install underground feeders.

Conduit. We have not discussed conduit (metal tubing) used for the SE conductors. Old style SE cable (unprotected) is seldom used now. The

FIGURE 93. To find the full-load current required by the 5-hp motor shown, multiply its rated load of 28 amperes by 1.25; thus, $28 \times 1.25 = 35$ amperes. Note: This example assumes that the 5-hp motor is the largest motor on the farm.

power supplier, or planner, should be consulted as to size of conduit needed. The N. E. Code attaches much importance to the maximum number of conductors allowed in a given size of conduit. Where large SE conductors are to be used, the conduit may have to be 4-inches or more in diameter. The type of wire used will affect the maximum number allowable, because the "over-crowding" of conductors in a given size of conduit can create a fire hazard. A weatherhead and other conduit fittings are necessary in conduit installation. It would probably be wise to have the power supplier install the conduit while the meter socket is being installed. See Figure 156 and other illustrations for details.

Underground Feeders. There is a trend toward the use of more underground wiring. Wire size would be selected by using Tables 9A, B, C, D for wiring "in conduit or buried in earth." See Chapter 8 for further details regarding the installation of underground circuits. Use Type USE, UF, or other "underground type."

Example Two, Table 11

At the beginning of this chapter we stressed the need for adequacy, efficiency, expandability, and economy in the exterior wiring system. But we also pointed out the fact that the exterior wiring frequently acts as a bottleneck for the entire electrical supply for a large farm. Therefore, in our example we will consider both *present* and *future* needs in working out the plan for Table 11. Hereafter, we will use the symbols "A-load" for the present, and "B-load" for the future.

Assumptions. In order to complete the plan, it is necessary to assume some distances. The owner could provide exact measurements as shown below:

	Length of Run (feet)	"A-load" (amperes)	"B-load" (amperes)
Service Drop to Meter Pole	125	320*	400*
Pole to Dwelling	150	101	175
Pole to Shop and Machinery Shed	200	132	135
Pole to Hog House	350	36	40
Pole to Irrigation Pump	300	50	50
Total (approximately)	—	320*	400*

* You will notice an allowance of 80 amperes for expansion. Most of this will be prorated for the dwelling. Totals have been "rounded" off to 320 amperes and 400 amperes.

What Equipment and Materials Are Used in Exterior Wiring?

We may be criticized for *over-wiring* in our plan for the farmstead in Table 11. But the rapid expansion of electrification, shortage of labor, and the types of loads involved will, in our situation, justify the use of *100-percent computed loads* as the "demand."

On the other hand, the committee that is responsible for producing the *Farmstead Wiring Handbook* recognizes a maximum-demand formula where several different loads are involved in the main service, one or more "with diversity." This method will allow somewhat smaller conductors for the service drop and SE conductors than the computation shown at the end of Table 11. That formula is as follows:

1. For the largest computed demand, 100 percent, including 1.25 percent of the largest motor
2. For the second largest demand, 75 percent
3. Third largest computed demand, 65 percent
4. All remaining loads, 50 percent of the total

For the farmstead in Table 11, this formula would yield the following:

	Amperes
1. One hundred percent of Farm Shop and Machinery Shed (131.2 amperes)	131.2
2. Seventy-five percent of Dwelling (101 amperes)	76.0
3. Sixty-five percent of Irrigation (50 amperes)	32.5
4. Fifty percent of all other loads (36.1 amperes)	18.0
Total	257.7

In like manner we could take the "B-load" (future needs) data and determine the minimum size of service drop and SE conductors for the master service entrance, using the preceding formula.

TABLE 11 COMPUTED LOADS AT FOUR FARMSTEAD LOCATIONS

Building or Use	Specifications	Connected Load (Watts)	Probable Demand (Watts)
	DWELLING (36' × 50')		
LIGHTING: General, Yard	3w per sq ft, 1,800 @ 3w 3 yard-poles, 2 150w Par-38 per pole (total, 6 @ 150w)	5,400 900	Code specifies several different methods of computing probable demands for dwellings.*

CIRCUITS:
General Purpose
  Kitchen
  Laundry
  Outdoor
	
	2 20-amp circuits for kitchen
	1 20-amp spare circuit
	1 20-amp circuit for laundry
	1 20-amp circuit for general outdoor use (total, 5 circuits @ 1,500w)
		
		7,500	

RANGE	1 range, 230v sp.p. circuit @ 12,000w	12,000	
WATER HEATER	1 double-element type, 230v sp.p. circuit @ 4,500w	4,500	
DISHWASHER	1 separate 230v-circuit, ¾ hp motor @ 6.9 amps	1,587	@ 125% 1,984
SPACE HEATERS	4 space heaters, one per circuit @ 115v, 1,600w @ 100%	6,400	
CLOTHES DRYER	1 230v separate circuit	4,500	
WINDOW AIR-CONDITIONERS	2 ½ hp units on sp.p. 230v circuits (not counted since total watts is less than space heating load)	xxx	
		42,787	43,184

* COMPUTATION: (One Code method)
1st 10,000 @ 100% = 10,000
Remainder (33,184) @ 40% = 13,274

Total Demand   23,274

$\dfrac{23,274}{230v}$ = Approximately 101 amps

TOTAL COMPUTED DEMAND, 101 amperes

FARM SHOP, 30′ × 30′, and MACHINERY SHED, 30′ × 40′

LIGHTING	1. General shop, 1 150w ref. lamp per 100 sq ft (Total, 9 × 150w)		1,350	The Code specification for farm-service buildings is taken as 100 percent, except for motors. (See final computation *)
	2. General machine shed, 1 100w ref. lamp per 150 sq ft (Total, 8 × 100w)		800	
	3. Special work, 2 150w ref. lamps per 10 ft bench, 3 benches in shop, 1 bench in machine shed (Total, 4 × 2 = 8 × 150w)		1,200	Note: In actual practice the power supplier may classify this as a "diversity" load and reduce load by a special formula, when several separate buildings are involved.
	4. Doorways, 1 100w ref. lamp @ each entrance (Total, 2 doors = 2 × 100w)		200	
	Total Lighting		3,550	
CONVENIENCE OUTLETS FOR GENERAL PURPOSE: Shop	2 115v outlets per work station (3 stations) = 6. 1 115v outlet per working wall (4 walls) = 4	10		
	Sub Total			
Machinery Shed	6 115v outlets for general use = 6 2 115v outlets per work station (1 work station) = 2	8		
	Sub Total			
	Total 115v Outlets	18		
	18 115v outlets (5 20-amp circuits @ 1,500w) Total Outlets		7,500	
OTHER LOADS: (Motors and Motor Outlets)	Largest motor, 3hp compressor, 230v @ 17 amps Drill, ½hp @ 230v Jointer, ½hp @ 230v Portable Saw, ½hp @ 230v Total "Other Loads"	1,127 1,127 1,127	3,910 3,381	@ 125% 4,888
ARC WELDER	1 50-amp, 230v circuit, 37.5 "limited input" type, 37.5 × 230v		8,625	
HEAT LAMP	2 300w infra-red type @ each work station (3 in shop, 1 in machine shed) = 4 4 sta. × 2 = 8 × 300w		2,400	
	TOTALS		29,366	30,344

* COMPUTATION: $\dfrac{30{,}344}{230\text{v}}$ = 131.9 amps

TOTAL COMPUTED DEMAND 131.9 amperes

* Data computed using *Farmstead Wiring Handbook*, National Electrical Code, and Table 12.

TABLE 11 (Continued) COMPUTED LOADS AT FOUR FARMSTEAD LOCATIONS

Building or Use	Specifications	Connected Load (Watts)	Probable Demand (Watts)
HOG FARROWING HOUSE			
(12 sows, 2 rows of pens, center walk)			
LIGHTING	1 100w lamp for each 12–15 ft located over pen lines (2 lines, 6 lamps)		
	2 over center walk, 1 at front door (Total 9 lamps @ 100w)	900	
HEATER CABLE	2 runs 1,750w heater cable, 1 run each side	3,500	
VENTILATOR FAN	½ hp fan, sp.p. circuit, 230v	1,127 @ 125%	1,409
WATER HEATER	1 230v circuit (2,500w size)	2,500	
	Totals	8,027	8,309

* COMPUTATION: $\frac{8,309}{230v} = 36.1$ amps

TOTAL COMPUTED DEMAND 36.1 amperes

IRRIGATION SYSTEM
(One 7½ hp, 230v single-phase motor)

MOTOR FOR PUMP	Full load, 7½ hp @ 230v @ 40 amps	9,200 @ 125%	11,500
	Totals	9,200	11,500

* COMPUTATION: $\frac{11,500}{230v} = 50$ amps

TOTAL COMPUTED DEMAND 50 amperes

FINAL COMPUTATION:

Using the total demand load at each of the four buildings, we can now determine the total demand load for the entire farmstead:

1. Dwelling   101.0
2. Shop and Machinery Shed   131.9
3. Hog Farrowing House   36.1
4. Irrigation   50.0

                                   319.0 amperes

This would be "rounded off" to 320 for easier computation.

* Data computed using *Farmstead Wiring Handbook*, National Electrical Code, and Table 12.

Which "Demand Load" Should Be Used?

Two different "total computed loads" were derived by the computation at the end of Table 10—(1) 320 amperes (100 percent of all computed demands for present needs), and (2) 400 amperes for the future demand. A third "total computed load" (258 amperes) was found by using the committee formula as an alternate method specified in the *Farmstead Wiring Handbook*. For our example, we have used the "B-load" figure, *400 amperes,* for reasons already discussed, mostly for expansion in the future. The committee formula would affect only the master service entrance, since each separate building (load) must be wired for a 100-percent load basis, except in buildings (loads) that operate with diversity. (By diversity is meant that *heating* and *cooling* will *not* be on *at the same time.* See *Farmstead Wiring Handbook* for details.)

After computing and selecting the master service conductors and equipment, the data for other buildings will be handled in a condensed manner, since you have already gone through the steps involved in the building in Table 10.

1. Service Entrance Conductors and Disconnecting Means. Again, we have used 2-percent, Aluminum Tables for 230 volt service. Refer to Table 9B and look for 400 amperes, using Type TW or THWN "In Conduit." *No size* for 400 amperes [10] is listed for *aluminum,* so we go to Table 9D, "2-percent, Copper Conductors," "In Conduit." The size for

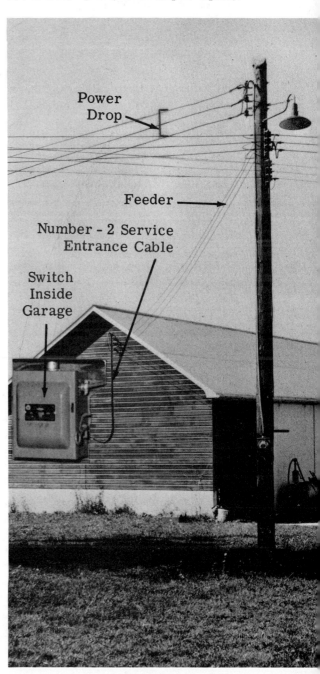

FIGURE 94. No. 2 service entrance cable is the size required for the garage-feed mill here. The service switch (inset) is 100-ampere capacity.

---

[10] The N. E. Code shows 1,250 (000) cir mils, which appears too large to be practicable for our purpose.

400 amperes is 750 (actually 750,000 cir mils) sometimes stated as "750 MCM."

The main disconnect switch would obviously need to be 400-ampere capacity and may be fused or fuseless. It could also be a combination type disconnect and fuse panel or breaker-panel.

2. Service Drop. Refer again to Table 9B and look for 400 amperes, 125-feet run. The size indicated is 400 (000) cir mils. Now check this size against the carrying capacity (data left of double vertical lines) and find the size for single conductors "Overhead in Air"; 350 (000) cir mils is indicated. Since we must use the larger of the two sizes, 400 (000) cir mils is specified. Types such as RH, RHW, THW, and others could be used for the service drop. Triplex is not available in this large size, but large size conductors can be twisted and installed in "Triplexed" fashion.

3. Ground and Conduit. The N. E. Code specifies 2/0 copper or 4/0 aluminum for the ground wire for 750 (000) cir mils, and specifies 3½-inch conduit for Type TW, copper conductors.

4. Power Poles, Auxiliary Equipment. We have already stated that the power supplier is responsible for selecting and installing line poles, guy lines, insulators, meter base, et cetera, for the service drop only. The owner generally pays for the conduit and fittings, the service entrance conductors, and ground wire.

This gets us to the main switch and take-off lugs for other disconnects. So we are ready to compute and to select wiring materials and equipment to serve the four loads detailed in Table 11. We have grouped wire sizes, types, and other items for all four buildings (loads) together as we proceed.

For the most part we have used Table 9B, "2-Percent Voltage Drop for 230v, 3-wire, Aluminum." We have arbitrarily used Type TW for SE conductors and Triplex for the feeders (same as service drop in the National Electrical Code). Service switches have been selected on the basis of equal capacity or the next standard size above.

1. Select SE Conductors, Type TW, or Triplex:

	Wire-size
For the Dwelling (175 amperes)	300 (000) cir mils
For the Shop and Machinery Shed (135 amperes)	No. 3/0
For the Hog House (40 amperes)	No. 6
For the Irrigation Pump (50 amperes)	No. 4

2. Service Switches, Nearest Standard Size:

For the Dwelling—200-ampere main
For the Shop and Machinery Shed—150-ampere main
For the Hog House—60-ampere main
For the Irrigation Pump—60-ampere main

3. Feeders as Indicated by "Length of Run": (Use Triplex).

	Wire-size
Pole to Dwelling (175 amperes—150-ft)	No. 4/0
Pole to Shop (135 amperes—200-ft)	No. 4/0
Pole to Hog House (40 amperes—350-ft)	No. 1/0
Pole to Irrigation Pump (50 amperes—300-ft)	No. 2/0

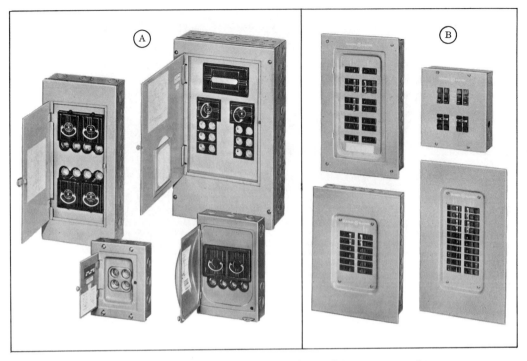

FIGURE 95. The four boxes at left (A) are fuse-type distribution panels. A circuit can be broken by removing the plug or block fuses. The four boxes at right (B) are circuit breakers. An automatic trip breaks the circuit if it becomes overloaded.

4. **Ground Wires:** The N. E. Code indicates the following sizes of grounds for loads as specified. Use bare copper:
1. For Dwelling—No. 2
2. For Shop and Machinery Shed—No. 2
3. For Hog House—No. 8
4. Irrigation Pump—No. 8

5. **Power Poles and Other Auxiliary Equipment:** Follow recommendations mentioned previously, making certain that poles are of adequate height and proper type. Some instructions concerning installation of feeders can be found in Chapter 8. For example, "clearance to highest point along feeder route must be at least 10 feet," "12 feet over driveways," et cetera.

As previously stated, you will probably want to have a competent lineman and/or electrician install your feeders and service entrances. Make certain that the ground wires and grounds are up to standard (Code). Also, the Code is very strict concerning the size of conduit relative to the size and number of conductors allowed. A newer type of wire, Type THWN, is smaller in diameter; so it is being used more for SE conductors, especially in the larger service entrances where conduit size is a problem.

6. **Fuses and Fusing.** You have previously studied the problems involved in getting *both* hot legs of the main switch properly fused. There will likely be some

FIGURE 96. Breaker units are available in ampere sizes comparable to fuses.

difference called "unbalanced load." The power supplier, or planner, should recommend fuse size for both switch legs and should check again when new equipment is needed.

Having computed the various wire sizes and having selected types of wire and other wiring materials for the master service entrance for four farm service buildings, it is desirable to make up one bill of materials for the exterior system. Since the power supplier will be involved, you should have this list of materials checked by the power supplier and ascertain what part of it they will pay for and install. Of course, you could use either of the other two "computed loads" or "computed demands" and select wiring materials for that load in the same manner as we have done, using the 400-ampere load as the ultimate-need load.

Why Are Some Wiring Practices Debatable?

In the preceding examples, we referred to such terms of "Type TW," "Triplex," "maximum demand" versus "100-percent computed load," et cetera. We also specified aluminum conductors in most cases, using Triplex and/or Types RHW, THW. These selections and specifications are *not intended as recommendations*.

The manufacturers of wiring materials and auxiliary equipment are constantly producing new and better materials, so authorities and organizations that publish recommendations are constantly having to revise their specifications as dictated by research and new materials. For example, the National Electrical Code has a long-term schedule for publishing new editions. Also, electrical contractors and local regulations affect actual wiring practices. And still another complicating factor is that the newest *Farmstead Wiring Handbook* cannot always be current with the latest edition of the N. E. Code.

Taking all these things into consideration, it is impossible to "satisfy" everybody. Even power supplier experts disagree on some aspects of exterior wiring. Hence, we have inserted brief explanatory notes, in various forms, throughout the chapter, wherever it appeared that confusion might arise.

One example of a so-called new problem is whether or not a smaller size neutral may be used in service drops and/or feeders. The answer has several ramifications: First, if Triplex, or any twisted-type span, is used, the Code specifies that the neutral shall have carrying capacity equal to the hot conductors. But if the conductors are spaced 12- or more inches apart, the neutral *may* be *one* or *two sizes smaller;* depending on the diversity of loads and the "unbalance" between the hot conductors; depending also on whether the conductors are considered by the Code to be *insulated.*

Concerning the use of copper versus aluminum, the major differences are that aluminum costs less and is lighter in weight, even though one size larger than copper is usually necessary for a given load. Local regulations may prohibit the use of aluminum for certain uses in particular locations.

Finally, the question of diversity of operation is one that can only be determined by studying the various uses in a building, then deciding whether certain appliances or equipment would or would not be in operation at the same time. It is only in the case of loads that operate with diversity that a maximum demand formula may be applied, as in "Alternate Method of Computing Maximium Demand." As previously stated, we used the formula of 100-percent of all computed loads for the total demand. Reasons for using this larger service have been discussed earlier in the chapter.

HOW CAN PROPER LOCATION OF METER BE DETERMINED?

The location of the meter for the farmstead building and demand loads, shown in Table 11, will be affected by several factors, such as the terrain of the farmstead, location of the main high lines in relation to the large loads on the farmstead, and other factors. The power supplier should be consulted before plans are finalized for locating the meter. We *assumed* 125 feet for the service drop and indicated the distance from the meter to each building to be served. In actual practice, these distances would be different, since the figures used are estimates.

The "x-y" Line Method [11]

One method of getting a close approximation of the load center (pole location) is as follows:

1. Lay off the farmstead, or home grounds, on a square or rectangular area

[11] This method was taken from the *Farmstead Wiring Handbook.*

to scale *in feet*. Designate the vertical line at left as x and the base line as y.

2. Multiply the ampere-load in each building by the shortest distance to x-line. Continue this procedure for each building to be served. Total the several results (in our case, four) to obtain a grand total.

3. Divide the *grand total* by *total amperes* (in our case, 400 for future demand). We would then have grand total $\div$ 400 = feet from x-line to meter pole.

4. Repeat steps 1, 2, and 3 for y-line, and again divide by 400 amperes. The result would be the distance of meter pole from y-line.

5. Using the two figures as an arc, the load center would be easy to locate.

If Meter Must Be in a Building

If the meter is to be located in a building, it should probably be placed in the building with the largest load, other things being considered. Other factors include the location of the high line; distance from the metering building to other buildings, especially if the latter are clustered or across the highway; terrain of farmstead; and other things. The power supplier usually has a voice in the location of the meter of the farm-

FIGURE 97. Main service switch (inside building) and five other "feeder" switches. The N.E. Code specifies "no more than six (total) disconnecting means for one service."

stead, so they should be consulted before plans are completed. Every case would be different, of course. (NOTE: Ordinarily metering from a building will result in somewhat longer runs of feeder circuit.)

SUMMARY

This chapter deals with the exterior distribution system, often the bottleneck in the electrical service for the home and farmstead. Many electrical suppliers employ expert planners to assist home and farmstead owners to plan their electrical needs for five to even ten years in advance.

Things stressed in planning the exterior system are safety, economy, efficiency, expandability, and convenience. Other important considerations are nominal first cost of the system, ease of making repairs, and ease of making additions.

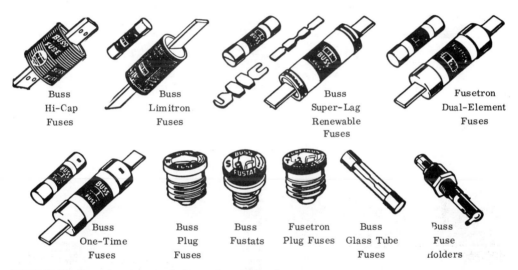

FIGURE 98. *Top:* Assortment of plug and cartridge fuses.

FIGURE 99. *Bottom:* To replace a blown link in a renewable cartridge fuse, unscrew the end caps, insert a new link, and replace the caps. A new link must have correct amperage capacity.

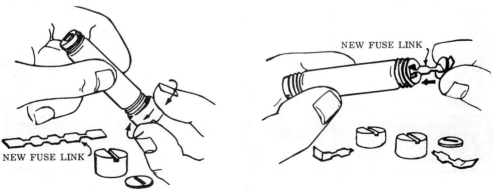

108 USING ELECTRICITY

A case of "fuse blowing" trouble frequently leads to the necessity of re-wiring the entire electrical system in order to obtain satisfactory service. This is the case when the fuse panel, or breaker, is already loaded and new appliances or machinery are to be added.

Major parts of the exterior system fall into two categories—namely, parts furnished and serviced by the power supplier and parts paid for by the owner. Included in the first category are the high line, transformer, service drop, and meter (socket and meter). The owner

FIGURE 100. Pole metering affords shorter runs of feeders. Note method of attachment to building and standard grounding (insets).

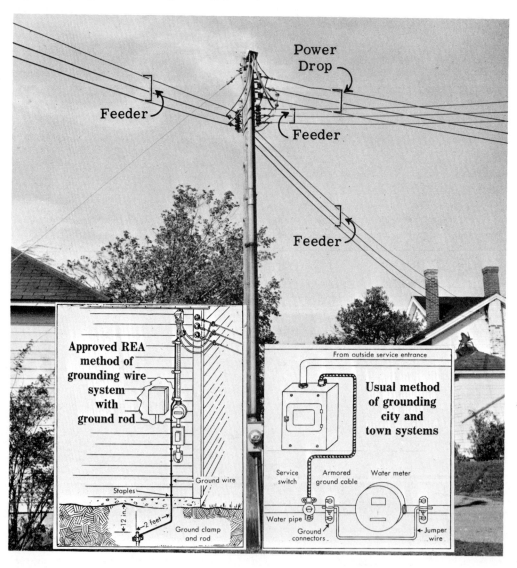

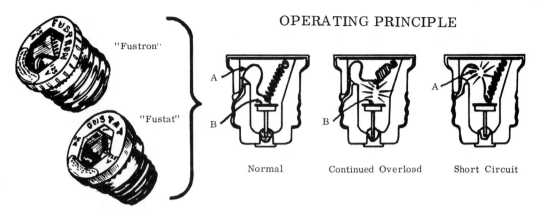

FIGURE 101. The Fusetron at upper left is designed for electric motors. The Fustat at lower left is said to be tamper proof because it will fit only a particular "sized" adapter.

pays for the service entrance (except the meter) including the weatherhead, all conduit and fittings, service entrance conductors (or SE cable), ground wire and accessories, and the "disconnecting means." The latter comes in varied forms —fused and fuseless disconnects, as well as combination types of disconnects and fuse panels or breaker panels. The usual and perhaps wisest practice is to hire the power supplier or a competent lineman to install the service entrance while the meter is being installed. The owner also pays for feeders and auxiliary equipment needed in feeder circuits.

Pole metering is rather widespread throughout the nation, especially on large farmsteads where several buildings are to be served. Advantages include lower first cost of the exterior distribution system, greater efficiency due to the usual shorter feeder runs, and convenience in cutting off the entire service at a central location. However, pole metering is not permitted in some areas (local regulations).

Wire-size tables and instructions for using them are included, the major authorities being the *Farmstead Wiring Handbook* and the National Electrical Code. Data for both "carrying capacity" and "length of runs" are included and explained.

The method used to plan the exterior system is two actual cases—(1) a single building and (2) a farmstead that has four buildings to be served from one master service entrance.

Major parts for the exterior system in both cases are selected, using wire-size table and the N. E. Code. The selection of certain types and brands of wiring materials is *not* to be regarded as *recommendations*. Several different types and styles of conductors would be equally suitable.

Three different "load" figures are illustrated for the farmstead service— the first two taking each "computed load" as a "demand"; the other obtained by using a formula in the *Farmstead Wiring Handbook*. The selection of materials in the example is based on the largest (future) demand, each building wired for 100-percent of computed load except the dwelling, which is computed by a special National Electrical Code formula.

Special attention is given to proper grounding because of the potential danger of faulty materials and workmanship.

Some areas of disagreement and related problems concerning exterior

FIGURE 102. A Fusetron installed in a water-pump circuit allows heavy overload for starting but only for a few seconds. Fuse blows if overload continues.

TABLE 12 FULL-LOAD CURRENTS, IN AMPERES FOR ELECTRIC MOTORS *

HP	115v	230v
⅙	4.4	2.2
¼	5.8	2.9
⅓	7.2	3.6
½	9.8	4.9
¾	13.8	6.9
1	16	8
1½	20	10
2	24	12
3	34	17
5	56	28
7½	80	40
10	100	50

* Taken from National Electrical Code, 1968.

wiring are discussed. The reason for this appears to be the rapid change in the electrical manufacturing industry, change of the National Code, and changes in local wiring practices. Formulas and practices that are acceptable today will no doubt become obsolete in a few years. Even the authorities in the electrical industry disagree quite frequently on exterior wiring.

Electric motors require special protection and draw four to six times the running amperage while starting. Table 12 shows motor amperage data.

QUESTIONS

1. How can you show that the wiring in a pole metering system is usually less expensive than metering from a building? More efficient?
2. In counting the total load in a farm building, why must the amperage of the largest electric motor be multiplied by 1.25?
3. Why is only 50 percent of the total amperage of a range counted in totaling the load at the residence?
4. Why will "enclosed" wire carry less amperage than wire in "open air"?
5. Why is it wise to choose the next standard size above that indicated for the entrance service for a farmstead?

ADDITIONAL READINGS

Bredahl, A. Carl, *Home Wiring Manual.* McGraw-Hill Book Co., New York, N.Y., 1957.

Brown, R.H., *Farm Electrification.* McGraw-Hill Book Co., New York, N.Y., 1956.

Davis, Hollis R., *Adequate Farm Wiring Systems,* Extension Bulletin 849. Cornell University, Ithaca, N.Y., 1956.

Edison Electric Institute, *Residential Wiring Handbook.* New York, N.Y. (Being revised).

Industry Committee on Interior Wiring Design, *Farmstead Wiring Handbook.* New York, N.Y., 1965.

Iowa Southern Utilities Company, *Farm Wiring Systems.* Centerville, Iowa (no date).

Mix, Floyd and Pritchard, E.C., *All About House Wiring.* Goodheart-Willcox Company, Homewood, Ill., 1959.

Montgomery Ward and Co., *Modern Wiring.* Chicago, Ill., 1955 (Being revised).

National Fire Protection Association, *National Electrical Code, 1968 Edition.* Boston, Massachusetts, 1968.

Richter, H.P., *Practical Electric Wiring,* Seventh Edition. McGraw-Hill Book Company, New York, N.Y., 1967.

Richter, H.P., *Wiring Simplified.* Park Publishing, Inc., Minneapolis, Minn., 1965.

Sears, Roebuck and Company, *Simplified Electric Wiring Handbook.* Local, 1964.

6 BRANCH CIRCUITS AND OUTLETS FOR THE RESIDENCE AND FARM SERVICE BUILDINGS

After planning the feeder circuit and service entrance for each farmstead building, the next step is to plan the *branch circuits* and *outlets* for each. The purpose of a branch circuit is to carry the proper amount of electric energy from a source, usually a fused switch or breaker, to the outlets where it is to be used. A branch circuit, then, is the final connecting link in the distribution system.

One farmer made the mistake of using No. 12 wire for a 250-foot branch circuit to operate his 1,750w pig brooder. He thought the two-wire, 115-volt circuit would provide 115 volts times 20 amperes, or 2,300 watts. However, he forgot to figure the voltage drop; so when the brooder failed to operate properly, ten pigs died from chilling. A test showed that the voltage at the brooder was 100, not 115; the amperage, at 100 volts, had dropped from 20 to 15 amperes. This was 100 volts $\times$ 15 amperes $=$ 1,500 watts, not 1,750 watts as the brooder required. This amounted to a loss of 800 watts between the fused switch and the brooder. Besides paying for that loss, the farmer lost his profit on the litter. The purchase of the No. 12 wire proved to be very uneconomical. There was only a difference of $10 between the cost of No. 12 and No. 6 electrical wires. (No. 8 is the smallest overhead span allowable according to the National Electrical Code.)

Before you plan your branch circuits, it will pay you to study your present and future needs for electric power. Also, very carefully study the most common troubles and faults that are usually associated with branch circuits as it is summarized in the table which follows on page 114.

TABLE 13 SUMMARY OF COMMON FAULTS AND RESULTS IN BRANCH CIRCUITS

Circuit Fault	Usual Result
Too few circuits	Overloading, poor service
Use of undersize wire	Loss of power; voltage drop; high cost of energy used
Use of 115-volt circuits where 230 volts are required	Loss of power; poor service; reduced life of equipment
Circuits too long	Voltage drop; high cost of wire; high cost of energy used; reduced life of equipment
Too few outlets	Inconvenience; restricted use of electrical equipment; overloading
Too few switches and other controls	Inconvenience; hazard to life and property
Improper controls	Hazard to life and property
Improper junction boxes and connections	Hazard to property; poor service
Miscellaneous unsafe wiring practices	Hazard to life and property; poor electrical service; reduced life of equipment; restricted use of equipment

WHAT RECOMMENDATIONS SHOULD YOU FOLLOW IN PLANNING ALL BRANCH CIRCUITS?

You can profit from the mistakes of others by applying the following recommendations:

1. Plan for a sufficient number of circuits and outlets for each building. Figure 104 shows a layout of a 200-ampere breaker with twenty-two branch circuits for a three-bedroom residence. A little later you will find a discussion of the types of circuits shown.

2. A common rule of thumb is to limit the number of lighting outlets on one circuit to eight. Five or six convenience outlets constitute a load for one branch circuit. Even then, the total load on a two-wire circuit of No. 12 wire should not exceed 1,840 watts. If you already have some circuits wired with No. 14 wire, the total load on any two-wire circuit should not exceed 1,400 watts. Remember, however, that No. 14 wire is *not* considered adequate for branch circuits in a residence. Always insist on No. 12 gauge or larger, as the load may require. (No. 14 is allowable for a small dwelling.)

3. With few exceptions, appliances rated at 1,200 watts or more and all electric motors ½ hp and over should be permanently connected. The exceptions are heavy-duty portable motors, room heaters, or other appliances that are used in several locations. See that the wiring for these appliances is adequate. For example, a 1,500 watt room heater ordinarily will operate on a general-purpose circuit, but should have a separate appliance circuit.

4. Your range, food freezer, water heater, and other high-wattage equipment should be wired for 230-volt (three-wire) service. Put your food

BRANCH CIRCUITS AND OUTLETS

FIGURE 103. The proper size of wire for a heat lamp insures correct operation.

freezer and brooders on individual circuits to reduce the danger of interruption of the current.

5. For high-wattage equipment requiring a circuit longer than 50 feet, select the wire size by reference to Tables 9A, B, C, D. No. 12 wire is usually adequate for most household appliances if the run is not over 50 feet. Your range, dryer, and water heater, however, will need a wire size up to No. 6.

6. Insist on standard wiring supplies and good wiring practices. For example, make certain that every splice is properly joined and is protected by junction boxes. Wiring supplies should carry the UL label. Also insist on having plenty of wall switches (two- and three-way, as needed) and plan for an adequate number of convenience outlets. Remember that it is less expensive to install adequate wiring at the time your wiring job is completed than it is to rewire later.

7. Determine the proper type and size of protection devices for each circuit and follow the strict rule of never over-fusing.

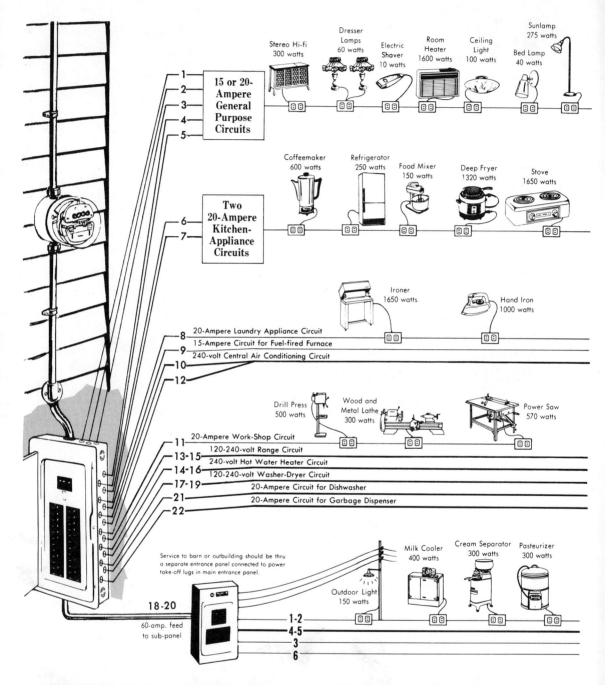

FIGURE 104. Layout of circuit arrangement for a three-bedroom residence. The diagram here shows twenty-two branch circuits grouped according to uses. Breaker panel is 200-ampere capacity. Circuits are "sized" according to circuit loads and fused the same.

BRANCH CIRCUITS AND OUTLETS 117

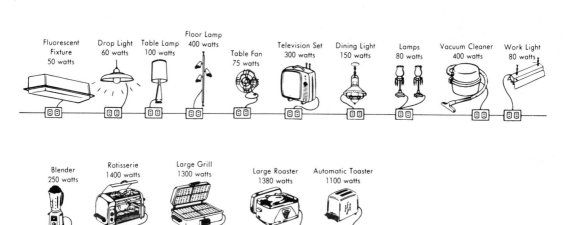
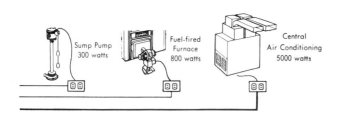
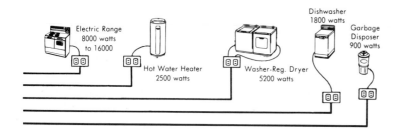
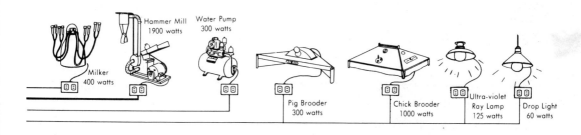

FIGURE 105. *Top:* This ¾-hp motor is wired in, not plugged in.

FIGURE 106. *Bottom:* Three-prong plug and 50-ampere receptacle for a 230-volt range circuit. Cable for this circuit should be at least No. 6, three-wire style.

WHAT TYPES OF BRANCH CIRCUITS WILL YOU NEED?

Branch circuits are classified according to (1) voltage rating, and (2) use or purpose.

How to Plan Your Branch Circuits to Meet Voltage Requirements

The voltage of a two-wire circuit may vary from 110 to 120 volts, and a three-wire circuit may vary from 220 to 240 volts. To simplify examples, however, this book uses 115 volts for two-wire circuits and 230 volts for three-wire circuits. This voltage is considered standard at the point of entry at the buildings.

Almost every home wiring system requires both 115- and 230-volt circuits. Two-wire circuits are adequate for lighting and low-wattage appliances. Three-wire service may be needed, however, for all single loads of 1,200 watts and higher. Also, three-wire, 230-volt service is needed for motors rated at ½ hp or more.

Protection Should Match the Type of Circuit. A two-wire circuit should be protected with a 15- or 20-ampere fuse if the wire size is No. 12. If you already have some No. 14 wire in your buildings, fuse these circuits with 10- to 15-ampere fuses. A range usually requires a 50-ampere fuse. Other equipment should be fused according to the amperage rating. Figure 109 shows common types of fuses for circuits and appliances for the home and farmstead.

Electric motors will draw from two to six times their normal amperage while starting; therefore, circuits that serve electric motors should be protected

FIGURE 107. *Top:* A Plugmold wiring system in the kitchen *(A)* provides plenty of plug-ins, one every 18 inches in the metal channel. A three-way switch at the head and foot of the stairs *(B)* reduces the falling hazard; the UL label *(C)* is a guarantee of safe wiring materials.

FIGURE 108. *Bottom:* One hot wire and a neutral = 115 volts; two hot wires with or without a neutral = 230 volts.

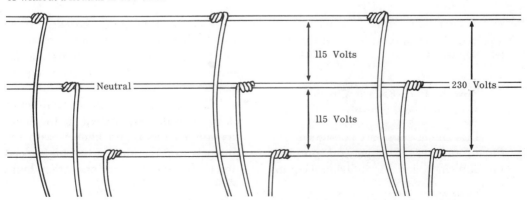

120 USING ELECTRICITY

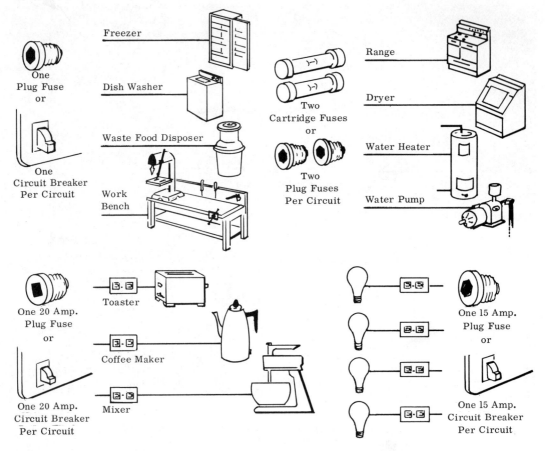

FIGURE 109. Sample of fuses and circuit breakers for the residence and farm-service buildings.

with a *time-delay fuse*. (See Chapter 5, Table 12, for data on protection for electric motors.)

How to Plan Branch Circuits According to Purpose

Branch circuits are designed for three purposes: (1) general purpose, (2) appliance, and (3) individual equipment.

1. *General-purpose circuits* serve outlets for lighting, vacuum cleaners, small fans, radios, table lamps, clocks, and other portable household equipment.

2. *Appliance circuits* operate most of the portable equipment in the laundry, dining room, and kitchen with the exception of the range, the high-speed dryer, the food freezer, and other high-wattage equipment.

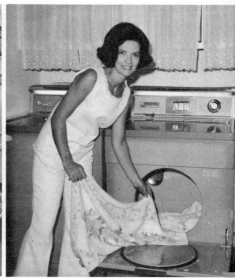

FIGURE 110. *Left:* This combination refrigerator-freezer is usually operated on a general-purpose circuit of No. 12-2 grounded wiring, but the trend is toward the use of a separate, individual circuit, as newer models get larger. Freezer holds 250 pounds; circuit fused at 20 amperes.

FIGURE 111. *Right:* High-speed dryer should have a special-purpose, or individual circuit, fused according to amperage rating of dryer. Washer at left is plugged into grounded-type laundry circuit.

3. *Individual-equipment circuits* can serve high-wattage equipment that is usually installed in a permanent location. The range, dryer, water pump, air conditioner, water heater, and food freezer must have individual circuits for satisfactory service.

The individual circuit may be two-wire, 115-volt, but it is usually three-wire, 230-volt. A special-purpose receptacle is usually installed to serve the range; however, individual circuits are usually wired directly to the piece of equipment being served.

Circuit accessories include wall switches, pull-chain switches, convenience outlets, ceiling outlets, special-purpose outlets, junction boxes, outlet boxes, and other miscellaneous items.

A two-way wall switch will control lighting outlets at one location, and a three-way * wall switch will provide control from two different locations. Three-way switches are needed to control outlets at two entrances to a room, to control a yard light or garage light, to control lights for a stairway, and to operate lights in out-of-way places.

---

* The term "three-way" indicates that a switch has three terminals on it and will control a circuit at two locations. A four-way switch has four terminals on it and will control a circuit at three locations.

FIGURE 112. *Top:* Dishwashing is child's work with a "wired-in" electric dishwasher. The N. E. Code specifies a separate, grounded circuit for dishwashers.

FIGURE 113. *Bottom:* Yard light controlled by a three-way switch inside the farm home and another three-way switch at the barn.

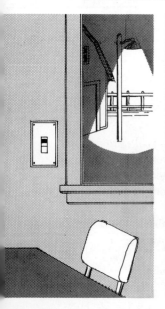

Use four-way switches wherever you may need to control a light from three different locations. A room having three entrances or a yard light would be an example of this need.

Use a pull-chain switch to control lights in a closet or bathroom. A pull chain must have an insulated, nonconducting link in it to protect against shock. Never install any pull chain or other electric fixtures in a location that can be reached from the bath tub or shower.

Most convenience outlets should be of the duplex grounded type, which provide two places for plugging in at one outlet. (See Figure 115.) Your range will probably be equipped with a special-purpose plug having three blades arranged in such a way that it can be plugged into the outlet in only one position. Your 230-volt air-conditioner, and perhaps other high-wattage equipment, may have a 230-volt polarized plug-in.

Special Note on Convenience Outlets. With the release of the 1963 National Electrical Code, it has been standard practice to use "grounded" receptacles in new wiring. This is the reason why appliances now come equipped with three-prong plugs (caps). The result on the wiring industry has been the manufacture and use of a wire style that contains a third (usually bare) grounding wire. It may be specified as "No. 12-2 grounded," "No. 10-2 with ground," et cetera. Old, two-wire circuits can be rewired for grounding (see Chapter 8).

Most receptacles, therefore, have three slots instead of two, the third being for the extra ground. Older, two-slot receptacles can be used by inserting an adapter, but this *does not ground* the

BRANCH CIRCUITS AND OUTLETS 123

FIGURE 114. *Left:* A pull-chain light fixture is suitable for use in closets, bathrooms, and other small spaces.

FIGURE 115. *Bottom:* A duplex receptacle has two plug-in positions. *(A)* grounded-type receptacle; *(B)* pilot light; *(C)* grounded-type switch; *(D)* cover plate (for receptacle only—need 3-hole plate if pilot light is used).

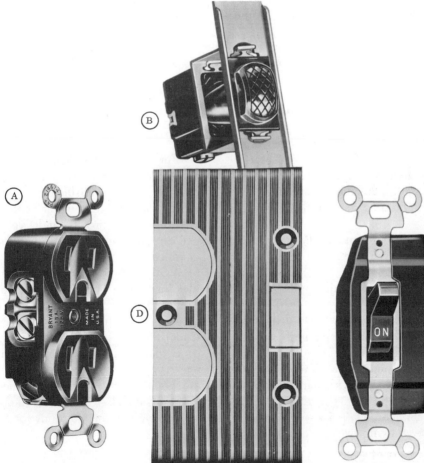

outlet unless the wiring system is conduit or flexible armored cable. Chapter 8 deals with wiring outlets with grounded-style wire.

HOW MANY CIRCUITS OF DIFFERENT TYPES WILL BE NEEDED FOR THE RESIDENCE? *

In a previous example, you selected a 200-ampere entrance switch for your residence and selected service entrance conductors and the feeder circuit for a maximum demand of 200 amperes. Using the same 1,800-square-foot residence, design the number and type of the branch circuits needed.

General-Purpose Circuits. Allow one general-purpose circuit for each 400 square feet of floor area as a minimum; for extra-heavy use of electric equipment, this should be increased by one circuit.
1,800 sq ft ÷ 400 = 4 (minimum) or 5 (heavy usage). **Total 5**

Appliance Circuits. Allow at least two 115-volt, 20-ampere circuits for kitchen and dining room appliances. Sub-total 2

Allow at least one appliance circuit for the laundry and one for outdoors. Sub-total 2

Total number of all appliance circuits: **Total 4**

Individual Circuits. Each of the following high-wattage appliances should have a separate circuit:

(a) Air-conditioner: Two 115-230 volt circuits. No. 10 wire minimum. Sub-total 2

(b) Range: One 230-volt circuit, using fuse and wire size as recommended by the manufacturer. This will be three-wire, No. 6 cable or larger, depending on the wattage of the range and the length of the circuit. Sub-total 1

(c) Water heater: One 230-volt circuit. Sub-total 1

(d) Washer-dryer: One 115-230 volt circuit. If you have a high-speed dryer, you will need No. 8 or No. 6 cable for the circuit. Sub-total 1

(e) Food freezer: Sub-total 1

(f) Shop equipment and other items in the basement: One; plus four space heaters: Sub-total 5

Total number of individual circuits: **Total 11**

Summary:
1. Total general-purpose circuits: 5
2. Total appliance circuits: 4
3. Total individual circuits: 11

Total All Circuits   **20**

Circuits for Smaller Homes

For a very small home having less need for electricity, a 60-ampere service entrance may be satisfactory. This will provide for a range, a water heater, two general-purpose circuits, and two appliance circuits. No major appliances can be added.

A 100-ampere service entrance provides for sixteen branch circuits, but the total capacity is 18,400 watts as compared to 36,800 for the 200-ampere size. (Code: 80 percent of total amperes.)

If there is a chance that you will increase your present use of electricity, you should select a service entrance capacity of not less than 100 amperes, leaving spare circuit positions for future use.

* Fig. 104 shows 22 circuits.

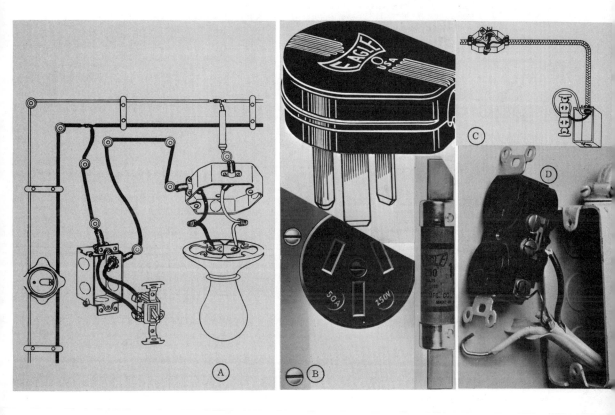

FIGURE 116. Common types of outlets for the home and farm. *(A)* Ceiling outlet, one for each 150-square feet of residence area; *(B)* parts of a 230v range-welder, outlet ("plug-in" style); *(C)* common "plug-in" for 2-wire cable; *(D)* grounded type "plug-in" (note bare ground wire at center).

WHAT OUTLETS AND ACCESSORIES WILL BE NEEDED FOR THE HOME?

The data contained in Table 14 will help you to complete the plan for the electrical system in your residence.

An *outlet* is a place in a circuit from which electricity may be taken and used for an appliance. It may be a box in which the wires for a ceiling light are terminated and connected to a ceiling lamp fixture, or it may refer to a receptacle where electric devices can be plugged in for using electricity.

For wiring the residence, three types of outlets will be needed: (1) ceiling or lighting outlets; (2) convenience outlets; and (3) special-purpose outlets such as range receptacles and other polarized plug-ins; also called "Individual-Equipment" circuits.

The general rule for determining the number of lighting outlets needed in a residence is one outlet for each 150 square feet of floor area. A 14 x 14 room would contain 196 square feet and would need two lighting outlets. Convenience outlets are specified on the basis of

GRAPHICAL ELECTRICAL SYMBOLS FOR WIRING PLANS*

These symbols have been extracted or adapted from American Standards Association Standard ASA Z32.9-1943, wherever possible. Adaptations and new symbols included in this list have been proposed for inclusion in the next revision of that standard.

General Outlets

- ○ Lighting Outlet
- Ceiling Lighting Outlet for recessed fixture (Outline shows shape of fixture.)
- Continuous Wireway for Fluorescent Lighting on ceiling, in coves, cornices, etc. (Extend rectangle to show length of installation.)
- Ⓛ Lighting Outlet with Lamp Holder
- Ⓛ$_{PS}$ Lighting Outlet with Lamp Holder and Pull Switch
- Ⓕ Fan Outlet
- Ⓙ Junction Box
- Ⓓ Drop-Cord Equipped Outlet
- ⊣Ⓒ Clock Outlet
- To indicate wall installation of above outlets, place circle near wall and connect with line as shown for clock outlet.

Convenience Outlets

- Duplex Convenience Outlet (Grounding type outlet required)
- Triplex Convenience Outlet (Substitute other numbers for other variations in number of plug positions, such as 1 for single convenience outlet).
- Duplex Convenience Outlet — Split Wired
- Weatherproof Convenience Outlet
- Multi-Outlet Assembly (Extend arrows to limits of installation. Use appropriate symbol to indicate type of outlet. Also indicate spacing of outlets as X inches.)
- Combination Switch and Convenience Outlet
- Combination Radio and Convenience Outlet
- Floor Outlet
- Range Outlet
- Special-Purpose Outlet. Use subscript letters to indicate function. DW-Dishwasher, CD-Clothes Dryer, etc.

Switch Outlets

- S   Single-Pole Switch
- S$_3$ Three-Way Switch
- S$_4$ Four-Way Switch
- S$_D$ Automatic Door Switch

*Farmstead Wiring Handbook

Switch Outlets (continued)

- S$_P$ Switch and Pilot Light
- S$_{WP}$ Weatherproof Switch
- S$_2$ Double-Pole Switch

Low-Voltage and Remote-Control Switching Systems

- Ⓢ   Switch for Low-Voltage Relay Systems
- Ⓜ$\overline{\text{S}}$ Master Switch for Low-Voltage Relay Systems
- ○$_R$ Relay—Equipped Lighting Outlet
- — — — —   Low-Voltage Relay System Wiring

Auxiliary Systems

- Push Button
- Buzzer
- Bell
- Combination Bell-Buzzer
- CH   Chime
- Annunciator
- D   Electric Door Opener
- Interconnection Box
- T   Bell-Ringing Transformer
- Outside Telephone
- Interconnecting Telephone
- R   Radio Outlet
- TV   Television Outlet

Miscellaneous

- ○$_{a,b}$
- ⊖$_{a,b}$ Special Outlets. Any standard symbol given above may be used with the addition of subscript letters to designate some special variation of standard equipment for a particular architectural plan. When so used, the variation should be explained in the Key of Symbols and, if necessary, in the specifications.
- ▲$_{a,b}$
- □$_{a,b}$
- ────── Feeder
- ────── Branch Circuit
- —·—·— Branch Circuit, Concealed in Floor
- — — — Switch Leg Indication. Connects outlets with control points.
- ▨▨▨▨ Service Entrance Equipment
- ■■■■ Distribution Panel for Feeders and/or Branch Circuits

> The electrical symbols appearing on this page provide a key to the wiring symbols used in the drawings appearing in this Handbook.

FIGURE 117. Graphical electrical symbols for wiring plans.

TABLE 14 SUMMARY OF ELECTRIC OUTLETS FOR THE HOME

Space	Lighting Outlets	Type of Circuit	Convenience Outlets	Type of Circuit	Special-Purpose Outlets	Type of Circuit
Living Room, Farm Office	General illumination; wall-switch controlled.	Gen.	No point at wall line more than 6 feet from an outlet; outlet in mantel shelf. Two or more outlets switch controlled. Outlet in any wall space 2 feet wide or greater.	Gen.	1 for air-conditioner.	Ind.
Dining Areas	1 outlet; wall-switch controlled.	Gen.	No point at wall line more than 6 feet from an outlet. Outlet in any wall space 2 feet wide or greater.	App.		
Kitchen	General illumination plus light over sink; wall-switch controlled. Work-area lighting.	Gen.	1 for every 4 feet of kitchen work-surface frontage. 1 at refrigerator location. 2 at table location. Two or more 20-ampere circuits to serve these outlets.	App.	1 for range. 1 for clock and home freezer. 1 for fan. 1 for dishwasher-waste disposal unit (if plumbing facilities are installed).	Ind. Ind. Gen. Ind.
Laundry	General illumination; wall-switch controlled. Work-area lighting.	Gen.	1 outlet, for general use.	App.	1 for washer. 1 for hand iron or ironer. 1 for clothes dryer. 1 for water heater.	Ind. App. Ind. Ind.
Bedrooms	General illumination; wall-switch controlled.	Gen.	No point at wall line more than 6 feet from an outlet. Outlet on each side, and within 6 feet of center line of each bed location. Outlet in any wall space 2 feet wide or greater.	Gen.	1 for room air-conditioner.	Ind.
Bathrooms Lavatories	Good illumination of face at mirror essential; wall-switch controlled.	Gen.	1 near mirror.	Gen.	1 for built-in space heater. 1 for built-in fan; wall-switch controlled.	Ind. Gen.

TABLE 14 * (Continued) SUMMARY OF ELECTRIC OUTLETS FOR THE HOME

Space	Lighting Outlets	Type of Circuit	Convenience Outlets	Type of Circuit	Special-Purpose Outlets	Type of Circuit
Recreation Room	General illumination; wall-switch controlled.	Gen.	No point at wall line more than 6 feet from an outlet; outlet in mantel shelf. Outlet in any wall space 2 feet wide or greater.	Gen.		
Hall	General illumination; wall-switch controlled.	Gen.	1 for each 12 feet of hallway. Halls over 25 sq ft at least one outlet.	Gen.		
Stairways	Outlets for adequate illumination of each stair flight. Multiple control at head and foot of stairway.	Gen.	1 at intermediate landings.	Gen.		
Closets	1 outlet.	Gen.	None.			
Exterior Entrances	1 or more outlets; wall-switch controlled.	Gen.	1 preferably near front entrance.	Gen.		
Porches	One outlet for each 100 square feet; wall-switch controlled. If enclosed, treat as a living room space.	Gen.	1 for each 12 feet of wall bordering porch. If enclosed porch, treat as living room.	Gen.		
Yard Lights	One or more post lights controlled by time-delay switch or photoelectric cell.					
Terraces and Patios	General illumination; wall-switch controlled.	Gen.	1 for each 15 feet of wall bordering terrace or patio.	Gen.		
Basement and Utility Space	General illumination of work areas, equipment, and stairways.	Gen.	2 outlets, one at work-bench location.	Gen.	1 for electrical equipment in connection with furnace. 1 for freezer (and kitchen clock).	Ind.

Garden and Christmas Lighting, and Garden Tools	Gen.	Provide weatherproof outlets at locations convenient for connection of lawn mowers, hedge trimmers, and Christmas or garden lighting.
Accessible Attics	Gen.	1 outlet; wall-switch controlled. 1 for each enclosed space.
	Ind.	1 for cooling fan, with switch control.
Garage	Gen.	1 per car near hood and one near back for trunk; wall-switch controlled. 1 for exterior lighting, multiple-switch controlled if garage is detached from house.
	Gen.	1 for one or two car garage.
	Ind.	If food freezer, work bench or automatic door opener is planned, provide appropriate outlets.

Notes on Preceding Table

Gen.—Outlets supplied by General-Purpose Circuits.
App.—Outlets supplied by Appliance Circuits.
Ind.—Outlets supplied by Individual-Equipment Circuits.
A convenience outlet to be at least of the duplex grounding type (two or more plug-in positions), except as otherwise specified.

All spaces for which wall-switch controls are required, and which have more than one entrance, to be equipped with multiple-switch control at each principal entrance. If this requirement would result in the placing of switches controlling the same light within 10 feet of each other, one of the switch locations may be eliminated.

* Taken from *Farmstead Wiring Handbook*, Edison Electric Institute, 1965.

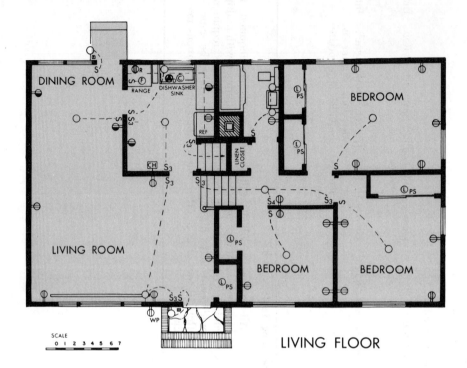

FIGURE 118. Typical wiring plan for three-bedroom farm home.

linear feet of usable wall lines. (See Table 14 for specifications.)

In addition to an adequate number of switches needed inside the dwelling, you should have three-way switches for your yard lights and garage lights so that these lights can be turned on or off at the two most convenient points along the circuit.

Before studying the information in Table 14, you will find it helpful to familiarize yourself with the electrical symbols in Figure 117. Then refer to the typical wiring plan for a farm home in Figure 118 and see whether you can read the symbols. Later, when you begin to make wiring plans yourself, you will need to use these symbols as a carpenter would use a blueprint.

WHAT BRANCH CIRCUITS AND OUTLETS WILL BE NEEDED FOR EACH FARM SERVICE BUILDING?

Most general recommendations for branch circuits and outlets covered earlier in this chapter apply also to farm-service buildings—number of outlets per circuit, wire size, et cetera. However, Table 15 gives more detailed recommendations for specific types of circuits for different service buildings. The diagrams that accompany Table 15 do not in every instance agree with specifications for circuits and outlets as listed in the table. The reason for this difference is that the diagrams shown are designed for larger buildings and possible expansion; Table 15 lists minimum requirements.

The Farmstead Wiring Handbook * contains a discussion of each diagram shown in connection with Table 15.

Some Specifics

The rapid and widespread use of electrification in agriculture is making very thorough planning of circuits and outlets for specific tasks even more essential. Above all, *safety* should come first. Convenience and expandability are also important. One of the main obstacles to needed expansion of agricultural operations continues to be under-sized wiring. The wiring of farm-service buildings will require more frequent reference to wire-size tables in comparison to residential wiring because of the increasing use of larger motors, more heating, expanding use of feed-handling machinery, et cetera, generally requiring wire size larger than the "old standard" No. 12.

Wire Type

In some kinds of farm work, the wiring is subjected to excessive moisture, fumes, mechanical damage, and the like. It will pay you to ask the power company planner to suggest both sizes and types of wire for specific circuits. Most of the non-metallic sheathed cable now on the market will withstand ordinary moisture and fumes but may require protection from mechanical damage. Conduit or flexible armored cable may be needed in some locations.

Convenience Outlets

In the discussion on circuits and outlets for the dwelling, we mentioned the national changeover from old-style "2-wire Romex" to a newer style, specified as "No. 12-2 grounded," et cetera. The same regulation applies to farm-service buildings—namely, use of grounded-type outlets. As previously stated, this three-wire style is for extra safety; you simply get an extra ground by using wire that contains the usual hot and neutral wires plus a bare ground for use with 3-prong plug-ins. We stated before that an adapter makes a 2-slot receptacle usable for 3-prong plugs, but this *does not* ground the circuit unless it is in conduit or flexible armored cable.

Convenience outlets in farm buildings should be located 5 to 6 feet above floor level in locations where farm animals are kept. Use non-metallic covers on convenience outlets.

Wall Switches

Most farm buildings should have wall switch control for the lighting throughout and 3-way switches at the residence and each farm building that is used much after dark. Since stairs are especially hazardous in farm buildings, their lighting should be controlled by a 3-way switch. Again, use non-metallic covers for the switches.

Grounding

Good grounding is frequently a problem on farms because they are usually not adjacent to a city water system. The new type community (rural) water systems often use "non-conducting" pipe for the main water supply line. Where

---

* *Farmstead Wiring Handbook,* Edison Electric Institute, 750 3rd Avenue, New York, N.Y. 10017. ($1.50 per copy)

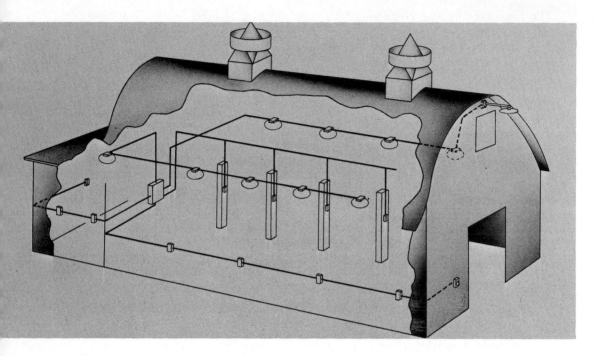

FIGURE 119. A modern dairy barn containing five circuits.

this is so, the home water system is generally *not* suitable for use as a ground. Driven grounds have to be used where the water pipes are not at or in the permanent moisture level. Size, type, and method of installation of grounds should conform to Code requirements. (See Chapter 8.) Also, large motors and other high-wattage loads should have a separate ground in addition to their regular grounding (service entrance at each of the farm buildings must have a standard ground).

Hazardous Locations

No electrical switch or other part of the wiring, nor any electrical appliance, should be located so that the operator can reach it while standing in water, or while any part of the body is immersed in water. Electric pumps and the inside of pump houses should be considered hazardous.

Electric Heating

Great expansion in electric heating is taking place in both rural and urban areas. Before completing your wiring plan for service buildings, consider whether or not you will use electric energy for heating.

In the following pages, Table 15 lists detailed information and recommendations for different kinds of farm-service buildings. Diagrams of wiring plans accompany the table.

TABLE 15* SUMMARY OF ELECTRIC OUTLETS FOR FARM BUILDINGS

Type of Outlet	Location of Outlet	Conditions to Observe
	DAIRY BARN	
Lighting	Every 12 feet in litter alley and 15 feet in feed alley	
Convenience	Every 20 feet in litter alley	For portable milkers, clippers, immersion heaters, infrared lamps, sprayers; grounded
	At certain windows and doors	For electric fly screens
	At ventilating fans	Less than ½ hp
Special Purpose	At ventilating fan	½ hp and larger
	At pipe-line milker	½ hp and larger (230 volts)
	At gutter cleaner	1 or 2 large motors required
	LOAFING BARN (COWS NOT IN STANCHIONS)	
Lighting	1 for every 150 sq ft open pen area; double height	Place outlets in feed alley, every 20 feet
Convenience	At convenient locations	For clippers, immersion heaters, sprayers, etc.; grounded
Special Purpose	At water pipes	For heating cable to protect water system; grounded
	BOX STALLS AND PENS	
Lighting	1 per stall	In low-partitioned stalls, one outlet located over separating partition may serve two stalls
Convenience	1 per stall	In low-partitioned stalls, one outlet may serve two stalls
Special Purpose	At water pipes	For heating cable to protect water system
	MILKING BARN OR MILKING ROOM	
Lighting	1 for every 3 cows, in passage front of cows	
	1 for every 2 cows, in passage back of cows	
Convenience	1 for every 5 cows	For portable milkers, clippers, immersion heaters, sprayers; grounded
	At ventilating fans	Less than ½ hp
Special Purpose	At ventilating fan	½ hp and larger
	At milking machine	½ hp and larger (230 volts)

* *Farmstead Wiring.* Pittsburgh, Pennsylvania, Westinghouse Electric Corporation.

TABLE 15 (Continued) SUMMARY OF ELECTRIC OUTLETS FOR FARM BUILDINGS

Type of Outlet	Location of Outlet	Conditions to Observe
	MILKING PARLOR	
Lighting	At rear of each cow on center line of pit	Light for cleaning cow's udder and for milking
	At each entrance and exit	Light for entering and leaving
Convenience	Every 2 cows on either side of pit	For clippers, immersion heater, lamp, clock, radio, portable milker; grounded
Special Purpose	1 outlet at one end of milking parlor or milker may be located in separate room	For operating milking machine
	1 outlet at window	For ventilating fan
	MILK HOUSE OR MILKROOM	
Lighting	1 for every 100 sq ft floor area, and at loading platform; at toilet	Over wash basin
Convenience	At each work area	For portable appliances, such as plug-in water heaters and motor-driven devices of less than ½ hp; grounded
	At ventilating fans	Less than ½ hp
Special Purpose	At ventilating fan	½ hp and larger
	At milk cooler	½ hp and larger (230 volts)
	At water heater	Omitted where 115-volt plug-in heaters are used
	At utensil sterilizer	
	At toilet	Quartz-type heaters as needed
	Milking station	

FIGURE 120A. *Left:* Wiring plan for dairy barn, face-out arrangement.
FIGURE 120B. *Right:* Wiring plan for dairy barn, face-in arrangement.

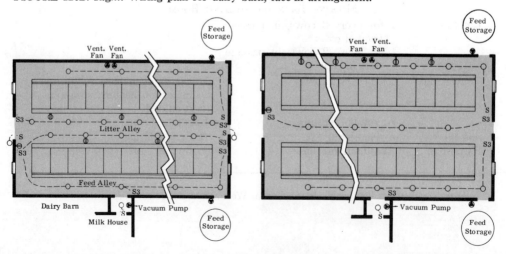

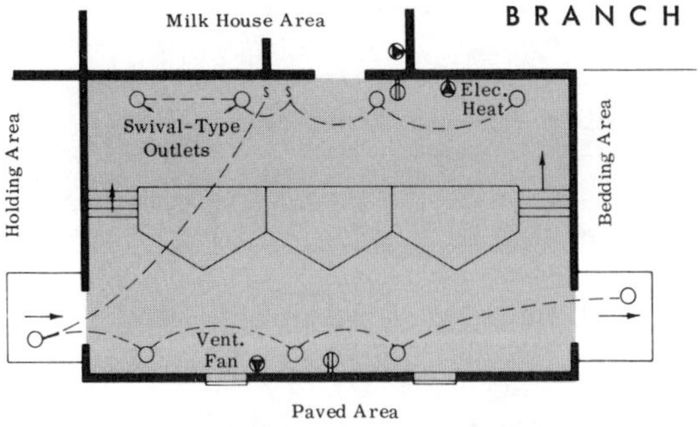

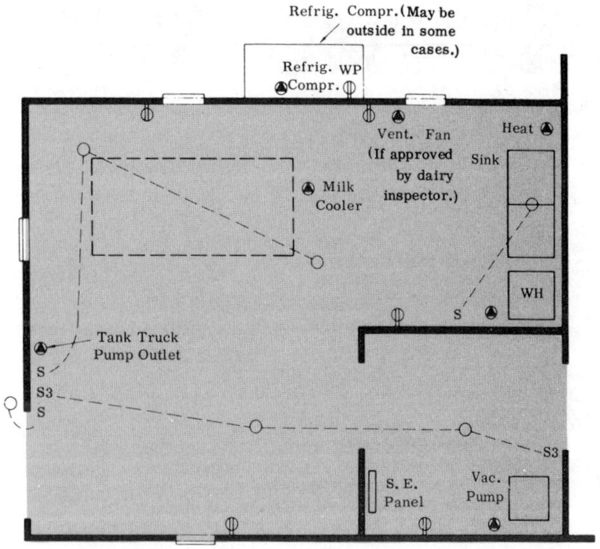

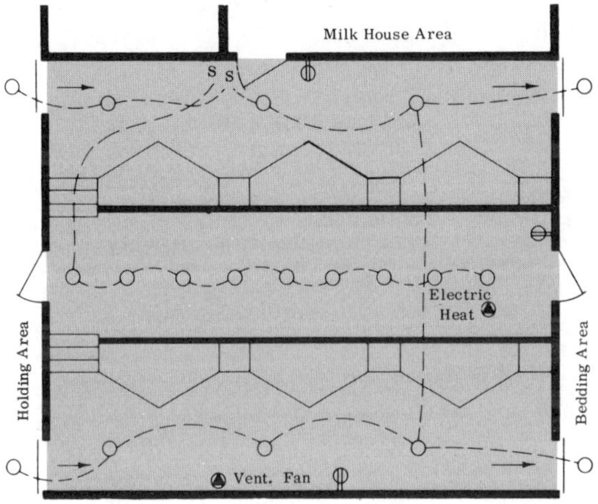

BRANCH CIRCUITS AND OUTLETS 135

FIGURE 121A. *Top left:* Wiring plan for milking room (single row of cows).
FIGURE 121B. *Top right:* Wiring plan for milk house.
FIGURE 121C. *Left:* Wiring plan for milking room (two rows of cows).

TABLE 15 (Continued) SUMMARY OF ELECTRIC OUTLETS FOR FARM BUILDINGS

Type of Outlet	Location of Outlet	Conditions to Observe
	LAYING HOUSE *	
Lighting	1 for every 200 sq ft floor area (a) 20 ft deep house—one row of outlets 10 feet apart on center line between dropping board and house front (b) Greater than 20 feet in depth: with roosts at back of pen; two rows of outlets on .10 foot centers, staggered. With roosts in center; two rows of outlets on 10 foot centers, one row on center line between dropping board and front of house, the other between dropping board and rear	Time switch controlled (Reflectors desirable for all lighting units)

* Special note: Laying houses vary greatly in size and type. Consult local authorities.

TABLE 15 (Continued) SUMMARY OF ELECTRIC OUTLETS FOR FARM BUILDINGS

Type of Outlet	Location of Outlet	Conditions to Observe
	Laying House (continued)	
Lighting (With Dimmers)	Bright lights: 1 for every 200 sq ft floor area (Locate same as outlined under Item 27)	Time switch controlled (Reflectors desirable for all lighting units)
	Dim lights: 1 for every 400 sq ft floor area. Placed in a row slightly back of bright light outlets, toward roosts	Time switch controlled
Lighting (All Night)	One 10–15w unit for every 200 sq ft floor area (Locate same over feeding areas)	Wall switch control outlets in each pen (Time switch controlled if desired)
Convenience	1 for every 400 sq ft floor area, but at least one in every pen (On post in center or on wall)	For portable poultry water heaters and for small manually-controlled ventilating fans; grounded
Special Purpose	At ventilating fan	Automatic control desirable
	(Note: See local authorities for wiring plans.)	
	Brooder House * (Portable or Individual Type)	
Lighting	On ceiling or wall	
Convenience	On ceiling in center of space	To serve brooder (also water warmer if needed—1,150 watt); grounded

* Special note: Brooder houses vary greatly. Consult local authorities.

FIGURE 122A. *Bottom:* Wiring plan for conventional-type laying house.
FIGURE 122B. *Right:* Wiring plan for poultry lighting circuit, including optional photoelectric control.

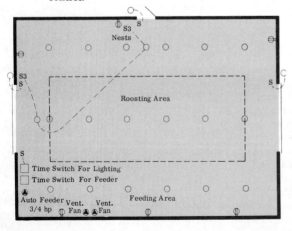
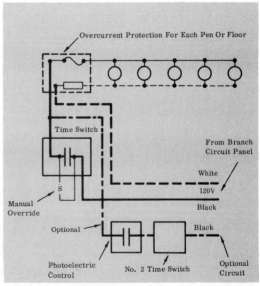

BRANCH CIRCUITS AND OUTLETS

TABLE 15 (Continued) SUMMARY OF ELECTRIC OUTLETS FOR FARM BUILDINGS

Type of Outlet	Location of Outlet	Conditions to Observe
	BROODER HOUSE (COLONY TYPE)	
Lighting	1 per brooder pen	Controlled individually or in groups
Special Purpose	1 in each pen for brooder	1,000w brooder and 150w for water warmer; grounded
	EGG STORAGE AND HANDLING ROOM	
Lighting	1 for every 150 sq ft floor area, and 2 over each working surface	
Convenience	At egg cooler or humidifier fan	Grounded
	At egg candler, egg cleaner and egg grader	Grounded
Special Purpose	At refrigeration unit	½ hp and larger (230 volts); grounded
	At egg grading unit	

FIGURE 122C. *Right:* Wiring plan for egg handling and storage.
FIGURE 122D. *Bottom left:* Wiring plan for time-lighting control (laying house).
FIGURE 122E. *Bottom right:* Wiring plan for multi-floor time-lighting control (laying house).

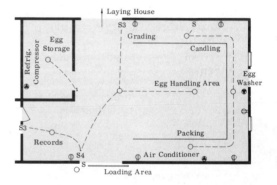

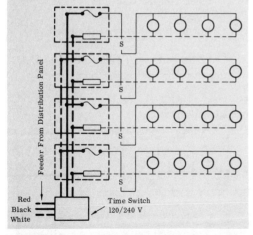

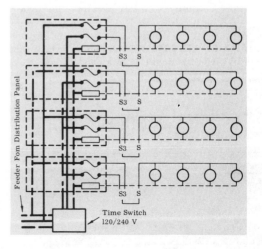

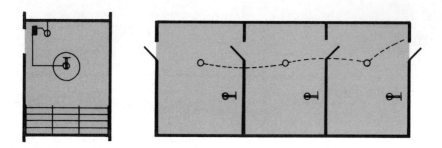

FIGURE 122F. Wiring plan for two types of brooder houses.

TABLE 15 (Continued) SUMMARY OF ELECTRIC OUTLETS FOR FARM BUILDINGS

Type of Outlet	Location of Outlet	Conditions to Observe
	POULTRY DRESSING ROOM	
Lighting	1 for every 200 sq ft floor area and 2 over each working surface	
Convenience	1 for every 400 sq ft floor area	For general-purpose use; grounded
Special Purpose	At poultry scalder	1,000 to 4,500 watts (230 volts)
	At waxer	1,000 watts (230 volts)
	At picking machine	¾ to 1½ hp (230 volts)
	At refrigerator	½ hp and up (230 volts)
	BEEF CATTLE BARN	
Lighting	1 for every 250 sq ft pen area	Place outlets in feed alley, every 20 feet
Convenience	At convenient locations	For clippers, sprayers; grounded
Special Purpose	At ventilating fan	½ hp and larger (230 volts)
	At feed mixer and conveyor	½ hp and larger (230 volts)

FIGURE 123. *Left:* Wiring plan for a beef cattle barn.
FIGURE 124. *Right:* Wiring plan for horse barn (individual stalls).

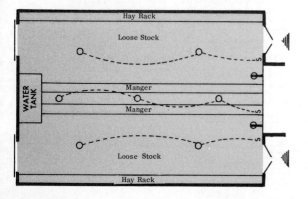
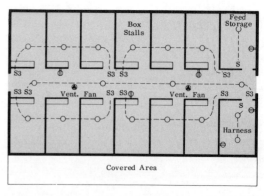

TABLE 15 (Continued) SUMMARY OF ELECTRIC OUTLETS FOR FARM BUILDINGS

Type of Outlet	Location of Outlet	Conditions to Observe
Horse Stables (Tie Stalls)		
Lighting	1 in back of each pair tie stalls 1 per box stall	Reflector type Where partition separates stalls, one outlet may be located over partition to serve two stalls
Horse Stable (Individual Stalls)		
Lighting	Every 20 feet in feed alley, and 1 for each pair of stalls (Locate outlet over partition)	If partition is high, provide one for each stall
Convenience	1 for every four stalls in service alley	Grounded, 6 ft above floor
Sheep Barn and Lambing Shed		
Lighting	Every 20 feet of feed alley and 1 for every 250 sq ft of pen area	Reflector type
Convenience	1 for each pair of pens (Located over partition) At shearing location	For lamb brooders For sheep shearing; grounded
Hog and Farrowing House		
Lighting	1 for each pair of pens (Located over partition)	One line over each row of pens
Convenience Special-purpose heater cable in concrete	At each farrowing pen	For pig brooders; where building structure permits, one outlet may serve two pens; grounded

FIGURE 125. *Bottom left:* Wiring plan for sheep and lambing shed.
FIGURE 126. *Right:* Wiring plan for 12-pen hog farrowing house.

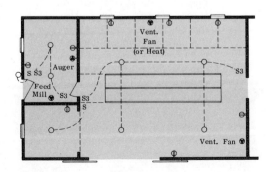

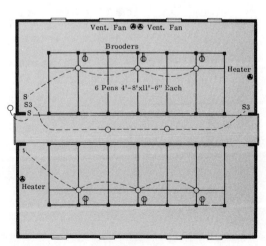

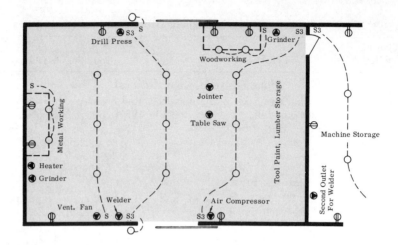

FIGURE 127. Wiring plan for farm shop and machinery shed (combination).

TABLE 15 (Continued) SUMMARY OF ELECTRIC OUTLETS FOR FARM BUILDINGS

Type of Outlet	Location of Outlet	Conditions to Observe
	FARM SHOP	
Lighting	1 for every 100 sq ft floor area 1 for each item of permanently placed equipment, such as drill press, saw, anvil; 1 for each 10 feet of bench length	Reflector units desirable
Convenience	1 for each 10 feet bench length (also on post in center of room, if desired)	Two at each work station plus one per wall; grounded
	1 for each permanently placed piece of equipment	Less than ½ hp
Special Purpose	At drills, grinders and forges	½ hp and larger (230 volts)
	Arc welder; 50-amp grounded outlet	Only special farm-size welders should be considered. Welders are being developed with electrical characteristics especially suited for use on rural electric lines. Also, most utility companies have regulations concerning the use of welders on rural lines. To assure the most satisfactory results, it will be advantageous for the planner of the wiring system to consult the local utility company

BRANCH CIRCUITS AND OUTLETS

TABLE 15 (Continued) SUMMARY OF ELECTRIC OUTLETS FOR FARM BUILDINGS

Type of Outlet	Location of Outlet	Conditions to Observe
	MACHINERY SHEDS	
Lighting	1 for every 150 sq ft floor area	Reflector type
Convenience	1 for every 500 sq ft floor area (at least 1 per building)	For portable tools for doing minor repair; grounded receptacles
	GARAGE	
Lighting	1 for every 200 sq ft and 1 over each workbench	Three-way switches
Convenience	1 at each workbench	
	At least 1 between each two vehicles	For charger, trouble lamp, etc.; grounded
	STAIRS AND PASSAGEWAYS	
Lighting	2 outlets at each stairway, 1 at head and 1 at foot (Locate outlets to avoid shadows and glare)	3-way switches where exit is normally made without retracing steps
	1 for every 25 linear feet of passageway (Locate outlets to avoid shadows and glare)	3-way switches where exit is normally made without retracing steps
	FEED-HANDLING ROOM	
Lighting	1 for every 200 sq ft floor area	Reflector units desirable; one over each work area

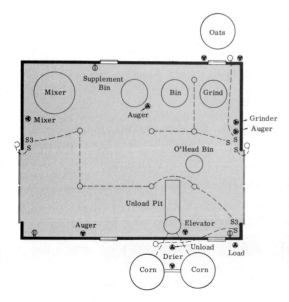

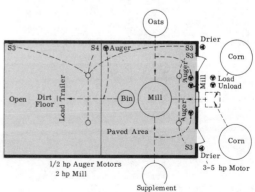

FIGURE 128A. *Left:* Wiring plan for feed-handling center.
FIGURE 128B. *Bottom:* Wiring plan for farm feed mill (bulk feed into trailer).

TABLE 15 (Continued) SUMMARY OF ELECTRIC OUTLETS FOR FARM BUILDINGS

Type of Outlet	Location of Outlet	Conditions to Observe
	FEED-HANDLING ROOM (continued)	
Convenience	At fan, small corn sheller, fanning mill, etc.	Less than ½ hp; grounded
Special Purpose	At feed grinder	½ hp and larger (230 volts)
	At feed mixer	½ hp and larger (230 volts)
	At corn sheller	½ hp and larger (230 volts)
	At grain elevator	½ hp and larger (230 volts)
	CROP STORAGE ROOMS AND CRIBS	
Lighting	1 for every 400 sq ft floor area. Spaces less than 20 sq ft need not be provided with outlet	One to each entrance, wall switch controlled
Special Purpose	Grain elevators and blowers	½ hp and larger (230 volts)
	BARN FLOOR AREA	
Lighting	1 for every 400 sq ft floor area	At each entrance, wall switch controlled
Convenience	1 for every 1,000 sq ft floor area	For general-purpose use; grounded
Special Purpose	At hay hoist	2 hp and larger (230 volts)
	At hay dryer	3 to 7½ hp (230 volts)
	At portable elevator	1 hp and larger (230 volts) (with portable motor, one outlet may serve various machines)

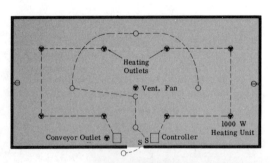

FIGURE 129A. *Left:* Wiring plan for 2,000-bushel sweet potato curing and storage house.

FIGURE 129B. *Bottom:* Wiring plan for potato cellar.

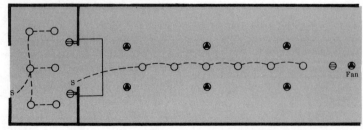

TABLE 15 (Continued) SUMMARY OF ELECTRIC OUTLETS FOR FARM BUILDINGS

Type of Outlet	Location of Outlet	Conditions to Observe
HAYMOW OR HAYLOFT		
Lighting	At least 1 for haymow	Glass or metal protector for bulb
Special Purpose	At hay hoist At hay dryer	2 hp and larger (230 volts) 3 to 7½ hp (230 volts)
SILO		
Lighting	1 at roof of silo; 1 at top; and 1 at bottom of chute	Outlets in chute to be controlled at foot of chute or at entrance to chute tunnel
Special Purpose	At ensilage cutter, blower At silo unloader	5 to 7½ hp (230 volts) 1 hp and larger (230 volts)
FRUIT AND VEGETABLE STORAGE		
Lighting	Every 20 feet of alley	
Convenience	1 for every 400 sq ft floor area	Grounded
Special Purpose	At refrigeration equipment At ventilating fans	½ hp and larger (230 volts) ½ hp and larger (230 volts)
ROADSIDE STAND		
Lighting	1 pair for every 50 sq ft display area and at floodlights and signs	
Convenience	At convenient locations	For space heater, refrigerator, etc.; grounded
Special Purpose	At convenient locations	For refrigeration, ½ hp and larger (230 volts)
PUMP HOUSE		
Lighting	1 centrally located	3-way switch, if needed
Convenience	At pump	Grounded
Special Purpose	For pump	½ hp and larger (230 volts)

SUMMARY

The wiring that extends from the fused switch, or breaker panel, to the final outlets where electric energy is used is called *branch circuits*. Most modern residences have at least a 200-ampere main switch and about twenty branch circuits. Branch circuits are classified as: (1) type of use; (2) voltage service.

According to use, circuits are further classified as:

(a) *General purpose*, serving the general lighting, and the convenience outlets except for the kitchen, laundry, and outside. General-purpose outlets should be fused at 15 amperes and convenience outlets should be of the grounded type—one for every 6 feet of wall space, not counting doorways. Use No. 12-2 grounded wire for these circuits (convenience outlets).

(b) *Appliance circuits* are used for the kitchen, usually two; for the laundry, usually one; for the basement and/or shop, usually one or two; and one or two for outdoors. No. AWG 12-2 grounded wire is needed for most of these circuits, fused at 20 amperes. Rule of thumb, install boxes 12 inches above floor level, except at the kitchen counter and for other special needs.

(c) *Individual circuits* should be installed for the range (50-ampere fuse), water heater, air-conditioners, permanently located space heaters, and for all motors ½ hp and larger. Individual equipment may be wired directly to the circuit or may be equipped with 3-prong, polarized caps. Size of the fuse should match the rated amperage (or wattage) of the appliance served; motor circuits should be protected with time-delay fuses. Wire-size for the range, water heater, and high-speed dryer may be as large as No. 6. The Code specifies that a dishwasher shall have a separate circuit and be separately grounded. (Polarized receptacles and caps are constructed so as to assure grounding in most instances.)

QUESTIONS

1. Why may a long branch circuit of small wire deliver less wattage than is required for an appliance to operate properly?
2. What is the advantage in using three-wire circuits for electric motors larger than ½ hp?
3. Why should a two-wire branch circuit be limited to eight outlets of any kind?
4. Why is No. 12 wire now recommended instead of No. 14 for both general-purpose and appliance circuits?
5. How many two-wire branch circuits could be wired from a 200-ampere service switch?

ADDITIONAL READINGS

Brown, R.H., *Farm Electrification*. McGraw-Hill Book Co., New York, N.Y., 1956.
Davis, Hollis R., *Adequate Farm Wiring Systems*, Extension Bulletin 849. Cornell University, Ithaca, N.Y., 1956.
Industry Committee on Interior Wiring Design, *Farmstead Wiring Handbook*. New York, N.Y., 1968.
Industry Committee on Interior Wiring Design, *Residential Wiring Handbook*. New York, N.Y., 1954.
Iowa Southern Utilities Company, *Farm Wiring Systems*. Centerville, Iowa (no date).
National Fire Protection Association, *National Electrical Code, 1968 Edition*. Boston, Mass., 1968.
Richter, H.P., *Practical Electrical Wiring*, Seventh Edition. McGraw-Hill Book Co., New York, N.Y., 1967.
Sears, Roebuck and Company, *Electric Wiring*. Chicago, Ill., 1955.

7 HOW TO GET READY FOR A WIRING JOB

Before purchasing any wiring supplies or electricians' tools you should check the following:

1. Understand the National Electrical Code and local code regulations that will affect your wiring plans. For example, will you be required to have conduit for the service entrance? Will you be allowed to use pole metering? Will the local inspector approve a wiring job done by you personally? Will the local power company serve your new job? May you use galvanized water pipe for your ground rod?

2. What type of wiring is best for your situation?

3. What tools are needed? Can you have them available?

4. What kind of wiring plans are needed? How can you use them? What will be the cost?

WHAT WIRING REGULATIONS SHOULD YOU OBSERVE?

Your wiring plans are controlled by certain regulations:

1. *The National Electrical Code.* The National Electrical Code is discussed in detail in Chapter 4. Its purpose is to protect human life and property by preventing dangerous wiring. This book conforms to the National Electrical Code and, if you follow it, your wiring plan should pass all inspections. However, it is advisable to check your final wiring plans with your power supplier to make certain that your diagram meets all Code requirements. Code compliance does not guarantee *adequacy*.

2. In some rural areas, especially in communities near large cities, a *local electrical code* may apply also. Local

FIGURE 130. A new transformer and larger powerdrop wire were required when this farm was rewired.

codes are often stricter than the National Electrical Code. If you comply with the local code, you will likely meet the specifications of the National Code also. Power suppliers in some areas will not serve a newly-wired building unless it is covered by a permit from an inspector showing that the wiring materials and installation are in compliance with the local code. Sometimes this procedure requires the payment of an inspection fee. Therefore, it is wise to determine the kind of code regulations that apply to you before you start a wiring job. Finally, determine whether you will be permitted to install your own wiring. In some areas, only licensed electricians are permitted to wire a building.

3. The power supplier serving your farm should be consulted in planning a large wiring job, because your plan may involve a new power drop or a new transformer; there is a possibility also that the present high line will not be adequate. If your power supplier cannot provide the new service you need, it would be foolish to go to the expense of buying and installing wiring that you cannot use.

WHAT KIND OF WIRING MATERIALS SHOULD YOU CHOOSE?

Of the hundreds of types and styles of wire on the market less than fifty are used in the home and on the farmstead. You will find that it pays to choose the correct wire for the job, because this is the best way to combine economy, good service, safety, and long life for your wiring. For example, it is false economy to wire a light-duty motor with expensive heavy-duty portable cable. Yet it is poor management to skimp on buying light-duty cord for a heavy-duty motor, since the result will be poor service and perhaps a "burned out" motor too.

To make certain that you always get the correct wire, select wiring supplies and equipment that are stamped with the UL label. Also, learn the characteristics of the types of wire commonly used in home and farm wiring.

You will need to become familiar with four classifications of wiring: (1) styles of wire, (2) types of wire for outside use, (3) wire for special purposes, and (4) wire for interior use.

HOW SHOULD YOU CHOOSE THE PROPER STYLE OF WIRE?

Seven styles of electric wire most commonly used on the farm are shown in Figure 132 and are described below.

1. Single-Conductor Wire. The first sample of wire in Figure 132 is a single-conductor style composed of small strands twisted together. This style also comes in a solid-wire conductor and is finished in numerous ways. The sample shown here is a well-constructed cord finished with (1) a fabric wrap on the wire (nylon, silk, cotton, etc.); (2) a course of heat-resisting rubber insulation over the fabric; (3) a second course of fabric over the rubber; and (4) an outer jacket of heat-resisting rubber.

2. Cable. A conductor containing two or more separate conductors is referred to as cable. The lighter weight wiring materials of this style are usually called cords or lamp cords instead of cable. Sample 2 shows a piece of heavy-duty rubber-covered cable (also referred to as cord) containing four conductors.

(NOTE: Refer to Glossary on symbols for wire types.)

FIGURE 131. The UL label on switches tells you that they meet safety standards.

3. Bare Ground Wire. Sample 3 shows a piece of bare copper ground wire. (It comes in different sizes.) This is used for grounding the electrical system, generally at the service entrance.

4. Parallel Lamp Cord. Sample 4 is a piece of No. 18 gauge, two-conductor lamp cord. Numerous small strands are twisted together to form the conductors. The sample shown is covered with plastic, but rubber covering is often used for this type of wire.

5. Braided Lamp Cord (Twisted). Notice the better construction of cord in comparison with sample 4. The additional courses of inner and outer fabric make this style more durable, but it is seldom used now for house wiring.

6. Portable Cord. The sample of cord shown here is widely used for extension

FIGURE 132. Seven styles of wire suitable for use on the farm.

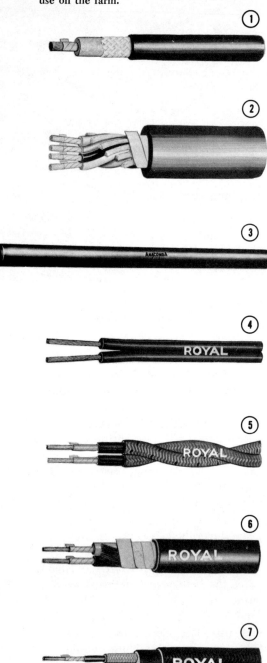

cord where the wire is not subjected to mechanical abuse. Notice the order of construction: (1) fabric wrap, (2) rubber insulation, (3) reinforcing fibers, (4) paper wrap, and (5) rubber jacket.

7. Reinforced Portable Cord. This cord is more durable than plain portable cord. Notice the two jackets of rubber in addition to the fabric wrap, rubber insulation, and braid.

WHAT KIND OF EXTERIOR WIRING SHOULD YOU CHOOSE?

Figure 133 shows nine samples of electrical wire (conductors) that are widely used for service drops, service entrances, and feeder circuits; and several of the samples have multi-purpose uses, such as underground circuits, interior feeders, et cetera.

1. Overhead Wiring. Item No. 7 is a polyethylene insulated (black) style constructed with a cross-linked neutral, making it especially useful for service drops. The manner of construction of this style relieves tension from the insulated wires. Item No. 6 called *Triplex* has a "rope-like" appearance—two hot wires and a bare neutral. This twisted style is also available as *Quadruplex* (not shown) where three hot conductors are twisted together with one bare neutral. Hot wires in both styles are insulated with polyethylene, adequate to meet Code specifications.

2. Service Entrance Wiring. Sample No. 4 shows service-entrance cable with stranded bare neutral. Unprotected SE cable is prohibited by local code regulations in some parts of the country. Where allowed, this style of wiring affords an economical service entrance directly to the range without additional protection, so it is less expensive than conduit installations. For conduit installations, Type TW, Item No. 1, is a popular choice. It has polyethylene insulation and comes in standard (red, black, and white) colors for polarizing (identifying) particular hot wires.

3. Feeder Conductors. The National Electrical Code considers feeders to be service drops. Therefore, the same wiring is recommended as for the main service drop.

4. Power Cables. Items No. 3, 8, and 9 are specially constructed to withstand high temperature, moisture, and chemicals. These physically tough conductors can be used underground, overhead, or for inside feeders. Polyethylene insulation is color coded; aluminum or copper.

WHAT WIRE SHOULD YOU CHOOSE FOR SPECIAL PURPOSES?

Figure 134 shows seven samples of wire for special purposes.

1. Heater Cord. This is a sample of heater cord, used for irons and other heating devices. The construction consists of (1) a fabric wrap, (2) rubber insulation, (3) an additional fabric wrap, and (4) a braided cover.

2. Kitchen-Unit Cord. This three-conductor cord is used for roasters and

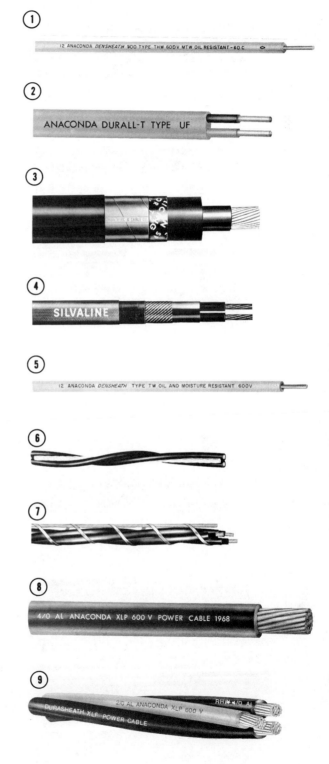

FIGURE 133. Nine widely used styles and types of exterior wiring materials suitable for residences and farm-service buildings; one type (UF) is used for both underground and interior wiring. Notice that wiring is labeled, in keeping with Code requirement.

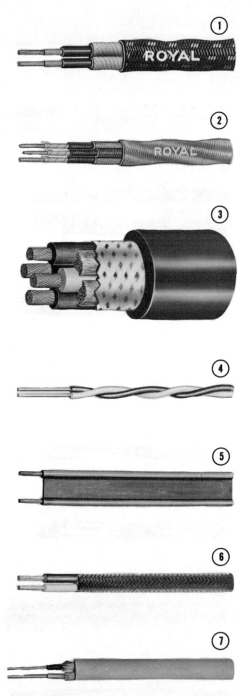

FIGURE 134. Seven special-purpose types of wire common to the farm.

other high-wattage equipment. Its construction is similar to heater cord. One bare ground is included.

3. Type-W Cable. This heavy-duty, rubber-covered cord is designed for welders and heavy motors. Rubber insulation is covered first with fabric and finally with a thick rubber jacket. Notice the bare, stranded neutrals. Outer covering may be neoprene.

4. Thermo-Cord. This type of wire is suitable for wiring thermostats. The covering is black and white plastic.

5. TV Cord. This special TV wire is used for a television antenna. High-voltage lead-in current is dangerous if improper wire is used.

6. Asbestos Heat-Resisting Cord. In locations where electric wires are subjected to excessive heat, this asbestos-insulated cord will stand up where rubber would soon wear out.

7. Plastic Cable. This plastic covered cable can be buried directly in the ground and Type UF can be used inside buildings. It is more expensive than ordinary types and grades of interior wiring. Note the solid conductors with plastic insulation followed by fiber filling and an outer jacket of tough plastic. Several new types of wire on the market are suitable for underground wiring.

WHAT KIND OF INTERIOR WIRING SHOULD YOU CHOOSE?

The three types of interior wiring that are commonly used for home and farmstead are (1) non-metallic sheathed cable, (2) conduit, and (3) flexible armored cable.

Another type of wire that comes "partially wired" is called Plugmold, Wiremold, and other trade names. Fig-

HOW TO GET READY FOR A WIRING JOB

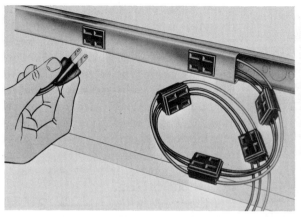 

FIGURE 135. *Left:* Plugmold wiring consists of a metal channel and pre-wired duplex plug-ins spaced 12 inches apart.

FIGURE 136. *Right:* Metallic raceway, outlet box, and plug-in. This type of wiring is easy to install since no carpentry work is required.

ure 135 shows a sample of this "pre-wired" type that is easy to install. Notice numerous plug-ins.

Metallic raceway with surface outlets, used in farm service buildings, is another type of wiring that is easy to install. Figure 136 shows this type.

When to Choose Non-Metallic Sheathed Cable

Non-metallic sheathed cable is widely used in residential wiring. It is easy to install, is not as expensive as conduit wiring, and is lightweight.

The sample non-metallic cable in Figure 137 has two insulated wires, one black and one white and a bare ground. (Or the ground may be insulated.) This three-wire cable is used for grounding various outlets and is largely replacing two-wire cable.

This covering is built to resist heat, moisture, and acid but is not suitable for use outdoors, underground, or in ma-

FIGURE 137. Sample of non-metallic sheathed cable, specified as "No. 12-2 with ground." Outer cover is tough plastic. Bare ground is center wire (connected to receptacle). White (neutral) wire is not connected.

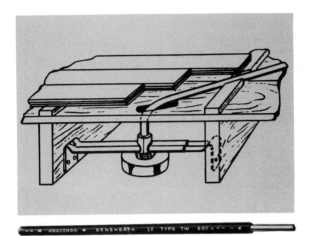

FIGURE 138. Sample of single-conductor wire and a conduit ceiling outlet. Conduit wire comes in black, red, and white covering. In wiring, *conduit* refers to both wire and tubing.

sonry. Non-metallic cable is also available in three- and four-wire styles.

More non-metallic cable is used in residential and farmstead wiring than any other type of wire. It is suitable for indoor use in homes, garages, and in most service buildings. It can be used in exposed runs as well as in concealed wiring. If exposed, it must be protected from mechanical damage by attachment to a baseboard or other rigid support. Local codes in some areas prohibit the use of this kind of cable; therefore you should ask before buying wire. For details on installation, see page 177.

When to Choose Thin-Wall Conduit

In the farm shop and other locations where your wiring is likely to be subjected to severe mechanical abuse, special protection for the wire is required. Of the three common types of interior wiring, thin-wall conduit is best, although it is the most expensive and the most difficult to install. The tubing serves the dual purpose of protecting the wire and grounding the electrical system. Conduit is usually galvanized but is sometimes finished with a special enamel. These special finishes make it suitable for use outdoors, indoors, or in wet locations. Conduit ordinarily comes in 10-foot lengths but is easily cut to desired dimensions.

Electrical wires must be "fished" through the conduit, and for that reason the size of the tubing must be large enough to fit the size of wire used. Connections to boxes are made with threadless fittings and the entire system is grounded when properly installed. Wires are usually red, black, and white, for color coding.

When to Choose Flexible Armored Cable

In permanently dry locations where your wiring is likely to be damaged, flexible armored cable is an excellent material and has two distinct advantages over conduit in that it is less expensive and easier to install. Flexible armored cable provides a grounded system when properly installed; therefore, a (third) ground is not needed as in NM cable.

The illustration in Figure 139 shows that armored cable is simple in construction. It consists of two or more insulated conductors encased in a metal shield. Armored cable may be imbedded in plaster or concrete inside buildings that are permanently dry, but should not be used in cellars, dairy barns, basements, or outdoors where dampness prevails. Armored cable is used to extend runs

HOW TO GET READY FOR A WIRING JOB

of conduit. This type of cable is more difficult to install than non-metallic sheathed cable, however. This is due largely to the special finishing job required wherever the armor is cut. Instructions for installing flexible armored cable are given on page 180.

WHAT TOOLS AND EQUIPMENT ARE NEEDED FOR A WIRING JOB?

A collection of tools needed in doing farm wiring is shown in Figure 140. You may already have most of the items in this collection in your farm shop. The entire collection can be purchased for about $50 to $60. Some of the more common uses follow.

A hacksaw is needed for cutting metallic armored cable, nails, bolts, and other metals encountered in a wiring job, and a brace and bit is necessary for boring holes. An extra-long bit, as illustrated in this collection, comes in handy for reaching difficult locations. A keyhole saw is indispensable for sawing openings for outlet boxes and for working in close places.

A pair of slip-joint pliers comes in handy for turning large locknuts on electrical fittings. You will also need a pair of electricians' pliers for cutting wire and

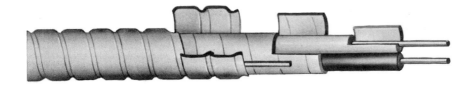

FIGURE 139. *Top:* Sample of flexible armored cable.

FIGURE 140. *Bottom:* Most commonly used electricians' tools. *(A)* hammer; *(B)* ⅝-inch bit or drill; *(C)* bit brace; *(D)* keyhole saw; *(E)* hack saw; *(F)* test light; *(G)* multi-purpose tool; *(H)* 6-foot folding rule; *(J)* chisel; *(K)* lever-jaw wrench; *(L)* linemen's pliers; *(M)* wire cutter, stripper; *(N)* jack knife; *(P)* fish tape and reel; *(R)* conduit bender; *(S)* screwdriver.

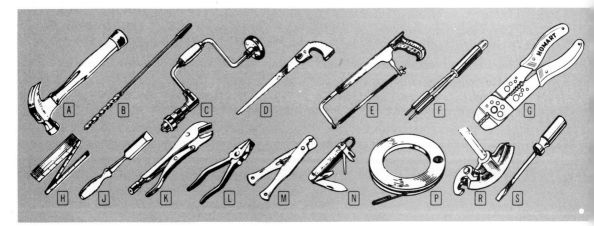

loosening insulation; for cutting insulation and doing numerous other wiring tasks, an ordinary pocket knife is essential.

A screwdriver is needed for tightening screws on electrical fittings and for other uses. A cold chisel comes in handy for cutting cable, nails, bolts, and other pieces of metal.

Notice the roll of wire at lower right in Figure 140. This is a special "fish tape," which is used for fishing wires into conduit. The conduit bender at upper right is used for bending and shaping tubing without crushing it. A folding rule is essential in measuring, and a hammer must be available for all types of pounding.

A neon test lamp, which costs about $2, makes it possible to trace circuits in both old and new work. This is an essential item in finding faults in the wiring system. You will find plans for a homemade test lamp on page 196.

Notice the cable stripper, which is needed for removing outer cable cover. This is an excellent, but not expensive, tool for stripping the insulation from wires. The rubber and friction tapes in this collection are needed for finishing splices and for repairing damaged wires. Plastic tape is being used more, for it serves the purpose of both rubber and friction tapes.

HOW CAN YOU WORK UP A WIRING DIAGRAM AND A BILL OF MATERIALS?

To be accurate, a bill of materials must be worked up from a wiring diagram made to scale. You can prepare your own diagram by following two steps: (1) draw to scale a floor plan of the building to be wired, and (2) sketch in the electrical system, using the proper electrical symbol for each item.

How to Draw a Floor Plan

A floor plan is easy to draw on graph paper. A size 30 x 36 inches, laid off in half-inch squares, is suitable for a wiring plan. Thus for a scale of 1 inch equals 2 feet, lay off 1 inch on the paper for each 2 feet of building. For example, if the width of your residence is 40 feet, lay off 20 inches on the graph paper. You can do this by counting off 40 half-inch squares.

Use standard building symbols for walls, windows, floors, stairs, closets, and other features of the building. Complete the floor plan. Be sure to check for accuracy before sketching in the electrical system.

How to Sketch in the Electrical System

Before attempting to sketch in the electrical system, you will find it helpful to study the wiring diagram in Figure 141. Locate circuits number 2 and 3 and refer to the bill of materials for these circuits. Take note of the name of each item and see that you can identify it by its symbol. You may also wish to review the list of electrical symbols in Figure 117.

Generally, the following steps are observed in sketching in the electrical system:

1. Determine the location of the service-entrance equipment and sketch it in, using the proper symbol of course. The size of the service cable and service switch will have been determined be-

forehand. Specify the number of circuit connections in the switch.

2. Sketch in the lighting outlets and wall switches throughout the building. Note that the symbols for two- and three-way switches are different.

3. Starting at the service switch, sketch in the lighting circuits, making certain that each wall switch is connected to the proper circuit. Remember that no more than ten outlets may be wired to one circuit. More than this number on one circuit ordinarily results in poor service as well as safety hazards.

4. Determine the location of the convenience outlets throughout the building. Sketch them in.

5. Starting at the service switch again, sketch in the circuits for the convenience outlets. Up to ten outlets may be wired to each general-purpose circuit but five is the maximum for appliance circuits.

6. Sketch in the special-purpose outlets (for range, food freezer, water heater, and the like) after determining their locations.

7. Sketch in each circuit for individual equipment. Indicate the type and size of wire to be used for each circuit.

8. When you have completed your wiring diagram, have it examined by a competent person. If it is approved, you may then work up the bill of materials.

How to Prepare a Bill of Materials

1. Starting at the service switch, measure the length of run of each circuit. Include vertical runs to wall switches and convenience outlets. Add 8 inches for making each connection.

2. Convert inches to feet and list the result. This is the amount of wire required for that circuit.

3. Count the number of junction boxes, switch boxes, "plug-in" boxes, toggle switches (two- and three-way), receptacles, cover plates, connectors, and other wiring supplies. List the totals for this circuit.

4. Likewise trace all other circuits and list the wiring supplies required.

5. Combine like items and list the total bill of materials. (See the sample bill of materials for a twelve-circuit dwelling on page 156.)

How to Use a Wiring Diagram

A completed wiring diagram and bill of materials serves two purposes. First, the diagram is a wiring plan or blueprint for the actual wiring job; second, the bill of materials can be submitted to dealers for bids or estimates of cost. You may realize considerable savings by obtaining two or more bids.

WHAT MATERIALS ARE NEEDED FOR WIRING A TWELVE-CIRCUIT HOME? *

The wiring diagram in Figure 141 includes twelve circuits. This is minimum for a three-bedroom home and is thought to be an economical and adequate wiring job. Many residences now have twenty or more circuits.

The following bill of materials is for the total wiring job in Figure 141.

---

* The drawing in Figure 141 and the bill of materials were prepared by Oran Lewellen, Engineer, Texas Power and Light Company.

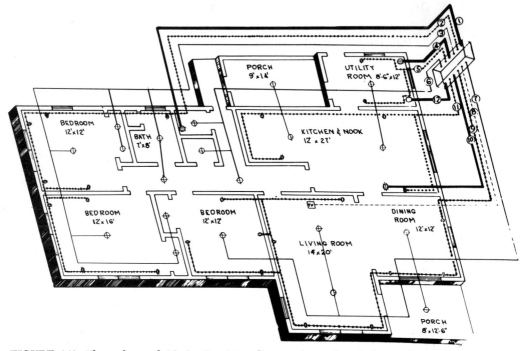

FIGURE 141. Floor plan and 12-circuit wiring diagram for a three-bedroom farm home.

Total Bill of Materials

ITEM	QUANTITY
200-ampere circuit breaker or fused switch	1
#6 solid copper ground wire	12 ft
Half-inch copper-coated steel ground rod	8 ft
Grounding connectors	2
Sill plate	1
#0 three-wire service cable	12 ft
20-ampere circuit breakers	8
30-ampere circuit breakers	6
50-ampere circuit breakers	2
#12 two-wire non-metallic sheathed cable, with ground	1,040 ft
#8 three-wire non-metallic sheathed cable	174 ft
#6 three-wire range cable entrance cable	40 ft
230-volt, single receptacle	3
230-volt, three-pole, flush-mounted range receptacle	1
Duplex receptacles, grounded type	27
Flush-mounted outlet boxes	30
Flush-mounted octagon boxes	21
Single receptacle flush-mounted cover plate	3

TOTAL BILL OF MATERIALS (continued)

Item	Quantity
Flush-mounted switch boxes	20
Flush-mounted toggle switches	20
Flush-mounted toggle switch plates	20
Duplex receptacle plates	27
Octagon boxes	17
Octagon box blank covers	17
Range receptacle cover plate	1
Lighting fixtures	20
Non-metallic sheathed cable connectors	177

Miscellaneous supplies: solderless connectors, plastic tape, cable straps, and a few other minor items.

Bill of Materials for Individual Circuits

The bill of materials for circuits number 2 and 3, in Figure 141, are presented here to illustrate the simplicity of working up the bill once the diagram has been completed. Refer to Figure 141 again and see whether you can identify each item listed under circuits number 2 and number 3.

CIRCUIT 2: CONVENIENCE OUTLETS

ITEM	QUANTITY
20-ampere circuit breaker	1
#12 two-wire non-metallic sheathed cable, with ground	115 ft
Flush-mounted outlet boxes	9
Duplex receptacles	9
Duplex receptacle plates	9
Octagon boxes	2
Octagon box blank covers	2
Non-metallic sheathed cable connectors	23

CIRCUIT 3: LIGHTING

ITEM	QUANTITY
20-ampere circuit breaker	1
#12 two-wire non-metallic sheathed cable, with ground	310 ft
Flush-mounted octagon boxes	8
Lighting fixtures	8
Octagon boxes	4
Octagon box blank covers	4
Flush-mounted switch boxes	8
Flush toggle switches, grounded type	8
Flush-mounted toggle switch plates	8
Non-metallic sheathed cable connectors	40

SUMMARY

Before beginning a wiring job, find out whether the local code requires that a licensed electrician do the job. Local power suppliers or inspectors furnish this information.

It is a good policy to ask your electrician (if you are hiring one) whether or not he is complying with all regulations of the local code and the National Electrical Code. Wiring methods and materials are covered by both codes.

For general use, wiring is classified in four ways: (1) style of wire; (2) exterior wiring materials; (3) special-purpose wire; and (4) interior wiring materials.

The styles of wire most often used on the farm include single conductor and cable; solid and stranded conductors; parallel and twisted cords; rubber, plastic, and braided covering; and plain and reinforced cord.

Exterior styles and types of wiring materials include insulated single-conductors; twisted style such as Triplex, with bare neutral for service drops and feeders; service entrance cable; conduit entrance conductors, color coded; and bare ground wire. Portable cords are used for operating portable electric machines and lighting.

Special-purpose wiring materials commonly used on the farm include heater cord, heavy-duty cable for welders and motors, thermostat cord, TV cord, asbestos heat-resisting cord, and acid-resistant plastic cable, which is suitable for burying directly in the soil. Many other special-purpose wiring materials are on the market although not very widely used on the farm.

The three most common types of interior wiring are (1) non-metallic sheathed cable, (2) thin-wall conduit, and (3) flexible armored cable. Non-metallic sheathed cable is the least expensive of the three types, is easiest to install, and is the most widely used type for residences and most farm service buildings. Thin-wall conduit is the best choice for locations where the wire may be damaged; however, this type is the most expensive and is the most difficult to install. Flexible armored cable provides protection for the inside wires and is easier to install than is conduit. The cost of flexible armored cable is between the range of conduit and non-metallic sheathed cable. Flexible armored cable must not be used in wet or damp locations.

A set of electricians' tools suitable for doing farmstead wiring can be bought for around $60 and should last a lifetime if given proper care.

The four steps required in preparing a bill of materials are as follows: (1) Draw to scale a floor plan of the building to be wired; (2) sketch in the electrical system, using electrical symbols to designate outlets, switches, circuits, and the like; (3) tabulate items in each circuit; and (4) list the total bill of materials.

QUESTIONS

1. For what kind of circuit would you use type-TW wiring?
2. Why is it dangerous to use flexible armored cable in wiring barns or cellars?

3 How much would it cost to buy a set of electricians' tools for doing farmstead wiring?

4 What is a "three-pole, flush-mounted, range receptacle"?

5 Why is it necessary to have a scale drawing of a building in preparing a bill of materials?

ADDITIONAL READINGS

Davis, Hollis R., *Adequate Farm Wiring Systems,* Extension Bulletin 849. Ithaca, N.Y., Cornell University, 1956.

Graham, Kennard C., *Interior Electric Wiring.* American Technical Society, Chicago, Ill., 1961.

Industry Committee on Interior Wiring Design, *Farmstead Wiring Handbook.* New York, N.Y., 1964.

Industry Committee on Interior Wiring Design, *Residential Wiring Handbook.* New York, N.Y., 1954. (Being Revised.)

Masterson, Robert H., *Study Guide for Interior Electric Wiring.* American Technical Society, Chicago, Ill., 1961.

Montgomery Ward and Co., *Modern Wiring.* Chicago, Ill., 1955.

National Fire Protection Association, *National Electrical Code, 1968 Edition.* Boston, Mass., 1968.

Sears, Roebuck and Co., *Electric Wiring.* Chicago, Ill., 1955.

8 HOW TO DO FARMSTEAD WIRING

The addition of a major electric appliance to the home or farmstead often requires some kind of wiring job. Because of this, people tend to delay buying equipment that is sorely needed. As a consequence, thousands of farm people are still using their backs and hands in competition with barn cleaners, silo unloaders, grain and hay elevators, wood saws, water pumps, automatic feeding and watering devices, and other electric machines. Perhaps you studied the example in Chapter 1 which shows that you earn as little as 2 to 5 cents per hour doing hand labor in competition with electric power.

Sometimes a new electric machine can be put into operation simply by installing a convenience outlet. More often, however, it is necessary to install a larger service switch and perhaps string a new feeder circuit for the new machine. As a matter of fact, there is a constant need on most farms for repairs, as well as additional branch circuits, wall switches, and so on. The farmstead wiring job is never done.

If you acquire the basic wiring skills covered in this chapter, you will have a decided advantage over many of your farm neighbors and other home owners.

SHOULD YOU UNDERTAKE TO DO A WIRING JOB YOURSELF?

Until you have had some experience in wiring, you should work on simple wiring jobs and not attempt to wire a major building without expert supervision. You should do a few simple wiring jobs first. An easy job for a beginner is to install a surface-type convenience outlet in your shop or garage. As a next step you might install a two-way switch, followed by a general-purpose circuit. Then, as you learn more about wiring, you can undertake more advanced jobs.

Many farmers, club boys, and housewives have shown by demonstration that farmstead wiring is not difficult. Proof of this can be seen in the thousands of electrification projects that are carried out each year in every section of the country. These range from small repair jobs to complete wiring of residences and other farm-service buildings.

The remainder of this problem-unit is devoted to instructions on *how to do* farmstead wiring. Instructions given in this section should enable you to do most of the wiring jobs on your farmstead.

HOW SHOULD WIRE BE CUT, SPLICED, AND CONNECTED TO TERMINALS?

The skills that are most often used in farm wiring are cutting, splicing, and connecting wire. Although these operations are rather simple, much faulty wiring can be traced to improper procedures in doing these jobs.

How to Cut and Connect Wire to Terminals

If you are working with cable, the first step is to remove about 8 inches of the outer cover. For instructions on how to do this, refer to page 179. Next, by holding your knife blade at a flat angle, or by using a stripper, remove about 1 inch of the insulation from the end of the wire. Take care not to damage the wire, especially if you are using a sharp knife. The traces of insulation left on the bare wire should be scraped off with a dull knife.

Next, form a right hand loop in the end of the bare wire as shown in Figure

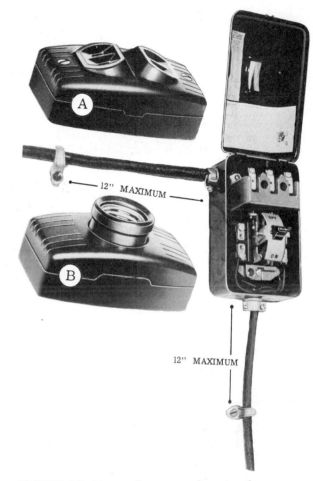

FIGURE 142. Two surface-type outlets, A and B, can be installed by cutting cable and connecting the wires to the proper terminals. Outlet boxes are easily anchored to the wall by two screws.

144-C (left). In Figure 144-C (right), a wire is connected to a terminal without being cut. Notice that the insulation has been carefully trimmed at an angle.

How to Splice Wire

Remove about 3 inches of insulation from the area to be spliced. Taper the

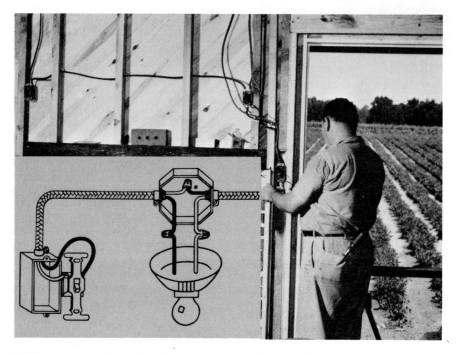

FIGURE 143. *Top:* A workman installing a two-way wall switch. The diagram at left shows a simple hookup.

FIGURE 144. *Bottom:* (A) Method of removing insulation with a knife. (B) A workman preparing wires for connection to a plug-in receptacle. (C) Two methods of forming loops for connection to terminals.

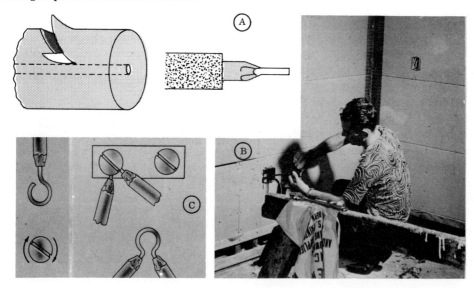

insulation cuts and scrape the bare wire clean as you did in preparing a wire for making a connection. A method of twisting the two wires together is illustrated in Figure 145-A. This type of splice is required where it will be subjected to strains.

In Figure 145-B, the splice has been soldered and is being wrapped with rubber tape. This will be followed with a course of friction tape. A dual-purpose plastic tape is easier to apply and can be used instead of rubber and friction tape.

The illustration at the bottom of Figure 146 shows a rat-tail splice, which is easier to make than the line splice above. In the rat-tail splice shown, three wires have been twisted together and soldered, and are being taped. This type of splice is excellent except for the fact that it will not withstand much strain. Still another kind of splice is called a *tap splice*. In this method, the end of one wire is spliced onto an existing electrical line.

How to Bond Electric Wires Properly

When two electric wires are twisted together, very little metal surface is in actual contact because the surfaces are round. Therefore, a splice restricts the flow of electricity unless it is bonded together with solder or by some other satisfactory method. To make matters worse, splices and connections tend to corrode with use and age. Therefore, careless splicing can create a fire hazard.

How to Solder a Splice. One of the methods that was formerly used to bond electric wires is soldering. First, clean the wire surfaces that are to be soldered, then apply a paste soldering flux. Have some resin-core solder on hand along with a hot, well-tinned, soldering iron.

Figure 147 shows the correct position of the soldering iron and wire while applying the solder. As soon as the wires become hot enough, quickly "feed" a

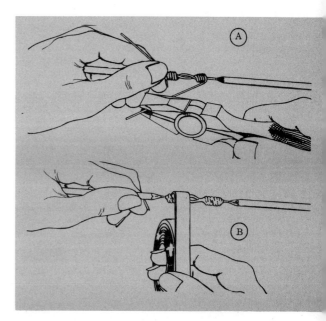

FIGURE 145. (A) Method of making an end splice. (B) The completed splice, soldered and being taped.

FIGURE 146. (A) Rat-tail splice (three wires), soldered and being taped. (B) Tap splice.

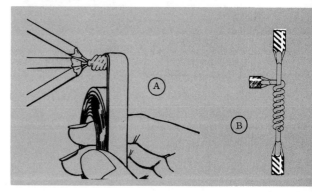

164 USING ELECTRICITY

small amount of solder into the splice and remove the soldering iron. You should barely cover the wires with solder, since too much produces a bulky joint.

Another method of soldering splices, especially if you are doing a large job, is to briefly dip the joint, which has been cleaned and fluxed, into a pot containing molten solder. Enough solder will stick to the wires to form a good bond. Wipe off excess flux.

How to Use Wire Connectors. The use of solderless connectors is illustrated in Figure 148. After the ends of the wires have been stripped of 1 inch of insulation and cleaned, the connector is screwed onto the ends of the bare wires. This forms a permanent bond on which tape is not needed.

In Figure 149 the two connectors at the left are the solderless type, while the two at the right are called split-bolt connectors. The latter are used to form permanent connections that require strong attachment; for example, connecting a ground wire to a ground rod.

HOW SHOULD THE EXTERIOR DISTRIBUTION EQUIPMENT BE INSTALLED?

It is essential to have the exterior distribution equipment properly installed. Generally, *the wiring of a yard-pole or the installation of metering*

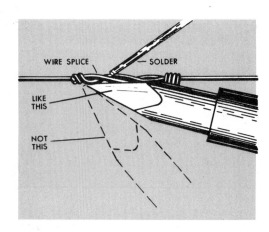

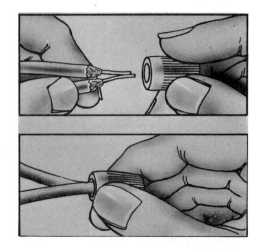

FIGURE 147. *Top left:* The correct position of soldering iron, wire, and solder for soldering a splice.

FIGURE 148. *Bottom left:* Steps in bonding wire with solderless connectors.

FIGURE 149. *Bottom:* At left, two solderless connectors; at right, two split-bolt connectors.

HOW TO DO FARMSTEAD WIRING

FIGURE 150. Metering installation at a farm building. *(A)* Conduit service entrance including meter. *(B)* Main service switch inside building.

equipment at a building is done by the power supplier or by an independent electrician. Even so, it will pay you to know the requirements of proper installation.

Metering at a Building

In Figure 150-A, the exterior part of a service entrance, including the meter, is installed on a centrally located barn. This is a conduit installation, but service entrance (SE) cable is also used in installations of this type.

The main service switch and fuse panels are shown in Figure 150-B. This part of the installation is located inside the barn.

Specifications for a standard metering service are shown in Figure 151. You

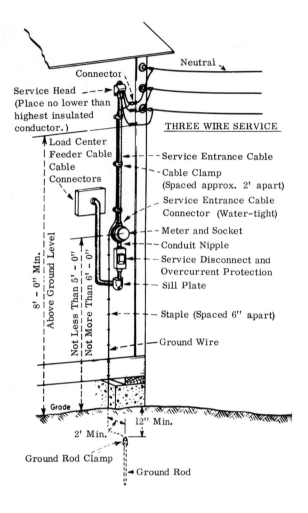

FIGURE 151. Specifications for a standard service entrance using SE (cable) and a meter installed at a farm building.

can refer to this diagram for information, knowing that it meets proper standards of farmstead wiring. Check it also to see whether or not your wiring job is being done properly.

Metering at a Yardpole

While there are many variations in the wiring of yardpoles for central metering, the two most popular types are shown in Figures 152 and 153. By referring to the appropriate installation you can determine whether or not your wiring job is being done properly.

Kind of Pole to Use. Insist on having a durable yardpole that will last many years without danger of coming down during a storm. Creosoted timbers are usually used for yardpoles. The height of the yardpole must be sufficient to maintain a 10-foot clearance of the lowest wire on the farmstead, except at driveways where at least 15 feet is required.

How to Install Feeder Circuits

Feeder circuits may lead from a yardpole to a building, from one building to another, or from one service switch to another in the same building. Feeder circuits may be three-wire, 230 volts or two-wire, 115 volts.

A farmstead will be a dangerous place if the feeders are not properly installed. The following "rules of thumb" will serve to make the service satisfactory as well as safe.

1. Feeders must touch nothing but their supporting insulators.
2. Wires should be at least 8 feet from roofs or other parts of farm buildings.
3. The lowest point of attachment of feeder to building must be 10 or more feet above ground level.
4. The bottom wire of feeder circuits should have a clearance of at least 15 feet over driveways; on the remainder of the farmstead, 10 feet is acceptable.
5. Feeder wires may be spaced 12 inches apart or twisted together and should not cross another building if possible.

HOW TO DO FARMSTEAD WIRING 167

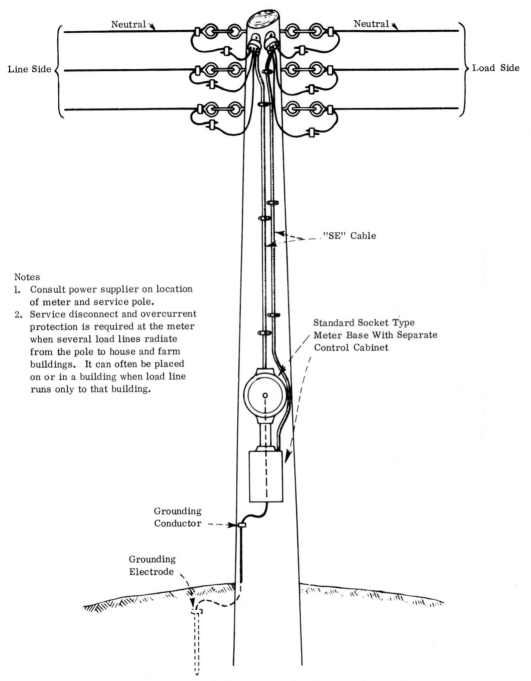

FIGURE 152. Specifications for a yardpole metering installation using service entrance cable.

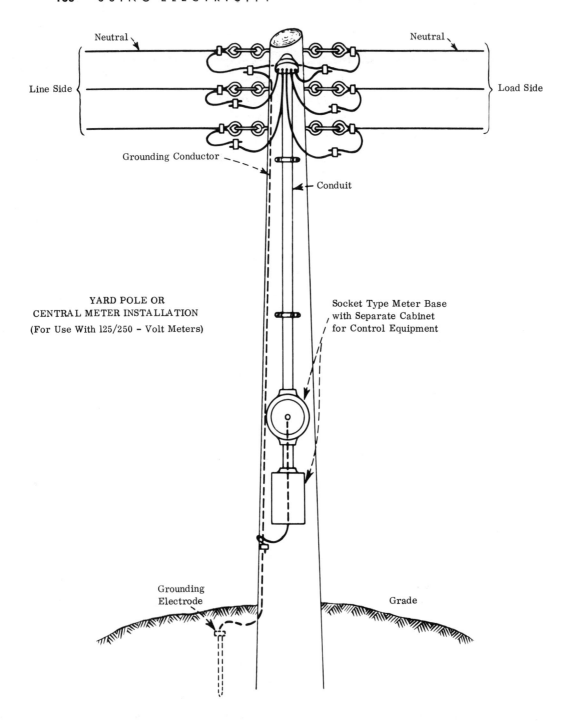

FIGURE 153. Specifications for a yardpole metering installation using conduit.

6. Feeders should be installed in such a way that persons and livestock will not normally come closer than 3 feet to them.

7. The National Electrical Code specifies that No. 10 copper wire or No. 8 aluminum wire may be used for overhead runs up to 50 feet; No. 8 wire or larger is required for runs that go over 50 feet.

8. Insulated wire shall be used for overhead circuits, but a special type of wire is required for underground feeder circuits.

Attachment of Feeders to the Power Source. By studying the diagram in Figure 154 you will see that the three feeder wires are attached to the yardpole in the same manner as to the power drop. The important points to observe here are (1) to use the proper type and size of insulators and make certain that they are securely anchored to the pole, (2) to see that each wire is correctly tied as shown in Figure 154-A or B, and (3) to make certain that the feeders are properly connected to the service wires leading out of the switch or meter. The top (neutral) feeder is connected to the neutral of the power drop by means of a jumper wire around the pole. (NOTE: *A qualified electrician should handle the yardpole wiring job.*)

Attachment of Feeders to Buildings. The diagram in Figure 155 shows one method of attaching wires to a building. The feeders are attached to the side of the building, each wire being tied to a screw-type insulator. (See Figure 154 for tying details.) The insulators should be anchored to 2-inch thick lumber or a double thickness of 1-inch lumber; otherwise the anchors may pull loose during storms and icing conditions. Feeder

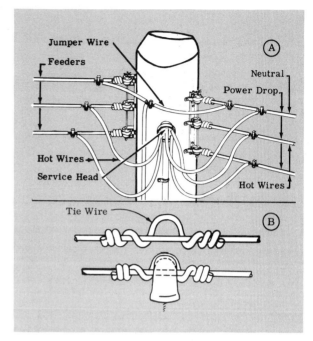

FIGURE 154. *(A)* Method of attaching feeders to a powerpole. Notice how the feeders are tied to the insulators. *(B)* Another method of tying, where the wire continues on to another attachment.

FIGURE 155. Method of anchoring feeders to the side of a building. Screw-type insulators are fastened to 2-inch lumber.

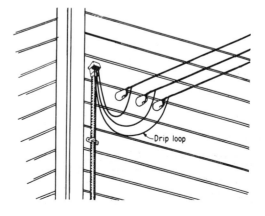

170 USING ELECTRICITY

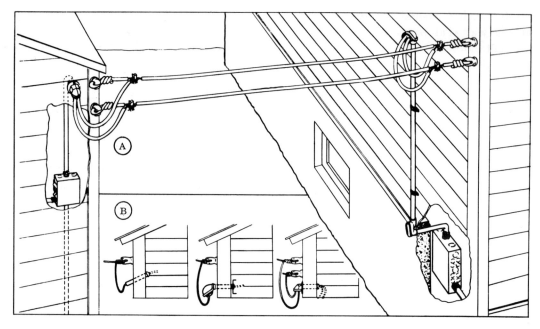

FIGURE 156. *Top:* (A) Method of wiring a two-wire feeder from the service switch at right to the building at left. (B) Three other methods of bringing feeder wires through a building wall.

FIGURE 157. *Bottom:* Underground feeder circuit wired with acid-resistant plastic-cable buried directly in the soil.

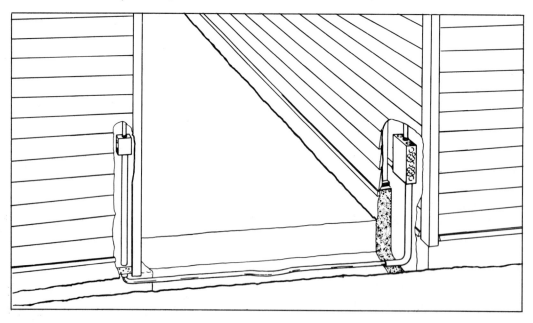

wires may be spaced 12 inches apart, or twisted together. (Figure 156)

If the weather is hot when you are installing feeders, tighten each wire so that about 6 inches of sag is present in a 50-foot run, with slightly more for a 100-foot run. In cold weather tighten each wire about as much as you can by hand.

How to Install a Two-Wire Feeder Circuit from One Building to Another

Figure 156-A shows a method of bringing a two-wire feeder circuit out of a service switch and stringing it to another building. The skills involved in this job are covered in the following section on installing a service entrance at a building.

Figure 156-B shows three methods of bringing feeder wires through a building wall.

How to Wire an Underground Feeder Circuit

In Figure 157, part of the wall sections has been removed so that you may see how an underground circuit is wired. In this installation, plastic cable is buried directly in the ground. The trench is 24 inches deep. You could use lead cable here, but this would require metal conduit in addition to the cable. The wires (either two or three) lead out of the bottom of the service switch in the building at right. Notice the length of conduit leading from the switch box and extending through the foundation wall. This tubing protects the wires from damage until they enter the ground outside the building.

The plastic cable enters the trench from the ends of tubing at both buildings. After the cable is laid and all connections have been made, the ends of the tubing are sealed with water-proof insulating compound. (NOTE: For instructions on how to connect wires to a service switch, refer to the section that follows below on installing a service entrance.)

In filling the trench, first lay a run of one-inch board over the plastic cable or other wiring, then cover with soil. Do not use cinders. Mark the course of the circuit with stakes or some other signals to follow.

How to Install a Service Entrance

Installing the service entrance for each farm building is one of the most important wiring jobs on the farm. You can learn to do this job by studying the diagrams and instructions that follow. Assume that a three-wire, 230-volt feeder circuit has been brought to the building and anchored to a 2-inch wall board.

Step 1: Anchor Service Switch in Place and Install Sill Plate or Conduit Ell. The service switch should be located near the point of attachment of the feeders, usually inside the building. Figure 158 shows a service switch and two distribution panels mounted on the surface of a wall. Another type, used in a residence, is called the flush-mounted switch because it is recessed into the wall and does not protrude.

Mounting a surface switch is a simple operation of fastening the anchor screws into the wall. Installing a flush-mounted switch, however, requires saw-

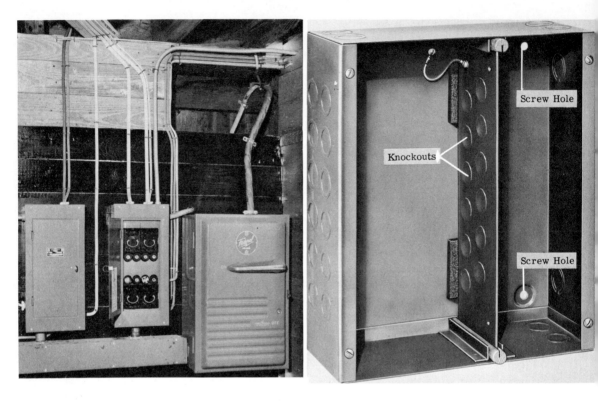

FIGURE 158. *Left:* The main service switch at right has a 200-ampere capacity and is fused at 150 amperes. It serves two sub-panels at left for the feed mill and the dairy barn.

FIGURE 159. *Right:* Metal case in which a fuse panel is to be mounted.

ing a proper size opening in the inside wall, then anchoring the box in place with screws or nails.

Notice the knockouts in the sides of the metal box in Figure 159. These can be removed as needed to provide holes for wiring. Cable connectors are installed at each hole where wiring enters or leaves the box.

To install a sill plate or conduit ell, bore a proper size hole in the wall and insert the device. Both devices prevent the seepage of water into the wall, when properly sealed with waterproofing compound.

Step 2: Prepare Service Wires for Installation. About 10 feet of SE cable or 10 feet each of red, black, and white conduit wire is sufficient for the average service entrance. (NOTE: Refer to Chapter 5 for instructions on how to figure wire size.)

(a) In preparing SE cable for installation, remove about 3 feet of the outer covering at one end to allow for drip loops. Also remove 18 inches of the outer covering at the other end to permit connections at the switch. The neutral wire in SE cable consists of numerous strands of bare wire (Figure 162)

FIGURE 160. Fuse panel mounted in a box.

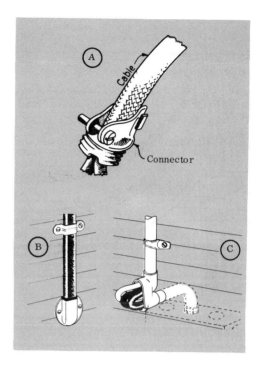

FIGURE 161. *Top:* (A) The cable connector fastens the cable to the switch box. (B) The sill plate makes a watertight connection for passage of the cable through the wall. (C) The conduit ell serves the same purpose as the sill plate.

FIGURE 162. *Bottom:* Numerous small strands of wire in SE cable are twisted together to form neutral wire.

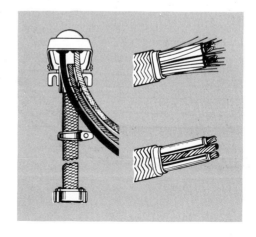

twisted about the two insulated wires. At both ends of the cable, these bare wires are twisted together to form the neutral. Remove 2 inches of insulation from both ends of the black and red wires and see that the bare wire is clean and ready for making connections.

(b) If conduit is used, remove 2 inches of insulation from both ends of all wires and see that the bare wire is clean and ready for making connections. (See, also, Figure 155-B.)

Step 3: Anchor Service Cable or Conduit to the Wall. Service cable installation is somewhat different from conduit installation.

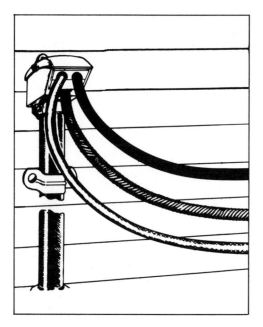

(a) In installing SE cable, anchor the service head in place at least 10 feet above ground level. Thread the service wires into the service head. To fasten the service cable to the wall, use one cable strap every 2 feet. Push the cable through the sill plate and pull the wires into the switch box. Make certain that there is ample wire for making connections to the switch. Be sure to tighten the cable connector where the cable enters the switch box.

(b) In installing conduit, the three wires (black, red, and white) are fished through tubing and threaded into the service head before the unit is anchored

FIGURE 163. *Top:* Service entrance cable threaded into a service head and anchored to a wall, with one cable strap every 2 feet.

FIGURE 164. *Bottom:* Conduit wires threaded into a service head and anchored to a wall with one conduit strap every 4 feet. The inset shows the connections inside the service switch.

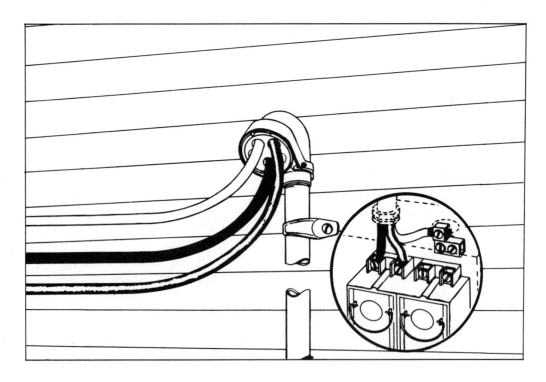

in place. Then fasten the conduit unit to the wall, one conduit strap every 4 feet. Push the service wires through the conduit ell, and pull them into the switch box, leaving sufficient wire to make connections to the switch. Tighten the connector where the wires enter the switch box.

Step 4: Install Grounding System. Have on hand a 12- to 15-foot length of No. 4 bare copper ground wire. Loosely anchor this wire to the wall alongside the service cable, using one staple every 6 inches. Allow about 3 feet of wire at the top for making a connection with the neutral feeder wire. Pound down the staples, taking care not to damage the ground wire.

Drive the ground rod. A ⅝-inch, copper-coated, solid steel rod, 8 feet in length, is suitable for a ground rod on the farm. Some electrical inspectors will allow ¾-inch galvanized water pipe to be used interchangeably for the ground. The rod, as indicated in Figure 165, should be driven to a depth of at least 12 inches below ground level; in the city, the ground wire may be attached to the city water pipes by special ground clamps. (See Figure 165.)

Attach the ground wire to the ground rod. Using a non-rusting, split-bolt connector, fasten the ground wire to the ground rod or to city water pipes. Do not cover the ground rod connection until the power supplier or inspector has examined it. A faulty ground is extremely dangerous.

Step 5: Wire the Service Switch. Refer to Figure 166 and note the order in which the service wires are connected to the upper terminals: (1) black to the extreme left, (2) red to the terminal on the right, and (3) bare neutral (or white, if conduit) to the center. Note also that the ground wire leads from the upper center terminal to the outside ground.

Notice that a three-wire, 230-volt range circuit leads from the knock-out hole at the upper right-hand part of the switch box. This circuit is protected by a 50-ampere cartridge fuse mounted at the back of the pull-out fuse blocks.

Another three-wire, 230-volt circuit is tapped from the "take-off" lugs located underneath the row of plug fuses. This 230-volt circuit can be used to operate a dryer, air-conditioner, or other high-wattage appliance, or it may be used as a feeder for another small service entrance. Four 115-volt circuits are wired from the plug-fuse panel, the wires leading from the bottom of the switch box. A modern all-electric residence requires a larger switch than this 60-ampere size—probably 200 amperes.

Step 6: Have Service Entrance Inspected and Have Service Wires Connected to Feeders. Always follow the practice of having a new wiring job inspected before turning on the current. Your power supplier will furnish this service for a nominal fee, if not free of charge.

The power supplier will furnish a lineman (or you may have to employ one) to connect your service wires to the feeders and to connect the feeders to the power drop. *Do not attempt to connect these wires yourself.* To do so is to invite death!

HOW SHOULD INTERIOR WIRING OPERATIONS BE DONE?

There are some half-dozen types of wire that are suitable for use inside farm

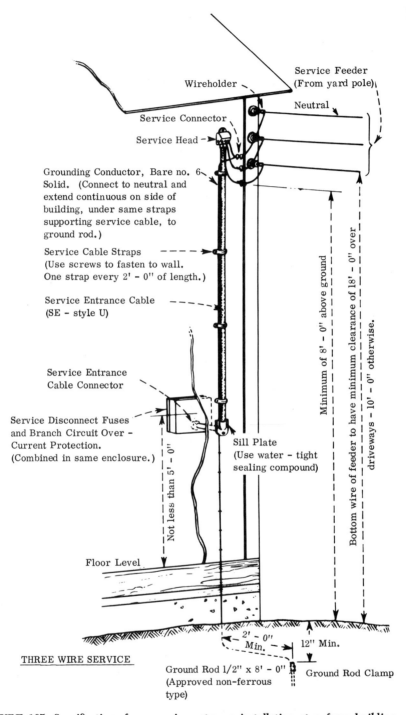

FIGURE 165. Specifications for a service entrance installation at a farm building.

buildings. The three most popular of these are: (1) non-metallic sheathed cable, (2) thin-wall conduit, and (3) flexible armored cable. These and other types of wire are described in detail in Chapter 7.

In this discussion, instructions are given for installing the three most common types.

How to Install Non-Metallic Sheathed Cable

Study the illustration in Figure 167 as you proceed. After laying out and cutting the cable into proper lengths, prepare the ends for making connections. Strip off about 8 inches of the outer covering to provide extra wire for making connections in the box, using a dull knife or wire stripper for the job. Take care not to damage the wire. After getting started, a curved knife of an electrician

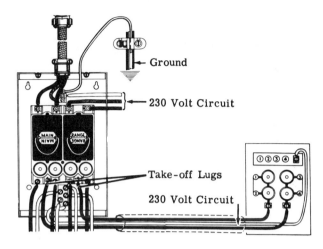

FIGURE 166. Combination service switch and fuse panel.

FIGURE 167. (A) Sample installation of non-metallic sheathed cable: (B) method of stripping off outer jacket; (C) cable fastened to supporting board with cable strap every 3 feet; (D) method of tightening locknut with screw driver; (E) sample of 3-wire cable.

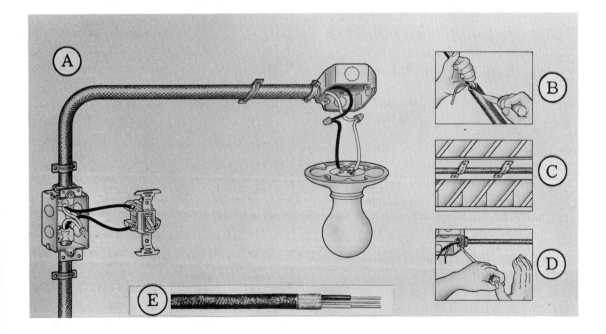

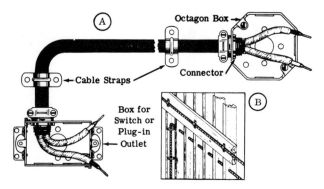

FIGURE 168. (A) Details of connecting non-metallic sheathed cable to outlet boxes. (B) Method of getting cable through studs.

(dull) provides an easy means of removing the outer cover. Remove about 1 inch of insulation from the end of each wire that is to be connected to a terminal, leaving the bare wire exposed. Refer to page 161 for more details on making connections.

Method of Connecting Cable to Boxes. The connector in Figure 167-D should be fastened to the outside covering of the cable in the manner as shown. Next, insert the wires into the box through the knock-out hole. Finally, fasten a locknut (inside box) onto the threaded end of the connector. The inset at Figure 167-D shows a method of tightening the locknut with a screwdriver.

Other types of boxes are equipped with built-in clamps and therefore do not require connectors and locknuts. With this type, a screw is provided for tightening the clamps.

Notice the sample of three-wire cable in Figure 167-E. The third wire is a bare ground wire. This extra ground is connected to the metal box and thus provides additional safety.

How to Fasten Cable to Supports. If cable is to be exposed, as in a barn or garage, it must be fastened with a cable strap every 3 feet. It must also be supported by a stud, beam, joist, or other solid member as shown in Figure 168-B.

In attics or other partially concealed spaces, non-metallic sheathed cable may be run across the edges of the rafters, provided that it is 7 feet or more above the attic floor. The cable should follow the contour of the roof or room and not cut across the spaces. In concealed spaces, cable straps should be spaced 4 to 4½ feet apart and 12 inches from each box. In old walls, cable may be fished through and used in these concealed runs without cable straps.

How to Install Thin-Wall Conduit

As stressed in Chapter 7, wiring that may be subjected to severe wear or possible damage should be protected by tubing. This method is referred to as "conduit wiring" and consists of a metal tubing and metal boxes connected so as to provide a grounded system. The wires used in conduit are not bound together as is true with cable, but rather are separate wires. So if you wish to run a two-wire, 115-volt circuit to a ceiling fixture, use one black and one white wire; or use two black wires to a two-way wall switch. For a three-wire, 230-volt circuit, use one black, one red, and one white wire. In short, conduit wiring provides stricter control of your colors (polarizing) than is true with non-metallic cable.

If you will study the illustrations in Figures, 169 and 170 as you proceed, you will find that it is not difficult to install conduit. You must, of course, have

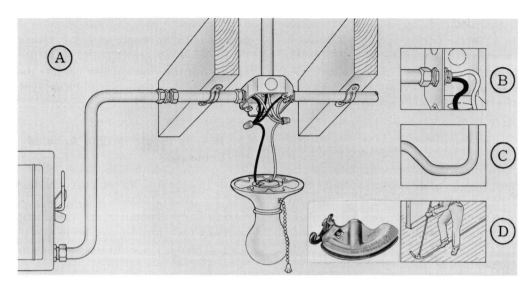

a few specialized tools for the job. For example, a bender, which is shown in Figure 169-D, is necessary if you are to do a very neat job of bending the tube. Directions for using this tool are furnished with it. You may be able to rent or borrow conduit tools from your wiring supply dealer or from your local department of vocational agriculture.

Step 1. Cut the conduit to proper lengths, rough in the boxes, and connect the tubing to the boxes. Since conduit is furnished in 10-foot lengths, you will have to cut shorter pieces as may be required in making runs from one box to another. Use a 32-tooth hacksaw blade to cut thin-wall conduit. The inside edges of these cuts should be smoothed with a pipe reamer or a rat-tail file so as to prevent damage to the wires.

Mount the empty conduit in place, using one conduit strap every 6 feet on exposed runs, or one every 10 feet on concealed runs, and connect to boxes before the wires are installed. Refer to the illustration in Figure 170 and notice that the tubing is fitted with a connector with

FIGURE 169. *Top:* (A) Details of a conduit lighting outlet. (B) Connection to a switch box. (C) Correctly bent tubing. (D) conduit bender.

FIGURE 170. *Bottom:* Details of conduit connection to an octagon box.

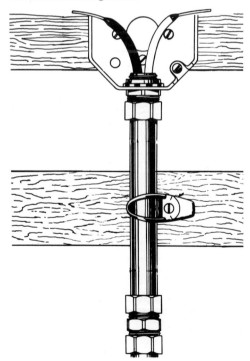

a threaded end. This is inserted into the box, then a locknut is tightened onto it. A screwdriver may be used to run the locknut down tight and thus give a good connection. This is important in grounding the conduit system. After all the boxes have been connected to the tubing and fastened to the wall or studs, you are ready to "fish" the wires into the conduit.

Step 2. Fish the wires into the tubing. Your wiring diagram should be followed when you insert the required number of differently colored wires for each run. The first step in fishing wires is to insert the fish tape from the outer end of a run until the hook protrudes at the inner box. Fasten wires to the hook and pull them through the tubing. Repeat this process for all runs in the wiring system, always leaving about 8 inches of excess wire for making connections at each box. You now have only

to prepare the ends of your wires, connect them to the proper terminals, mount the switch, convenience outlet, or fixture, and mount the cover plates. (NOTE: Remember to use only metal boxes with conduit, never bakelite or porcelain.)

How to Install Flexible Armored Cable (BX)

In Chapter 7 it was stressed that flexible armored cable should not be used in barns, basements, or other damp locations and should never be used outside. Remember also to use only metal boxes with armored cable. Bakelite or porcelain boxes with this type of wiring will destroy the "grounded" effect and result in a dangerous wiring system. The usual type of connector for fastening armored cable to boxes is shown in Figure 172-D.

FIGURE 171. Tubing and outlet boxes ready for fishing wires through. Notice the fishtape hook at right center.

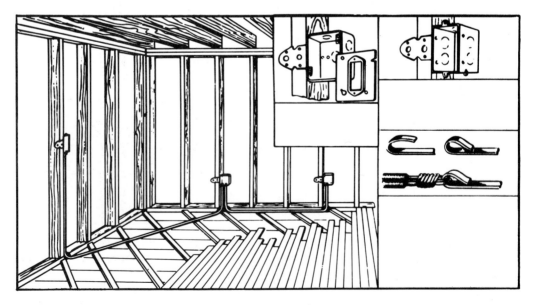

HOW TO DO FARMSTEAD WIRING

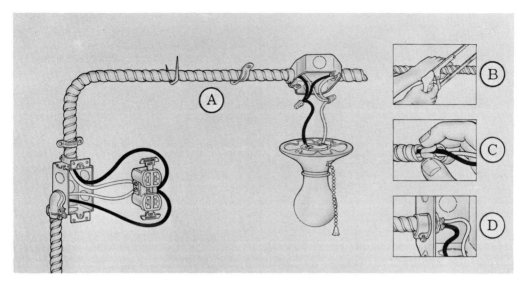

FIGURE 172. (A) Flexible armored cable installation, including lighting outlet and duplex convenience outlet. (B) Method of sawing armor. (C) Method of inserting bushing. (D) Cable connector.

As you proceed with the job of installing armored cable, study the illustration in Figure 172.

Step 1. Cut the armor and the wires and finish the ends of the armor. In order to get a good connection and a proper ground, it is necessary to give the cut ends of armored cable a special finish. First, after sawing through the armor with a 32-tooth hacksaw and cutting the wires, insert a fiber bushing between the insulation and the armor. This bushing is necessary to protect the insulation from damage by the sharp edges of the metal.

Next, cut off the bond wire, leaving about 3 inches, which is then bent back over the armor. Remove the paper covering from the wires and slip a connector (with locknut removed) onto the wires. Be sure that the fiber bushing is in contact with the connector. Tighten the set screw securely after winding the bond wire around it. This insures a sound connection between the bond wire and the metal box and thereby provides a good ground for the wiring system.

Step 2. Connect the armored cable to the boxes. Armored cable must be

FIGURE 173. Sample of flexible armored cable and fittings.

FIGURE 174. Surface-type switch in a barn, wired with flexible armored cable.

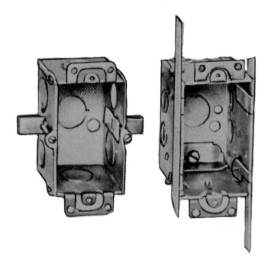

FIGURE 175. Switch or convenience outlet boxes. The screw-type box at left will hold in any wall.

supported by straps or staples every 4½ feet and at a point 12 inches from each box. This support is not required in concealed runs where it is not practical to use a fastener. Insert the finished end of the armored cable into its proper box so that the threaded end of the connector protrudes into the box. Insert a locknut inside the box and tighten with the aid of a screwdriver.

The 8 inches of excess wire is ready for connection to the receptacle, switch, or other fixture that is to be installed in the box. See later sections for further information.

A surface-mounted box and switch are shown in Figure 174. In barns, garages, and other locations where surface fixtures are not objectionable, this type of wiring is desirable because of its ease of installation.

How to Install Wall Switches and Convenience Outlets

The skills involved in installing wall switches and convenience outlets are very similar, except for the wiring connections.

Mark the Location of and Install Boxes. For recommendations on the number and location of switches and convenience outlets for each room and building, refer to Chapter 6. Generally, switches are placed 48 inches above the floor, and convenience outlets are placed 12 inches above floor level. In the kitchen, convenience outlets are installed at table height. Placement may vary depending upon the situations encountered.

In new construction, the installation of boxes is a simple matter of nailing or screwing each box to a stud or other strong member of the wall, after which the wall is finished out around the boxes.

In old work, where additional switches and outlets are to be installed, it is necessary to cut openings for the boxes.

How to Cut Openings for Boxes. An easy method of sawing openings for outlet boxes is shown in Figure 176. The cardboard pattern, as shown at upper left, makes the layout job simple. After marking the area and boring four holes as indicated at upper right, you can quickly saw out the opening. A brace bit and a keyhole saw are needed for this job. A sabre room is faster.

At lower right, the method of connecting the cable to the box is shown. Tightening the locknut will make this box ready to anchor in place.

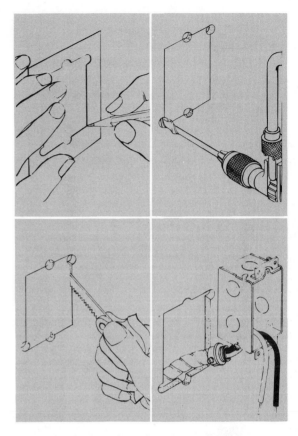

FIGURE 176. Steps to follow in installing a switch box.

Caution on Cutting Plaster or Wall Paper. Before beginning to saw through wall paper, using a sharp knife, cut the paper and canvas from the area along the pattern lines. This will eliminate the danger of tearing the paper when you begin to bore or saw holes.

In cutting plaster, first locate the laths by probing into the plaster at the desired location of your box. An ice pick, in conjunction with a 1/8-inch hand drill, will help you to find the exact location of three laths.

184 USING ELECTRICITY

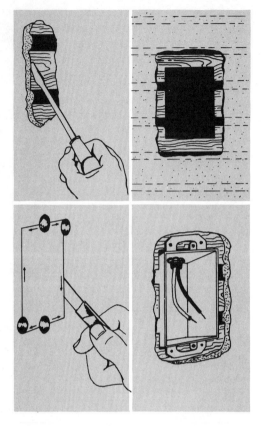

FIGURE 177. *Top:* Steps in installing a switch box in a plaster wall.
FIGURE 178. *Bottom:* A two-way wall switch (in conduit) beyond the lighting fixture it controls. One switch-leg wire would be white if non-metallic sheathed cable were used.

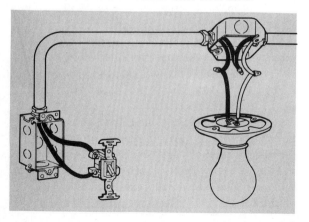

After laying out the pattern, saw through one lath and "dap out" portions of two others, as shown in Figure 177. In cutting plaster, support it with one hand and saw on the pulling stroke.

The final step in installing a switch or convenience outlet is to mount a cover plate. This is not done, of course, until the wiring connections have been completed and the circuit has been tested.

How to Wire Wall Switches

The next step is to wire and install the switches. Before connecting any wires, one basic principle should be reviewed; all wires connected to a switch are hot wires. Therefore, in using plastic or other cable, which contains one white wire, paint the ends of this wire black to show that it is hot; use rubber paint, not oil paint. In wiring switches with conduit, use black wires to the switch; however, it is sometimes necessary to make a white wire function as a hot wire in connecting three- or four-way switches. Again the end of the white wire may be painted black, but this may be confusing, so is not always done.

1. The diagram in Figure 178 illustrates the use of a run of wire beyond a ceiling fixture to a wall switch that controls this fixture. In the junction box, the incoming black wire is connected to a black wire of the switch cable, while the other black switch wire is connected to the black fixture wire. The incoming white wire in the ceiling box is joined to the white fixture wire. The use of solderless connectors eliminates the necessity of soldering. At the switch receptacle, attach the two wires to the two terminals. It doesn't matter which way, since both wires are hot. Remember that

one of the switch-leg wires will be white if non-metallic sheathed cable is used; then the end of the white wire should be painted black. (Usually not done.)

2. The hookup in Figure 179 illustrates switch control for one light fixture, with a convenience outlet on the circuit remaining hot whether the switch is on or off. For this job you will need a run of three-wire cable or three wires in conduit from the switch to the light fixture.

FIGURE 179. Wall switch (A) controls ceiling-light fixture (B) and leaves duplex plug-in hot all the time. (C) Shows grounded-type duplex outlet now specified by the Code. (Note bare ground wire at center. Neutral, white wire, is not connected.)

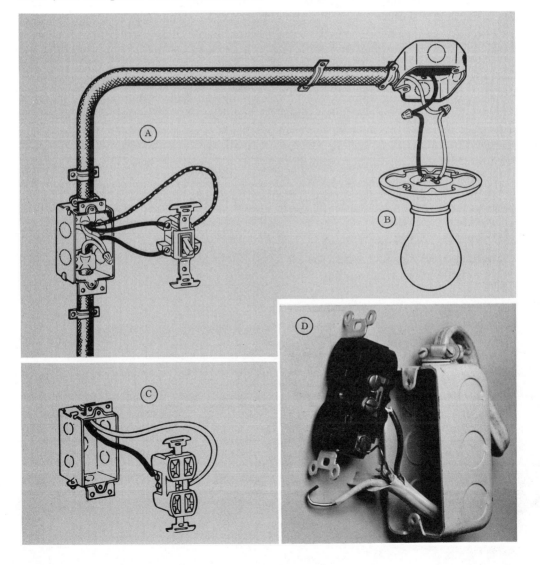

Proceed with connections as follows:

(a) Using solderless connectors, join the two white wires in the switch box; then, in the ceiling box, join the two white wires together and connect them to the white ceiling fixture wire. Notice that one white wire continues on to the convenience outlet, bypassing the switch.

(b) Join the two black wires in the ceiling box and connect the red wire to the black fixture wire.

(c) At the switch, connect both black wires to one terminal and connect the red wire to the other terminal.

(d) You can test your connections by using a home-made test lamp.

(e) Usually, a toggle switch or light fixture is mounted in its box by two screws. Certain types of ceiling fixtures require a different method of mounting. When the lighting fixture or switch is being installed, first tuck the surplus wire into the box, taking care not to damage it.

3. Figure 180 shows a hookup for controlling a ceiling light when the circuit continues on to other outlets.

The connections for this hookup should be made as follows:

The white wire from the source is attached to the ceiling fixture and then continues on to other outlets. The black wire from the source is attached to the black wire leading to other outlets and also to the white wire leading to the switch (when cable is used). The black wire to the switch is then attached to the ceiling fixture.

The connections to the switch itself are simple. Attach each of the wires to a terminal—it doesn't matter which way since both are hot. The white wire that is attached to the switch box should be painted black, using rubber paint. If conduit is used instead of cable, use a black wire here instead of white. The manner of making the connections depends upon the location of the switches in relation to the fixtures and to the source of electricity.

4. Figure 181 shows the simplest arrangement of a three-way hookup. Notice that both three-way switches control a light fixture that is between them. You may have several fixtures here instead

FIGURE 180. *Left:* The wall switch beyond the fixture is wired to control that fixture where circuits continue on to other outlets.
FIGURE 181. *Right:* Ceiling fixture controlled by two three-way switches.

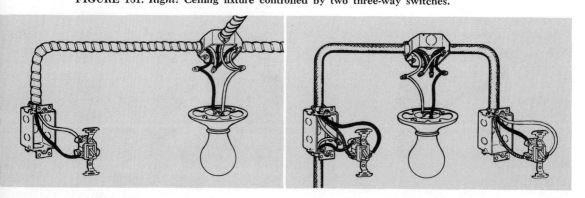

of one; for example, a large living room may require three or four lighting outlets. Notice that the source comes through one of the switches.

Proceed as follows in making connections. (NOTE: For this job you will need a run of three-wire cable from one switch to the other and to each outlet. Figure 182 will show you why two traveler wires are necessary.)

In Figure 181, the white wire is run through the first switch box, without being attached to the switch and continues on to the light fixture. As usual, the white wire is joined to other white wires in the ceiling box and continues on to all fixtures controlled by these two switches. All wires connected to both switches are hot, as is true with all wall switch hookups. At the first switch, the black wire from the source is connected to the dark-colored terminal, sometimes referred to as the *common* terminal. All three-way switches have one of these. (Not always in the same place.)

At the second switch, connect the black wire to the dark-colored terminal, and connect the other end of this black wire to the black fixture wire. At this point, unused terminals remain at both switches. Finish the wiring by connecting a red wire to the lower left terminal at each switch; join the red wires together in the ceiling box.

At the first switch the remaining black wire is connected to the unused terminal there, while the other end is joined to the unused white wire in the ceiling box. At the second switch, the white wire is connected to the unused terminal there and may be painted black, using rubber paint. The two wires that run from one switch to the other are the traveler wires.

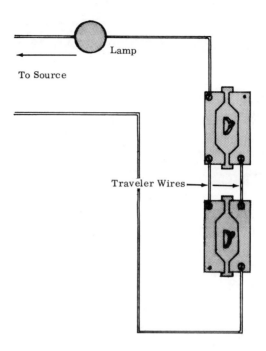

FIGURE 182. Diagram of two three-way switches showing two traveler wires between the switches. One traveler is hot all the time.

FIGURE 183. Three switches controlling a lighting fixture. The switch in the center is four-way, the other two are three-way.

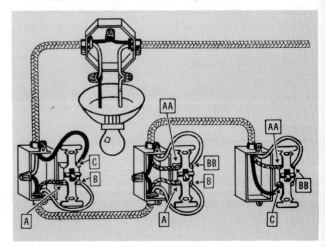

188 USING ELECTRICITY

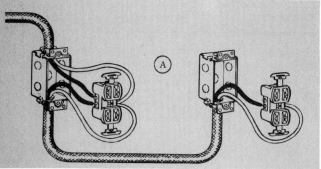

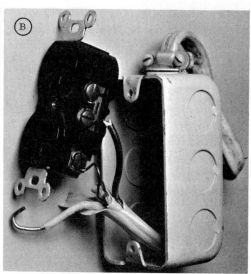

FIGURE 184. An additional convenience outlet (A) can be added to convenience-outlet circuit by extending run of cable to desired location. (B) shows grounded type of duplex outlet now specified by the Code. Note bare ground wire at center; neutral, white wire, is not connected.

A four-way hookup is shown in Figure 183. The connections here may appear altogether different from the previous illustration. A close study of this wiring arrangement, however, will show that it is essentially the same as the previous hookup. You have two traveler wires running from one switch to the others (both hot wires) and the white wire from the source runs directly to the ceiling fixture.

How to Wire Convenience Outlets

An easy and valuable job that you can do in your spare time is to add convenience outlets, often referred to as plug-in outlets. Often, it is possible to tap several plug-ins off your present circuits. Take care, however, not to overload your wiring as one homeowner did. (See story on page 80.)

In wiring convenience outlets, remember that they must always be hot; that is, no switch should be wired in such a way that it can interrupt the circuit for ordinary plug-ins, living room excepted.

1. The arrangement shown in Figure 184 illustrates a method of extending a run of wire from a convenience outlet to operate another outlet of this type. (NOTE: Instructions on how to install non-metallic sheathed cable and boxes can be found on page 177.) The wiring connections for this job are simple:

(a) Attach the black wire of the extension to the copper-colored terminals of both plug-in receptacles as shown in Figure 184.

(b) Likewise, connect the white wire to the silver-colored terminals of both receptacles.

(c) Using the screws provided for this, mount each receptacle in its respective box. Then, after testing has been completed, install the cover plate.

2. Another practical hookup is illustrated in Figure 185. A convenience out-

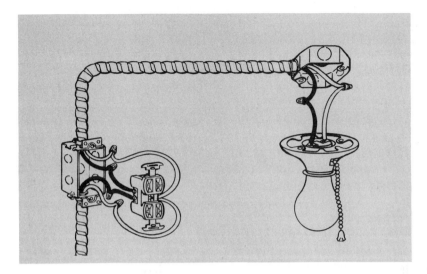

let is tapped into a circuit between two outlets. These may be plug-ins, pull-chain lighting fixtures, or a combination of the two. The wiring connections, again, are very simple and can be done easily.

(a) Cut short lengths of wire, two black and two white, and connect these to the plug-in receptacle. The blacks go to the copper-colored terminals and the whites go to the silver-colored side.

(b) Using solderless connectors, join the short black wires to the incoming and outgoing blacks. Likewise, join the white wires.

(c) Mount the receptacle, test your connections, and install the cover plate as previously instructed.

3. Figure 186 shows a method of tapping a plug-in off a wall switch. To make this outlet hot all the time you will need a run of three-wire cable from the switch to the ceiling fixture. Two-wire cable is used between the switch and plug-in.

Make connections as shown in Figure 186.

FIGURE 185. A convenience outlet tapped into a circuit between an existing plug-in and a pull-chain fixture.

FIGURE 186. Convenience outlet (A) tapped off wall switch (B) that controls ceiling outlets (C) plug-in (D) grounded type is now specified by Code. Note ground wire at center in (D).

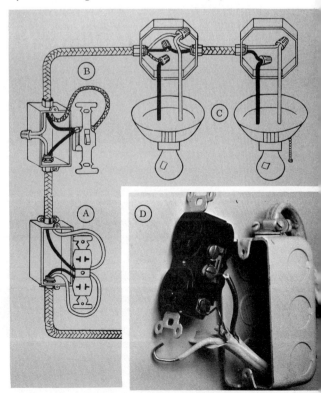

(a) Connect the white wire of the two-wire cable to the silver-colored terminal on the plug-in receptacle. Join the white wires in the switch box and connect the other end of the white run to the ceiling fixture. (NOTE: The white wire from the source is joined at this point.)

(b) Connect the black wire to the dark-colored terminal of the plug-in receptacle and also to one terminal of the switch. The two black wires in the ceiling box are then joined.

(c) This leaves one unused terminal at the switch. Connect the red wire to this and attach the other end of the red wire to the black wire at the ceiling fixture.

(d) Install the switch in its box, test it, and install the cover plate. Note that ground wire is now used, Figure 186-B.

4. The hookup shown in Figure 187 illustrates a method of adding a convenience outlet and a pull-chain light fixture from a junction box.

(a) In the junction box join the three black wires together. Also join the three whites together.

(b) At the plug-in receptacle, attach the incoming and outgoing black wires to the dark-colored terminal; attach the white wires to the other side of the receptacle. Connect ground wire.

(c) Connect black to black and white to white at the ceiling fixture.

(d) Install the receptacle and the fixture in the box, test, and install the cover plate.

How to Install and Wire a Light Fixture

In the preceding discussion, you observed several wiring hookups of ceiling fixtures. The connections for a particular fixture depend upon the situation there. Fixtures controlled by two-, three-, and four-way switches require special connections. Also, the location of the fixture in relation to the power source will affect the connection.

FIGURE 187. A convenience outlet and pull-chain fixture added from a junction box.

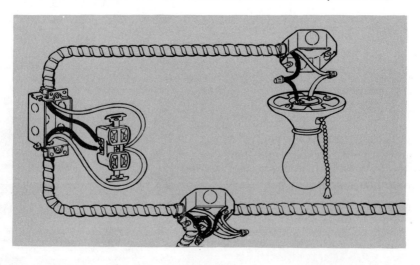

HOW TO DO FARMSTEAD WIRING

The basic principle to remember in wiring lighting fixtures is that you always have one white and one black wire attached to them; and the white wire runs an uninterrupted course throughout each circuit. Also, a ground wire is now used with non-metallic cable.

How to Mount Fixture in Outlet Box. In Figure 188 the workman is completing the installation of an outlet box. This is mounted on a metal rail, which, in turn, is anchored to the ceiling joists. (NOTE: There are several types of fixture boxes available to fit different types of fixtures.) In old work, it is necessary to fasten the box to the ceiling by means of a special hanger. Ask for this at your wiring supply house. Figure 189 shows three different methods of mounting light fixtures.

(a) In method A, the fixture is attached to a metal strap that is anchored to the ears of the box.

(b) In method B, the fixture is mounted to a metal strap that is then anchored to the box by means of a threaded center stud.

(c) In method C, the use of a hickey and a threaded stud is illustrated. By studying the illustration at right you can see how this is installed. Wall bracket fixtures can be mounted as shown in the two illustrations in Figure 190. These are self-explanatory.

How to Connect Wires to a Ceiling Fixture. The method of wiring light fixtures is very simple. With only one or two exceptions, all that is necessary is to connect a black wire to the dark-colored terminal on the fixture and a white wire to the silver-colored terminal. If your fixture is pre-wired with one black and one white stub, you can join the source wires to the corresponding stub by use of solderless connectors. An exception is that two blacks and two whites are necessary if the fixture is equipped with a plug-in that stays hot. Bond ground wires as needed.

For instructions on different methods of connecting wires in ceiling boxes, refer to the section on installing switches and convenience outlets. Several common wiring hookups are illustrated in those sections.

FIGURE 188. A ceiling box being prepared for installation of the light fixture.

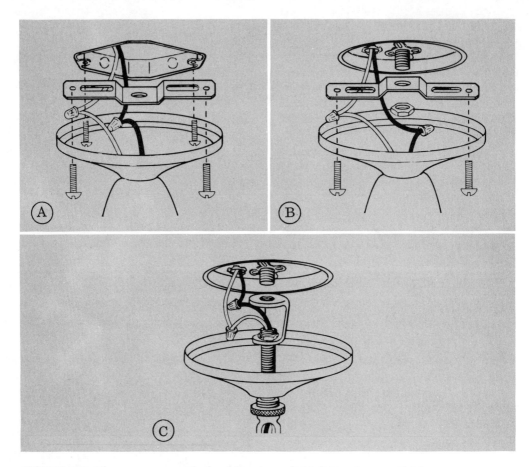

FIGURE 189. Three common methods of hanging a light fixture in an outlet box.

FIGURE 190. Two methods of hanging a wall fixture to a box.

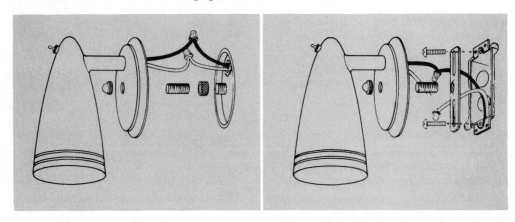

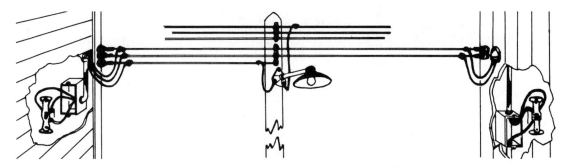

FIGURE 191. Method of wiring a three-way yard light between two buildings.

How to Wire for Three-Way Switch Control of a Yard Light

You can have three-way switch control for your yard lights by wiring them according to the diagram show in Figure 191. This arrangement is for overhead wiring, but an underground circuit can be substituted without changing the basic wiring connections. You can find instructions for installing overhead wiring on page 166; for underground wiring, instructions are on page 171.

Notice that three-wire cable is only required from the light fixture to one switch; a white wire to the fixture being tapped off the neutral line of the feeder circuit. The two wires that are continuous from one switch to the other serve as the travelers.

The connections at the three-way switch (at left) are made exactly as described on page 186. At the other switch, connect the black and red wires to the terminals at the top of the switch. The black wire from the source is attached to the bottom terminal, as shown. No white wire is necessary from the source since this is provided at the yardpole.

(NOTE: This method of wiring a yard light is dependent upon having a feeder circuit available as illustrated; otherwise you must run three wires from one switch to the other. In that event, both switches would have to be three-way. In the method as shown in Figure 191, one of the switches is an ordinary two-way type.)

How to Wire 230-Volt Circuits

Throughout this book, 230-volt circuits have been referred to as having two hot wires and one neutral. Three wires are not necessary to operate certain 230-volt appliances; for example, water heaters. For motors and appliances that are handled however, it is dangerous to operate them on two hot wires without a neutral. Cable may also have an extra ground.

Range Circuit. If you will refer back to page 175, you can find a diagram for tapping a three-wire circuit off the top right position of a service switch. The wiring may be three wires in conduit or three-wire cable.

The three wires shown leading from the switch provide both 115- and 230-volt service. The low voltage is used for low heat, high voltage for high heat.

194 USING ELECTRICITY

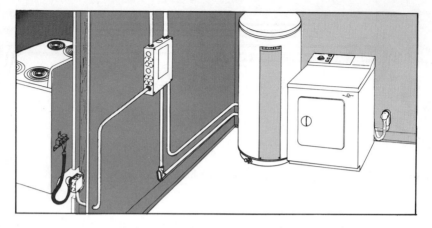

FIGURE 192. Appliances wired for 230 volts.

Fusing for the range circuit is also discussed on page 103.

Figure 192 shows the circuit layout for three 230-volt appliances. At the left you can see the wiring for the range. The wire size for a high-wattage range should be No. 6; and for lower-wattage models, No. 8 is adequate.

You can see how a "pig tail" is connected to the three terminals at the back of the range, while the other end is plugged into a 230-volt receptacle. Three types of 230-volt receptacles are shown in Figure 193. The flush-type at center is used in the installation in Figure 193.

The blades of the male plug shown in Figure 193 are arranged so that the neutral blade must connect with the neutral at the source. Other 230-volt appliances are equipped with the same type of plug. Many 115-volt portable machines are being adapted to a three-wire hookup for added safety.

Water Heater. Only two wires (both hot) are run to the 4,000w water heater shown in Figure 192. No. 10 wire is satisfactory for this installation, but you should use a larger wire for a run over 50 feet. The two hot wires are tapped off the bottom of the service switch and are protected by a 30-ampere cartridge fuse. These two hot wires are connected to the two terminals of the heater with each wire connected to a terminal.

(NOTE: The method of wiring your water heater may be governed by your power supplier; for example, if you are

FIGURE 193. Three styles of 230-volt receptacles. Plug-ins at left and right are the surface-mounted type; center one is for flush mounting.

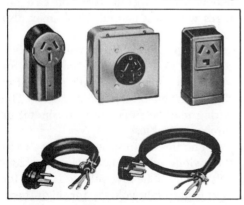

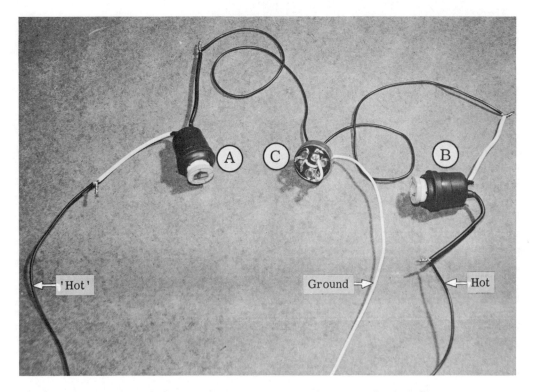

FIGURE 194. Simplified method of protecting a motor wired for 230 volts. Weatherproof sockets (*A* and *B*) are wired into the two hot lines; two Fusetrons (totaling motor amperage) are inserted in the sockets. A three-prong plug *(C)* automatically polarizes the circuit.

on an off-peak rate, you may have to use a separate meter or other special wiring.)

Clothes Dryer. Your clothes dryer, like your range, must have a three-wire circuit. The hookup for this is the same as for a range. The frame of the clothes dryer must be grounded. Ordinarily, No. 10 wire is sufficient for this installation. A "pig tail" plug may be used, as is customary with a range. Sometimes, however, a clothes dryer is connected permanently.

Other 230-volt circuits may be installed for air-conditioners and similar high-wattage appliances. Use two hot wires and one neutral, tapped off the service switch and fused according to the wattage of the appliance.

How to Protect an Electric Motor Circuit

The wiring necessary to install a Fustat or Fusetron for motor protection is very simple: (see Figure 194).

(a) Cut the hot wire of the two-wire, 115-volt circuit serving the motor. Prepare the ends of the wire for splicing.

196 USING ELECTRICITY

(b) Using solderless connectors or solder, join the two legs of a weatherproof socket to the two loose ends of the wire, either leg to either end of the cut wire. Anchor the wiring, including the socket, to a support.

(c) Insert a Fusetron of the proper size.

(d) For a three-wire, 230-volt motor circuit, install two weatherproof sockets and fuse both hot wires; divide the rated motor amperage by two and insert Fusetrons of that size, one in each socket. Fustats can be used for motor protection but they require a special adapter base.

(e) In connecting motor protectors (magnetic breakers, and the like) for large motors, follow the wiring diagram furnished with the device.

How to Construct a 115-Volt 230-Volt Test Lamp

To construct the inexpensive test lamp shown in Figure 195, obtain two weatherproof sockets and prepare the ends of the four wires for splicing. Splice two legs together and finish by soldering and taping in the usual manner.

Prepare two sharp pointed probes (nails will do) and solder one to each unused leg of the socket assembly. Finish by taping the soldered joints, leaving 1 inch of the nails exposed. Insert two

FIGURE 195. A test lamp for 115- and 230-volt circuits can be made from two weatherproof sockets. Splice and solder B and C. Solder pointed probes to the two unused wires. D shows taped probe with 1 inch exposed. Tape A and BC before using.

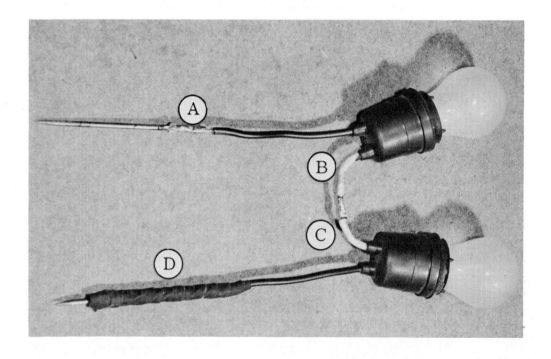

15w bulbs and you are ready to test circuits and electrical devices of all types.

To use this tester, touch one prong to the hot side of a circuit, fuse panel, motor, or other electrical item to be tested. Touch the other prong to the neutral side. If the circuit or device is hot, the bulbs will glow dimly on 115 volts, brightly on 230 volts.

If the bulbs fail to come on, the circuit or device being tested is dead. A little experience with this test lamp will enable you to trace and locate troubles in your electrical system in a few minutes.

SUMMARY

Thousands of homeowners throughout the United States are doing all types of wiring, from simple jobs to the complete wiring of a residence.

In learning to do wiring skills, the simplest jobs should be tried first. These should include splicing wire, repairing lamp cords, and, perhaps, installing convenience outlets. Afterwards, the wiring of two- and three-way switches and lighting fixtures will be in order. Following that, the installation of a small service entrance and branch circuits for a farm building will be a reasonable undertaking.

The first rule in wiring is to obtain a wiring diagram and instructions and follow these carefully.

Exterior wiring involves wiring a meter pole or service entrance at a building. The essentials here include securing the wires firmly to the pole or building so that there will be no danger of their coming loose. Insulators must, of course, be used at the points of attachment.

A good ground rod (or proper connection to city water pipes) is necessary for a safe electrical system. An improper ground may result in injury or even death.

The holes through which wires enter a building should be properly sealed, either by a sill plate or by sealing compound (for underground circuits).

In some areas the service entrance must be a conduit installation; in others, service entrance cable (type SE) is allowed.

Another essential of exterior wiring is the proper installation of feeder circuits. A feeder that runs too close to a building, a tree, or other object may cause a fire or an accident.

The essentials of interior wiring are good splices and connections, as well as properly polarized wires. A poor splice (one not soldered or otherwise bonded) may eventually cause a fire or other trouble. After soldering, a splice must be taped to provide insulation.

In making connections, all white wires should run an uninterrupted course throughout the system. That is, when a white wire runs to a fixture, it must be spliced so that a main white wire continues on to all other fixtures in the system.

A neutral wire is never attached to a switch except when a white wire is allowed to *function as a hot wire* in certain switch connections. Where this is so, the white wire may be painted black to show that it is hot, but this is not generally done.

Between two three-way switches there will usually be three wires. One of these is called a traveler wire.

With the exception of wiring for heating devices, all 230-volt circuits should have three wires. Usually one of the two hot wires will be black and the other red. If a neutral wire is not used in a 230-volt circuit, a dangerous (ungrounded) situation will be present. Thus, it is especially important to ground a television set, a washing machine, and all other high-wattage appliances that are handled.

Grounded type wiring (1 bare wire) is now a Code requirement for receptacles.

QUESTIONS

1 How does a solderless connector form a permanent bond in splicing wires?
2 Why is it important to drive a ground rod at least 8 feet into the earth?
3 Can you explain the principle that all wires connected to two- and three-way wall switches are hot wires?
4 Why must plug-in receptacles and lighting outlets be connected to both hot and neutral wires?
5 Why is a traveler wire necessary in three-way wall switches?
6 Why is it necessary always to use metal boxes when wiring with thin-wall conduit or flexible armored cable?
7 Why is it important that the bond wire in flexible armored cable be properly fastened to the metal casing?
8 Why must a washing machine always be grounded?
9 Why does a range "pig tail" have three wires?

ADDITIONAL READINGS

American Association for Agricultural Engineering and Vocational Agriculture, *Maintaining the Home Lighting and Wiring System*. Agricultural Engineering Center, Athens, Ga., 1965.

Anderson, Edwin P., *Wiring Diagrams for Light and Power*. Theodore Audel and Company, Indianapolis, Ind., 1965.

Brown, R.H., *Farm Electrification*. McGraw-Hill Book Co., New York, N.Y., 1956.

Davis, Hollis R., *Adequate Farm Wiring Systems*, Extension Bulletin 849. Cornell University, Ithaca, N.Y., 1956.

Industry Committee on Interior Wiring Design, *Farmstead Wiring Handbook*. New York, N.Y., 1964.

Larson, R.H. and Turner, C.N., *Handbook of Wiring Specifications, for Electrical Farm Equipment*. Department of Agricultural Engineering, Cornell University, Ithaca, N.Y., 1957.

Montgomery Ward & Co., *Modern Wiring*. Chicago, Ill., 1955.

National Fire Protection Association, *National Electrical Code, 1968 Edition*. Boston, Mass., 1968.

Richter, H.P., *Practical Electricity and House Wiring*. Frederick J. Drake and Co., Wilmette, Ill., 1952.

Sears, Roebuck and Company, *Simplified Electric Wiring Handbook*. Chicago, Ill., 1964.

Texas Education Agency, *Farm Electrification Lesson Plans Wiring and Safety*. Vocational Agricultural Education Division, Austin, Texas, 1967.

SUGGESTED PROJECTS FOR PROBLEM-UNIT THREE

1. Easy-to-Make Trouble Lamp. You can reduce expense and gain experience by constructing your own trouble lamp. Parts are listed in the diagram in Figure 196. Details for installing the plug are shown in four steps in Figure 196-B.

FIGURE 196. Easy-to-make trouble lamp.

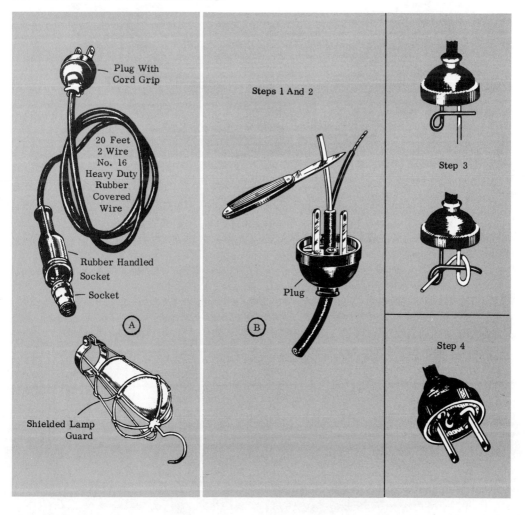

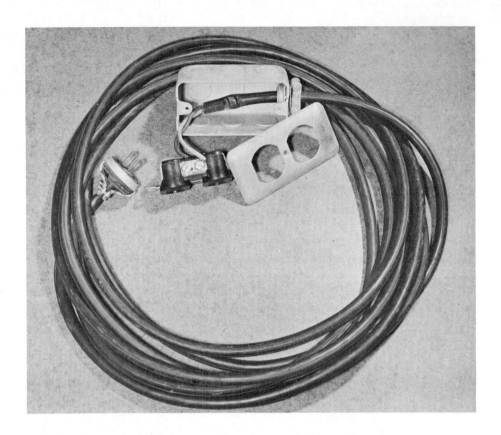

FIGURE 197. Heavy-duty extension cord for the farm shop. The plug-in consists of a metal outlet box, a duplex receptacle, a metal cover plate, and a cable clamp. The receptacle mounts in the box and is held in place by two screws. Note the durable male plug also equipped with a cable clamp. Wire is No. 12 gauge, rubber-covered heavy portable cable.

FIGURE 198. Underground circuit for a post lantern and outside plug-ins. The wire is acid-resistant plastic cable buried directly in the soil.

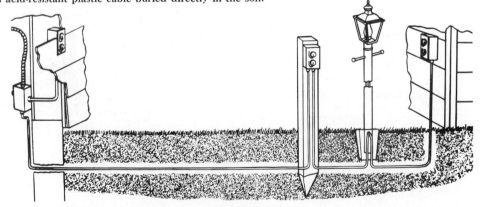

2. Heavy-Duty Extension Cord. A dependable extension cord for operating portable electric machines is shown in Figure 197. Parts needed include a metal outlet box (surface type) and cover, a duplex receptacle, and 50 (or more) feet of No. 10 reinforced portable rubber cord.

Lightly solder the ends of wires to be connected to the plug and receptacle. Make connections as previously instructed.

3. Outdoor Plug-ins and Post Lantern. Figure 198 shows a layout for an underground circuit (using plastic cable) to operate a post lantern, post plug-in, and two outside wall plug-ins. Make certain that boxes and receptacles are of the outside type. Plastic cable is wired to the source in the building at left and leads to (1) a box on building wall, (2) a post-mounted plug-in, (3) a post lantern, and (4) a surface-type plug-in at the other building. Instructions for installing plastic cable for an underground circuit are given on page 171.

4. Wiring Demonstration Board. The board shown in Figure 199 was constructed by a student in Mississippi. It involved most of the basic wiring skills, including the installation of a meter, a service switch, two-, three-, and four-way switches, circuits, convenience outlets, and lighting outlets.

FIGURE 199. Wiring demonstration panel constructed by a student in Mississippi.

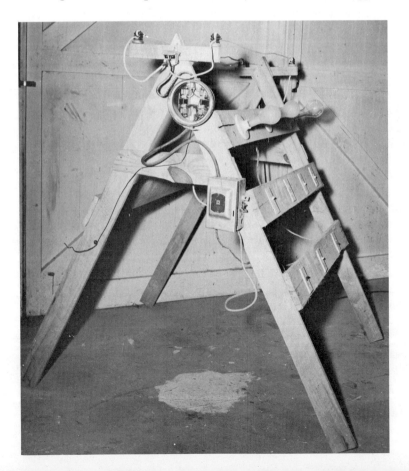

GLOSSARY FOR PROBLEM-UNIT THREE

Lighting Outlet   A wired box to which various types of lighting fixtures may be fastened and supplied with electricity.

Convenience Outlet   A wired box and plug-in receptacle supplying electricity for operating appliances. Receptacle outlets should be duplex type (two plug-in positions) except as otherwise specified by codes. Plug-ins for a dairy barn may be mounted on a heavy-duty cord suspended from the ceiling.

Special-Purpose Outlet   A wired box and receptacle supplying electricity for a particular appliance; for example, a range or water heater. A special-purpose outlet may be a plug-in receptacle, or it may be an outlet to which equipment is permanently connected.

Service Entrance   Includes the service head, the entrance cable, the main switch, the distribution panel, and all accessories such as the sill plate and the cable straps. The ground wire, the ground rod, and the connections are considered a part of the service entrance.

Feeder   Wires extending from a distribution panel in one building to the service entrance of another building, or from a distribution panel to a branch circuit panel in the same building. Also, conductors which connect a meter-pole installation to the service-entrance conductors of the various buildings served from the meter pole.

Branch Circuit, General Purpose   That portion of the wiring system extending from the final fuse or circuit breaker to the outlets for general use; for example, lighting and convenience outlets.

Branch Circuit, Individual Equipment   A circuit that is intended for supplying a single motor or appliance. In general, all stationary appliances over ½-hp or 1,000 watts should be permanently connected to an individual circuit.

Appliance Circuit   A 115-volt circuit consisting of one hot and one neutral wire designed to supply electricity to appliances in the kitchen, dining room, laundry, and outdoors. Appliance circuits should be wired with No. 12 wire (up to 50 feet) and protected by a 20-ampere fuse.

Voltages   In this book, two-wire circuits are treated as supplying 115 volts; three-wire circuits, 230 volts.

Usable Wall Space   All portions of a wall, except that occupied by a door opening or a fireplace opening. Window width is considered usable wall space. The minimum usable wall space requiring a convenience outlet is three feet in length at the floor line.

Floor Area   Refers to area computed from the outside dimensions of the building and the number of floors. In computing the floor area of a farm home, open porches, garages, and unfinished spaces in the basement and attic are not included, unless adaptable for future use.

Closed Circuit   A circuit that is carrying current is a closed circuit. It is sometimes referred to as a live or hot circuit.

Open Circuit A circuit that has been disconnected by a switch, fuse, or circuit breaker. An open circuit is sometimes referred to as being dead or cold.

Short Circuit An improper or accidental contact between two or more electric wires, or between one wire and a path to the ground.

Switch A device for controlling the flow of electricity. A switch opens or interrupts the circuit to stop the current.

Entrance (Service) Switch A wiring device for breaking the connection between the farmstead wiring system and the wires leading from the power company's lines, or for interrupting current to a building or separate service.

Ground A safety precaution consisting of an electrically sound connection to moist earth. A ground conducts into the earth currents and short circuits that sometimes develop in the wiring system.

Fuse A ribbon of soft wire or metal mounted in a container connected to an electric circuit. A fuse limits the amount of current in a circuit. When the circuit is overloaded, the ribbon of wire melts or blows, thereby disconnecting the circuit.

Circuit Breaker A device for protecting the wiring and appliances against too much current (overload). A circuit breaker may be used instead of a fuse. A breaker operates on the principle of tripping by spring tension when an overload occurs.

Wire Type Symbols

R	Rubber insulation.
RU	Latex rubber.
H	Heat resistant insulation.
W	Suitable for use in wet locations.
T	Thermoplastic.
N	Nylon jacket.
NM	Non-metallic cable (sheated or plastic coated).
NMC	Non-metallic cable, flame and moisture resistant.
UF	Underground cable, flame resistant.
USE	Underground cable, not suitable for inside use.
A	Asbestos covering.
MI	Mineral insulation.
L	Lead covered.

Note: Combination "THW" indicates *thermoplastic* insulation, *heat* and *moisture* resistant. Over a hundred such combinations are possible. The Code specifies that wire type must be marked on the outside of cables and cords.

PROBLEM-UNIT IV

HOW TO SELECT AND CARE FOR ELECTRIC MOTORS

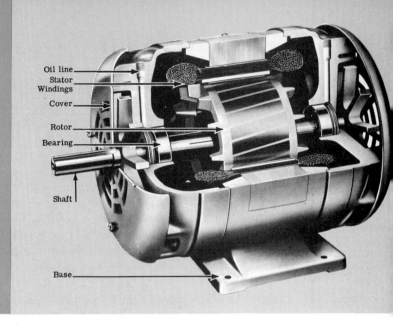

FIGURE 200. Electric power from this motor can do the work of a grown man for less than 15 cents a day.

About 90 percent of all stationary farm power jobs in the United States are done by electric motors. The popularity of electric power is mainly due to its convenience and low cost.

In Chapter 1, it was shown that a homemade elevator powered by a ¼-hp motor could be used in some situations to replace a hired hand. The total cost of electricity and depreciation on this equipment would be less than 60 cents per eight-hour day.

Studies have shown that you can earn the following amounts in competition with electric power:

1. Milking cows by hand, 1 cent per 100 pounds of milk.

2. Handling grain with a hand scoop, 2 or 3 cents per hour.

3. Pumping water by hand, 5 cents per 1,000 gallons.

4. Cleaning a barn and moving manure by hand scoop and cart, 3 cents per hour.

5. Grinding and mixing feed by hand mill, 35 cents per ton.

In addition to providing low-cost power for the farm, electric motors offer other advantages: up to 90 percent efficiency in comparison with 25 to 30 percent for engines; ease of starting in cold weather; quiet operation with no fumes; adaptation to automatic operation; long life with very little upkeep; less danger of fire with no inflammable fuel used; and little or no supervision necessary, permitting operation by unskilled farm help.

FIGURE 201. Illinois beef cattle producer and power company representative watch "endless chain" feeding system in operation. System mixes concentrate with silage, dispenses feed, and cuts itself off.

9 HOW TO SELECT ELECTRIC MOTORS AND MOTOR DRIVES

Most electric appliances come from the factory equipped with the proper type and size of motor. However, you may need to convert a hand-operated or engine-powered machine to electric power. Or, you may need to exchange the motor on a machine for a different type or size. If so, you must make a selection from a number of different types and sizes on the market.

The first electric motor to operate entirely by electric current was made by Thomas Davenport in 1837. He used it for drilling and other light work in his blacksmith shop. By 1850, Davenport's motor had been improved and was being used to do a greater variety of jobs. The one serious flaw in these early motors was the fact that they were operated by batteries and were therefore too expensive to be practical.

In 1886, an alternating-current system was installed at New Barrington, Massachusetts. This made available a plentiful supply of economical electric energy, so when Nikola Tesla produced the first workable a-c motor in 1888, it met with almost immediate success.

How to Identify the Parts of Your Motor. Today, the farm motor is a vastly superior machine in comparison with the early models. Yet the operating principle is practically the same as that used in 1888. Moreover, the parts of a modern electric motor are quite similar to those in Tesla's first a-c motor.

The cutaway section of the motor in Figure 202 shows the following parts: a stationary set of windings called a stator; a rotating unit of insulated metal sections called a rotor; a set of bearings; a capacitor (condenser) for giving greater starting torque; and a lubrication system. This motor also contains a

ELECTRIC MOTORS AND MOTOR DRIVES

wiring box with electric leads for connection to an extension cord. The motor shown has a drip-proof cover and solid metal base.

Before going further, you should refer to page 51 and review the explanation of how an electric motor runs.

WHAT POINTS SHOULD BE CONSIDERED IN SELECTING ELECTRIC MOTORS?

The things that you will be most concerned about in selecting a motor are the size in horsepower, the motor type, the starting torque, the type of motor enclosure, and the type of bearings.

In getting the most from your motor you will also need to select a suitable drive. This job involves figuring the types and sizes of pulleys as well as the types and sizes of belts needed.

The first cost of a motor is of less importance than the necessity of getting the right motor for the job. A cheap motor often turns out to be the most expensive by burning out in a short time.

You can find the speed and horsepower rating of a motor by checking its

FIGURE 202. Cutaway section of ½-hp motor with the major parts labeled.

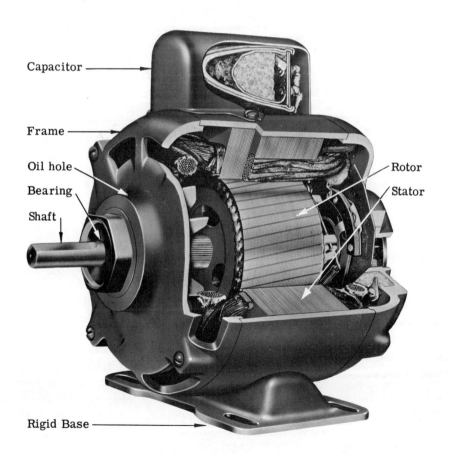

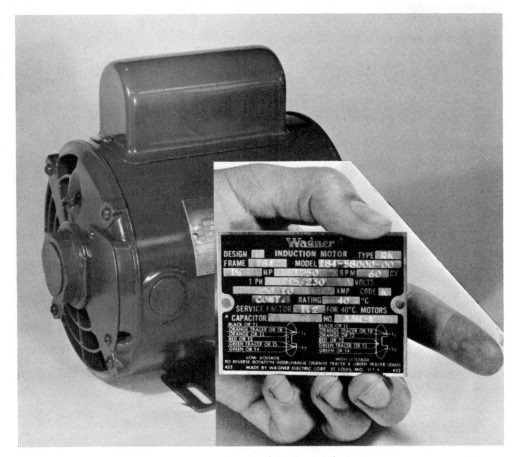

FIGURE 203. The nameplate of an electric motor specifies such information as hp rating, cycles, amperes, rpm, safe operating temperature, voltage service, and type. This model has a smaller frame than older models.

nameplate. Other information appearing on this plate includes cycles, voltage, phase, and allowable heat rise. All these facts should be carefully noted before you purchase a motor.

How to Determine Size of Motor Needed

Motor sizes in the range of ⅛ through ¾ hp are referred to as fractional-horsepower motors. Sizes from 1 hp up are called integral-horsepower motors.

The largest motor normally permitted on a single-phase line is 7½ hp, but 10-hp motors are allowed in some areas. There is no limit to the size of a motor that can be operated on three-phase service, provided, of course, the wire size is adequate. Where irrigation equipment is operated on three-phase service, it is common to find motors up to 50 hp and larger. There may be a rather heavy demand charge for electric service of this type. Phase converters are now making possible the use of large motors on single-phase current.

How Motor Power Is Related to Time and Speed

What is 1 horsepower? A motor that would lift 33,000 pounds 1 foot high in 1 minute or any equivalent of this would be doing work at the rate of 1 hp. This should expend 746 watts of electricity. Due to losses caused by friction, heat, and wind resistance, however, it takes about 1,400 to 1,800 watts to produce 1 horsepower. Motors in the larger sizes usually give higher efficiency than do the fractional-horsepower sizes. Thus, a 5-hp motor expends only 6,440w, whereas ten ½-hp motors together would expend almost double this amount—11,270 watts.

A ¼-hp motor can be used to replace a 1-hp motor provided the smaller motor is given four times as long to do the same amount of work. It would be necessary, of course, to give the ¼-hp motor a 4 to 1 mechanical advantage over the larger size; otherwise the smaller motor could not pull the load. This can be done by using a 4 to 1 pulley ratio. See page 233 for instructions on figuring pulley size.

There is an important principle involved here. For example, farmers have found that a 5-hp motor will operate a combination hammer mill and feed-handling system just as well as a 10-hp motor will. The 5-hp unit simply runs twice as many hours to do the same amount of work. Since the system can

FIGURE 204. Timers shown here not only control lights but also control motors that operate ventilation and heat for large greenhouse in Massachusetts.

be controlled by a time switch, it makes no difference how many hours are required to grind the daily feed supply so long as it can be done in 24 hours or less. The advantage is obvious. The original cost of wiring, motor, and equipment is much less for the 5-hp system.

(NOTE: Using a smaller motor and a slower speed for a water pump is not recommended. Stick to the formula on page 266.)

Common Rules of Thumb on Motor Size. A machine that is being converted from hand operation to electric power should be operated at its original speed. If this is done, a ¼-hp motor should be adequate. A faster speed requires a larger motor for the same load.

A ¾-hp electric motor will replace a 1-hp engine. A 3-hp motor should be adequate to replace a 5-hp engine.

(CAUTION: If a motor becomes too hot to handle with your bare hands, you will know that it is either too small for the job or is being operated improperly. Sometimes an overloaded motor will operate properly if its speed is reduced.)

If a motor is to be replaced, the new one should be the same size unless the load on the machine is to be increased or the speed of the machine is to be increased.

If a motor overheats because of low voltage, a change to 220-volt service may correct the trouble.

WHAT TYPE OF MOTOR SHOULD YOU CHOOSE?

The majority of electric motors today are built to operate on single-phase, 60-cycle current. Nearly all motors ½ hp and larger can and should be operated on 230 volts.

There are seven types of single-phase motors used on the farm: (1) split-phase, (2) capacitor-start, (3) capacitor-start capacitor-run, (4) repulsion-start, (5) repulsion-start induction-run, (6) repulsion-start capacitor-run and (7) universal.

Since three-phase service is now being made available in many localities, three-phase motors should be used more in the future. Two types of three-phase motors that are used on the farm are: three-phase, general-purpose, and the high starting torque. Converters make it possible to use these motors on single-phase current.

Why Starting Torque of Motors Varies

Torque refers to the thrust that is produced by a motor or an engine. The major weakness of electric motors is their low starting torque. Once a motor gets up to three-quarter speed it has very high efficiency. The problem is to get it up to that speed. The starting problem may be especially serious when a motor must start under a heavy load.

In general, three "extras" are found in electric motors to give them greater starting torque. An extra winding is added in the split-phase motor; other types are equipped with brushes and special windings; still others are equipped with capacitors (condensers) that give additional starting power.

Several combinations of these "extras" are recommended for various jobs, according to the difficulty of starting and running. The three windings of the three-phase motor give it greater natural starting torque.

You will likely find that most jobs for electric motors fall into three classes according to difficulty of starting: (1) easy starting loads, (2) moderate starting loads, and (3) heavy starting loads.

As you would perhaps expect, motors that are built for heavy starting loads have extra equipment on them and, therefore, cost more than motors having low starting torque.

Types of Motors for Light Starting Loads

If you need a motor for operating fans, power saws, small grinders, portable concrete mixers, or other easy-to-start machines, you should choose the *split-phase motor.* This type is the least expensive, simplest in construction, and easiest to maintain of any electric motor. Its weakness is its low starting torque. Then too, it is not available in sizes larger than ⅓ hp.

The split-phase motor gets its name from the two windings in the stator. The starter winding in this motor gives added torque to the rotor until it attains about 75 percent full speed, at which point a centrifugal switch disconnects the starter winding. The motor then runs on the running winding only.

Split-phase motors are available in fractional horsepower of ⅙ to ⅓ hp. This type of motor uses six to eight times the normal running amperage while it is starting. Such heavy draft of current makes it impractical to use large split-phase motors on rural lines.

The split-phase motor operates on 115 volts and will burn out if plugged into a 230-volt circuit for more than a few seconds. Its direction of rotation can be reversed, and it will operate at a uniform speed up to full load.

For light starting loads up to ⅓ hp, you can save money by selecting a split-phase motor.

Types of Motors for Moderate Starting Loads

For starting a pump jack, a corn sheller, a large shop grinder, a milker, a feed mixer, an air compressor, or other moderate starting loads, you should

FIGURE 205. This ⅓-hp split-phase motor is the least expensive farm type on the market.

choose a capacitor-start motor or a three-phase, general-purpose motor (if three-phase service, or converter, is available).

1. The Capacitor-Start Motor. This type differs from a split-phase motor in that it is equipped with a capacitor and has more starting windings and larger wire size. A capacitor is a condenser that has the ability to store electricity for a few seconds and use it when the motor needs an added charge of power for starting. This gives the capacitor motors about three times greater starting ability than split-phase motors.

A capacitor motor can be operated on either 115 or 230 volts. Its speed will be uniform up to full load and its direction of rotation can be changed by a simple adjustment. (See motor manual.)

The amount of current for starting a capacitor motor will be about four to six times the requirement for running. Sizes vary from $1/6$ to 7.5 hp, and the cost is about 20 percent more than the same size split-phase motor. Due to the higher

FIGURE 206. Cutaway section of a capacitor-start induction motor. The capacitor (a condenser mounted in a cylinder on top of the motor frame) gives the motor excellent starting ability.

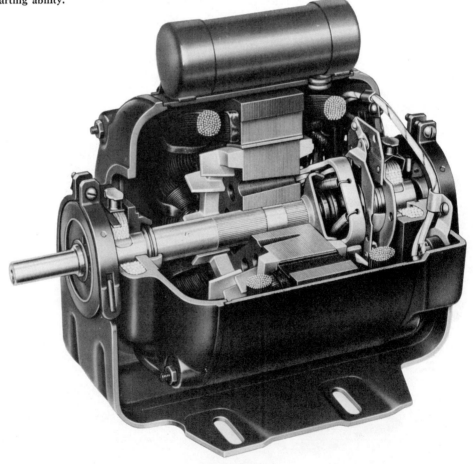

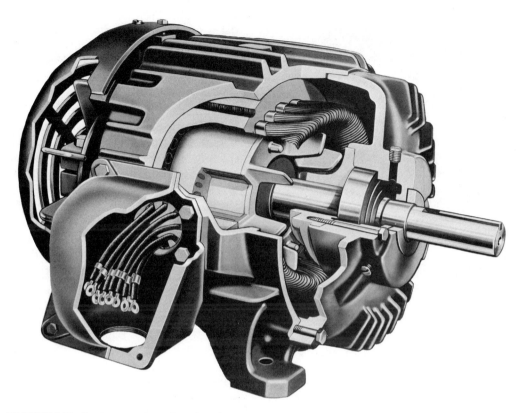

FIGURE 207. Cutaway section of a 5-hp, three-phase motor. Three separate windings give this motor excellent starting ability.

cost, you should not choose a capacitor-start motor where a split-phase type can be used. But, for moderate starting loads, a capacitor-start motor is desirable.

Although capacitor-start motors are made in 3- to 7.5-hp sizes, the cost of this type is about one-third greater than that of an equal size motor of the three-phase, general-purpose type. Therefore, you will probably choose a three-phase motor in preference to a capacitor type if you have three-phase electric service available. Also, you may use a converter for large motors.

2. The Three-Phase, General-Purpose Motor. This type motor is also for moderate starting loads. It differs from the split-phase and capacitor-start types principally in construction. The name "three-phase" comes from the three windings in the motor instead of the one that is usual for single-phase motors. Each of these is fed by a "separate" current from a three-phase high line. These three windings produce a rotating magnetic field, giving the motor a natural high-starting torque. This type of motor draws about three to four times the normal current while starting, which is about one half the starting draft of split-phase motors. The three-phase, general-purpose motor is available in sizes from one-third to several hundred hp. In comparison with the same size capacitor-start

214 USING ELECTRICITY

FIGURE 208. Cutaway section of a repulsion-start motor. Brushes and other special parts give this motor good starting ability.

motor, the three-phase, general-purpose type costs about two thirds as much. The cost of wiring for three-phase service, however, will be considerably more. The three-phase motor can be operated on 115 to 440 volts and the direction of rotation can be reversed.

Because of its ruggedness and simplicity of design, the three-phase, general-purpose motor gives long, trouble-free service; and there are not many parts to maintain.

If three-phase service or a phase converter is available, the best choice of a motor for medium-starting loads (1 hp and above) is the three-phase, general-purpose motor.

Types of Motors for Heavy Starting Loads

If you need a motor to operate a hammer mill, a heavy-duty compressor, a silage cutter, a grain elevator, a hay hoist, or other heavy-starting loads, you will need a *repulsion-start motor*, a *repulsion-induction motor*, a *capacitor-start capacitor-run motor*, a *three-phase, high starting torque motor*, or a *repulsion-capacitor motor*.

1. The Repulsion-Start Motor. For heavy starting loads with single-phase current, the *repulsion-start* motor is satisfactory. It has high starting torque with a minimum amount of current. Its direction of rotation can be changed, and the motor will operate on 115 or 230 volts. It has brushes and other parts that will require regular servicing.

Repulsion-start motors are made in sizes ranging from ⅙ to 10 hp. The smaller sizes cost about 10 to 15 percent more than the same size in a capacitor-start type. Larger sizes cost about the same for these two types.

2. The Capacitor-Start Capacitor-Run Motor. This type has good starting ability and runs on single-phase power. In addition to its capacitor for extra starting power, this motor has a second capacitor which operates in both starting and running.

This type of motor comes in 5- to 10-hp sizes and costs about the same as the repulsion-start motor. It draws three to five times the normal running current for starting. This type motor requires less maintenance than the repulsion-start type. Capacitors are more likely to need repairs than the other parts of the motor. The direction of rotation can be changed. (See operator's manual.)

3. The Repulsion-Induction Motor. When heavy starting loads are encountered and operating loads vary, the repulsion-induction type is needed. This motor is also equipped with brushes. Voltage can be 115 or 230. In addition to having good starting power, this type of motor draws a minimum of current while running. It has the advantage of being less complicated in construction and of requiring less maintenance than repulsion-start motors. The cost is about the same as for other high starting-torque loads. The repulsion-induction motor draws only 15 percent more current for starting than for running. The direction of rotation is easy to change in this motor. (See operator's manual.)

4. **The Repulsion-Capacitor Motor.** This type motor is a newcomer to the field. It is highly efficient while running. It combines the repulsion-start feature, which gives high starting torque, with the capacitor feature for running. The capacitor provides more efficiency in use of current after the motor has attained full speed.

The principal advantages of the repulsion-capacitor motor are (1) high starting torque, (2) low starting current, (3) efficiency, and (4) excellent construction features. Original cost runs about 10 percent more than a capacitor-start capacitor-run motor.

5. **The Universal Motor.** Motors used in drills and other variable speed machines are wound so that the load controls the speed of the motor. The motor itself is not easy to repair but is relatively inexpensive and has long life.

HOW WILL THE TYPE OF ELECTRIC SERVICE INFLUENCE MOTOR SELECTION?

The type of electric service available varies from one community to another. It is necessary, therefore, to check with the power supplier before buying and installing large motors. For instance, if you have a 115-volt service drop to

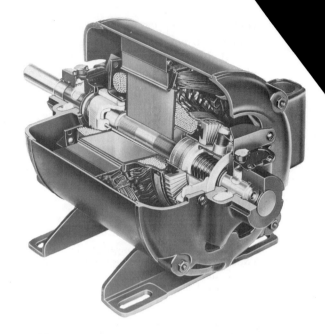

FIGURE 209. Cutaway section of a repulsion-induction motor. This type has excellent starting ability and a low draft of current while running.

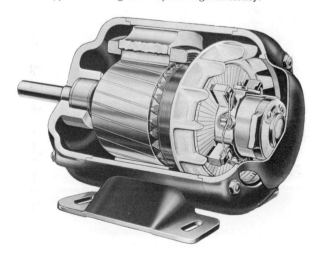

FIGURE 210. Cutaway section of a repulsion-capacitor motor, featuring excellent starting ability, low running current, and high efficiency.

...you cannot use a 230-volt... three-wire service drop is... same holds true for three-... ...rrent, which requires three hot wires and one neutral. If three-phase service is available, it will probably pay you to use three-phase motors for heavy work. Also, check the possibility of using phase converters on single-phase service.

Proper size service entrance must be provided for large motors. For example, if you have a 30-ampere switch now, you would probably need to install a larger service entrance to take care of an additional 5- to 10-hp motor.

WHAT MOTOR ACCESSORIES SHOULD YOU SELECT?

When choosing a type and size of motor, you should consider the accessories that will make your motor give satisfactory service over a long period of time. You should consider the type of bearings and the kind of motor enclosure needed.

Type of Bearings to Choose

Electric motors are equipped with two types of bearings: (1) sleeve bearings and (2) ball bearings. Small motors are often equipped with sleeve bearings and usually cost less than ball-bearing motors. Motors larger than 1 hp usually come equipped with ball bearings. Also, totally enclosed motors are of the ball-bearing type. Motors having ball bearings operate with less friction and may be mounted in any position. With one or two exceptions, sleeve-bearing motors must always be mounted with the motor shaft parallel to the floor. Otherwise, the shaft and bearings would soon wear out for lack of lubrication. Examples of the need for vertical mounting are lawn mowers or power drills.

Sleeve bearings require regular oil lubrication, whereas ball bearings, which are lubricated with grease, may be operated for long periods of time before new lubrication is required. Some ball bearings are sealed for life.

Excessive belt tension will wear out your motor bearings. See page 234 for further information on this.

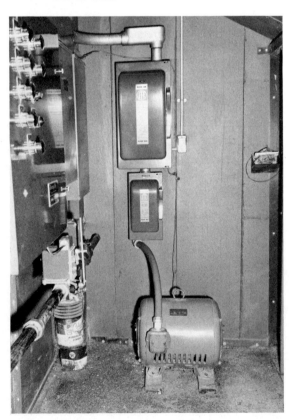

FIGURE 211. Phase converter shown here installed in single-phase line makes it possible to use this large (above 10 hp) three-phase motor. Used here in large grain drying operation in Ohio.

TABLE 16 COMPARISON OF FARM MOTORS

Name	Horse-power Size	Electric Service (volts & phase)	Starting Current % *	Starting Ability	Maintenance Cost	Relative Cost
Split-Phase	1/20 to 1/3	Usually 115, Single-phase	600 to 800	Poor	Low	Low
Capacitor-Start	1/3 to 7½	115–230, Single-phase	300 to 600	Good	Moderate	Moderate
Repulsion-Start	1/6 to 10	115–230, Single-phase	200 to 300	Excellent	High	Moderate
Capacitor-Start Capacitor-Run	5 to 10	115–230, Single-phase	300 to 400	Excellent	Moderate	Moderate
Repulsion-Induction	1 to 15	115–230	150 to 200	Excellent	Moderate	Moderate
Three-Phase, General-Purpose	1/3 and up	115–440, Three-phase	200 to 400	Good to Excellent	Low	Low
High Starting Torque, Three-Phase	1/3 and up	115–440, Three-phase	100 to 200	Excellent	Low	High
Repulsion-Capacitor	3 to 10	230, Single-phase	150 to 300	Good to Excellent	Moderate	Moderate
Universal	1/8 to 1	115, Single-phase	200 to 400	Good	Low	Low

* Starting current based on normal running current.

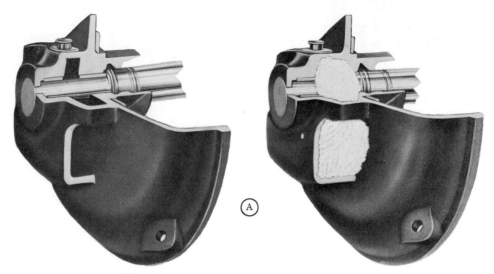

FIGURE 212. *Top:* (A) Shows cutaway of sleeve bearings. Yarn acts as oil reservoir and distributes oil. *Bottom:* (B) Ball bearings operate in a raceway and are lubricated with grease.

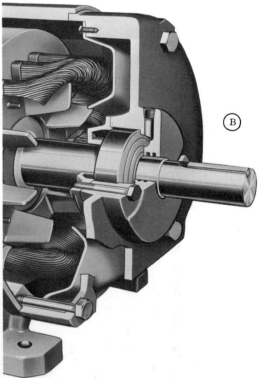

ELECTRIC MOTORS AND MOTOR DRIVES

FIGURE 213. *Left:* Farm motor with an open drip-proof cover. Not for use in wet or dusty locations.

FIGURE 214. *Right:* A splash-proof cover makes this motor suitable for use in locations where liquids may splash against the sides.

Type of Motor Enclosure Needed

The housings or covers for electric motors are designed to meet a variety of needs. Three types are adaptable to the farm. These are (1) the open, drip-proof, (2) the splash-proof, and (3) the totally enclosed.

If your motor is to be operated where the surrounding air is clean and dry, choose a drip-proof enclosure since this is the most economical kind.

Choose the splash-proof cover if liquids may splash against the motor from the sides; this kind costs more but the cover protects the motor windings.

Choose the fully enclosed cover if excess dust is present. This type costs still more but will save on bearings, shafts, and motor windings.

WHAT TYPE AND SIZE OF MOTOR DRIVE SHOULD YOU HAVE?

The most common method of transmitting electric motor power is by V-pulleys and V-belts. A few farm jobs require a flat drive or a combination flat and V-belt drive. For hookups where the motor is mounted directly on the machine, a flexible coupling should be used. For low-power applications in the farm shop, a flexible shaft is a valuable piece of equipment since it gives portable power in any position.

How to Determine the Size of a Pulley

Pulley problems will be easy for you if you think of them as a simple ratio. First, you will already know the speed of your motor (see label), and you will need to know the correct speed of the machine involved. In the case of a feed grinder, assuming the common speed 1,750 rpm for the motor and 350 rpm for the grinder, a 5 to 1 ratio is obtained. Next, select a pair of pulleys that are likewise in the proportion of 5 to 1. Example, if a 3-inch pulley is used on the motor, then a 15-inch pulley would be needed on the feed grinder. The motor

FIGURE 215. This totally enclosed cover is required in very wet or dusty locations. Note the amount of dust in the inset at lower left.

shaft will then turn over five times to each one revolution of the feed grinder. Remember that the machine will operate at the same speed as the motor if both pulleys are the same size, provided there is no slippage.

The easiest method of determining the proper size pulley is to use the chart in Figure 216. Simply lay a straight edge along the two columns and the correct size pulley will be shown.

Types of Pulleys for the Farm

Pulleys are made of cast iron, steel, or aluminum alloy. The larger, three-groove types are usually made of cast iron. Aluminum-alloy pulleys are used with smaller motors and for light loads.

The V-pulley is the most common type used with electric motors and the machines they operate. V-pulleys may be purchased in several styles; namely, A-section, B-section, C-section, two-groove, three-groove, four-step cone, and variable-speed pulleys. The standard V-pulley, single groove, is usually found on motors up to 2 hp unless the load is unusually heavy. For motors larger than 2 hp, a two-groove V-pulley with B-belts may be needed. The C-section type is for still heavier power transmission.

By referring to Table 17, you can find the type and number of V-belts required for various combinations of pulley sizes and horsepower. You will note that pulley size as well as horsepower influences the number and size of belts required.

Four-Step Cone Pulleys. Because of the four different speeds you can get,

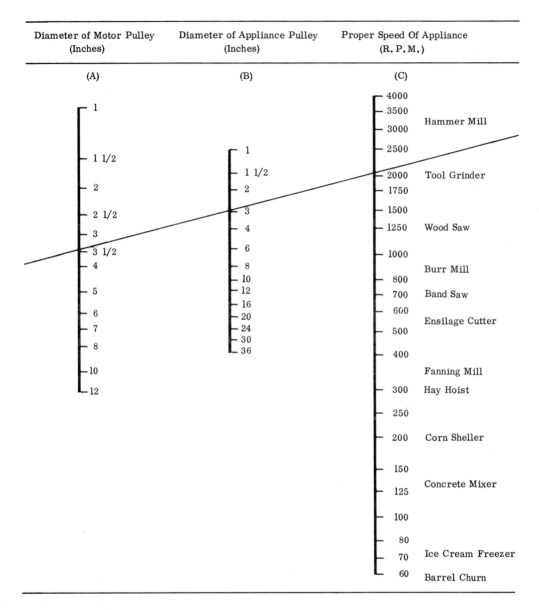

FIGURE 216. Pulley selector chart for 1,750-rpm motors.
HOW TO USE THE SELECTOR CHART:
1. Determine the proper speed of the appliance—for example, 2,000 rpm—and place straight edge at this point under C.
2. Determine the diameter of the motor pulley—for example 3½ inches—and place the straight edge at this point under A.
3. Read the proper pulley size for the appliance where the straight edge crosses the chartline under B. Answer: the appliance pulley should be 3 inches.
NOTE: If C and B are known, A can be determined in the same manner.

222 USING ELECTRICITY

FIGURE 217. Two styles of V-pulleys. At left, a B-section groove pulley fitted with an interchangeable arbor unit. At right, an A-section groove pulley with a ¾-inch arbor.

the four-step cone pulley works well with a portable motor that is used for operating more than one kind of machine. It is valuable for use with a drill and with other machines that operate at several different speeds.

How to Select the Proper Type of V-Belt

V-belts are usually made of rubber and fibrous materials. Some contain metal fibers also. The size of a V-belt refers to two dimensions; namely, its cross section and its length. For farm use, four cross-sectional sizes of V-belts are available: (1) the A-section, which is ⅜ inch wide at the top; (2) the B-section, ½ inch at the top; (3) the C-section, ⅝ inch at the top (actually $2\frac{1}{32}$ inch); and the FHP-type for motors un-

TABLE 17 NUMBER AND TYPE OF V-BELTS NEEDED FOR 1,750-RPM MOTORS *

Diameter of Pulley in Inches	½ or smaller	¾	1	Motor H.P. 1½	2	3	5	7½
2	FHP**	FHP**	..	..	..	..	..	..
2½	FHP**	FHP**	..	..	..	..	..	..
3	1-A	1-A	1-A	2-A	2-A	3-A	5-A	8-A
3½	1-A	1-A	1-A	2-A	2-A	3-A	4-A	7-A
4	1-A	1-A	1-A	1-A	2-A	2-A	3-A	5-A
4½	1-A	1-A	1-A	1-A	1-A	2-A	3-A	5-A
5	1-A	1-A	1-A	1-A	1-A	2-A	3-A	4-A
5½	1-A	1-A	1-A	1-A	1-A	1-B	2-B	3-B
6	1-A	1-A	1-A	1-A	1-A	1-B	2-B	2-B
7	1-A	1-A	1-A	1-A	1-A	1-B	2-B***	2-B
8	1-A	1-A	1-A	1-A	1-A	1-B***	1-B	2-B

* Adapted from *Electricity on the Farm Magazine* (New York, The Reuben H. Donnelley Corp., April, 1956), p. 18.
** FHP (Fractional horsepower) refers to a V-belt that is designed for small motors using small pulleys. Available in ⅜-, ½-, and ⅝-inch widths.
*** The same number of A-section belts could be used instead of B-section.
(NOTE: 2- or 2½-inch pulleys should not be used on motors larger than 1 hp.)

der ¾ hp in size. The larger cross sections are used for transmitting heavy loads; however, these larger, thicker belts will not bend to as small a pulley as will the ⅜-inch size or the FHP-type. Figure 219 shows cross sections of the major farm types.

How to Determine the Length of Belt Needed

If your motor can be installed in any location, it is best to locate it at a distance from the machine as determined by the following:

> Four times the diameter of the largest pulley plus 1.6 times the diameter of the motor pulley plus 1.6 times the diameter of the machine pulley equals length of belt for correct distance between motor and machine. Buy the standard size nearest to this figure.

In measuring the length of belt for a machine and pulley already installed, the best procedure is to take the old belt to the supplier and have it matched. If this is not possible, tie a string around both pulleys where the belt normally runs. Cut the string and measure with a tape or ruler and get this size. Most motors are mounted so that some adjustment in length can be made. This may make it possible to use a size above or below the old belt size.

Some jobs require a flat pulley on the machine and a V-drive on the motor. The size of a flat pulley is figured on the diameter at the largest point. V-belts are used to operate this combination.

In a few instances, it may be satisfactory to use a flat pulley on the motor

FIGURE 218. The four-step cone pulley at upper left gives four different speeds of machine. The B-section, three-groove V-pulleys and belts at right are for extra heavy duty.

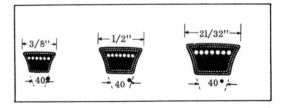

FIGURE 219. *Top:* Cross-sectional sizes of V-belts for farm use.

FIGURE 220. *Bottom:* Farmer's method of measuring the length of a V-belt.

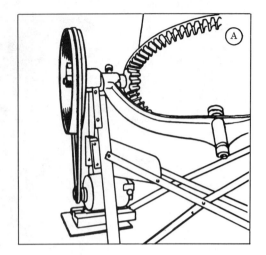 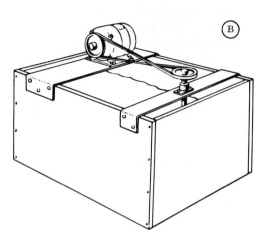

FIGURE 221. *Top:* (A) The drive unit consists of a V-pulley on the motor and a flat pulley on the machine. The large flat pulley costs less than the same size V-pulley yet does the same job. (B) The quarter-turn drive is valuable for operating machines in a vertical position where the motor must be mounted horizontally.

FIGURE 222. *Left:* Flexible hose connecting a small motor to a water pump is easy to install and easy on the motor and machine.

as well as the machine. A flat leather belt with the ends secured together with steel lacing can be used for flat drives; rubber fibrous belts are also satisfactory for flat drives.

Types of Positive and Direct Drives for Motors

Some jobs require that a motor be connected directly to its load. For this you may use a flexible hose, a rigid flange, or a flexible shaft.

1. A piece of flexible hose can be used for driving light loads, usually 1/3 hp or less. Hose will soon come apart if it is coupled to loads heavier than this. For light work, however, a piece of hose makes an inexpensive, convenient, and easily installed drive.

2. A rigid flange is sometimes needed where pulleys and gears are not feasible. Such a coupling eliminates all slippage. Usually one half of the flange is bolted to the motor and the other half is fastened to the machine. The two parts are secured together with pins. Rubber or leather fillers are used in the space between the two parts of the flange to serve as shock absorbers. This protects the motor bearings against damage from vibration. With the exception of fans, motors should not be attached directly to a machine without a proper flange.

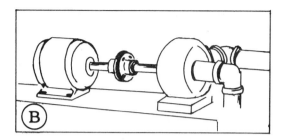

FIGURE 223. Rigid flanges for connecting a motor directly to a machine: (A) chain-drive flange; (B) a motor connected to a water pump by means of a rigid flange; (C) rigid flange with rubber filler.

Fans produce little or no shock and therefore may be attached directly to the shaft of the motor.

3. A flexible shaft is excellent for operating a grinder, a buffer, a sander, and other similar equipment. Almost every farm shop should have this piece of equipment for use with a ⅓- to ½-hp motor. This type of drive can be bought for motors up to 2 hp. The advantage is apparent. You can operate the equipment in any position and it can be moved about while in operation. The cost varies from $15 up.

SUMMARY

In all situations where electric motors can be used to do farm work, savings on power costs can be effected. Electric motors, on the average, are almost twice as efficient as engines. Also, motors that are given proper care should last up to 30 years with very little repair and upkeep.

The two main parts or units of an electric motor are (1) a rotor, or rotat-

FIGURE 224. A flexible shaft makes it possible to use power equipment in all positions.

ing part, and (2) a stator, or stationary part. All other parts act as accessories to these two main units. These other parts include bearings, lead wires, and covers. Some types of motors have capacitors and brushes, which are used for giving the motor added starting torque.

A rotor, when turning, is a spinning magnet. The current flowing through the stator coils creates a magnetic field and a current flowing through the rotor conductors also magnetizes the rotor. Magnetic forces in the stator, acting upon the magnets in the rotor cause it to turn.

The common size of single-phase motors for farm use ranges from 1/8 hp to 7½ hp. In some areas, up to 15-hp sizes are allowed. Three-phase motors are available in sizes up to a 100 hp or more. Most farm motors have a speed of 1,725 rpm; some operate at 3,450 rpm. Phase converters are becoming popular.

Single-phase motors for farm use come in seven types: (1) split-phase, (2) capacitor-start, (3) capacitor-start capacitor-run, (4) repulsion-start, (5) repulsion-induction, (6) repulsion-capacitor, and (7) universal. Three-phase motors come in two farm types: (1) general-purpose, and (2) high starting torque.

For easy starting loads, the split-phase motor (up to 1/3 hp) is satisfactory and is the least expensive. For moderate starting loads, the capacitor types or the three-phase, general-purpose motors should be used. For heavy starting loads, the repulsion-start types are best. Also, the three-phase, high starting torque motor is excellent for heavy loads, where it can be used.

If a motor must be operated in dusty or wet surroundings, a totally enclosed cover should be used. This type of cover is accompanied by sealed ball bearings. Sleeve bearings are less costly but are not suitable for operation out of level. Also, sleeve bearings must be lubricated periodically.

The most widely used drive for electric motors is the V-belt and V-pulley. These come in a variety of sizes and types suitable for most farm needs. Also, the flat belt, rigid flange, flexible hose, and flexible shaft are useful drives for a variety of needs on the farm.

QUESTIONS

1. Why do electric motors have such high running efficiency in comparison with engines?
2. Why do electric motors have low starting torque? How can starting torque be increased?
3. Why does a 1-hp motor draw more than 746 watts (1 hp)? What happens to the 1,000 or so watts (total draft 1,800 watts) that are normally used by a 1-hp motor?
4. Why is it economical to buy the proper type of motor, even if it costs more?
5. For what special combinations of farm jobs is the new repulsion-capacitor motor adapted?
6. What effects will excess tension in a belt have on the motor?

ADDITIONAL READINGS

Brown, R.H., *Farm Electrification*. McGraw-Hill Book Co., New York, N.Y., 1956.

Edison Electric Institute, *Electrical and Basic Controls Used in Agricultural Production*. New York, N.Y.

Fairbanks Morse and Company, *Electrical Machinery Catechism*. Freeport, Ill.

Industry Committee on Interior Wiring Design, *Farmstead Wiring Handbook*. New York, N.Y., 1964.

National Fire Protection Association, *National Electrical Code, 1968 Edition*. Boston, Mass., 1968.

Potomac Edison Co., *Introduction to Electric Motors, Selection, Operation, and Care*. Hagerstown, Md., 1964.

Promersberger, William J.; Frank E. Bishop; and Donald W. Priebe, *Modern Farm Power, Second Edition*. Prentice-Hall, Inc., Englewood Cliffs, N.J., 1971.

South Dakota Rural Electric Association, *Electrical Textbook for Vocational Agriculture Students*. Huron, S.D., 1956.

University of Illinois, Vocational Agriculture Service, *Electric Motors for Farm Use*. College of Agriculture, Urbana, Ill., (no date).

10 HOW TO CARE FOR AN ELECTRIC MOTOR

Despite the fact that an electric motor is built to last thirty years or longer, some motors have been known to burn out in a few minutes after installation. It is possible to select the proper type and size of motor and still get short life and unsatisfactory service from it. Whether or not your motors give good service and last as long as they should will depend upon the way you install and care for them.

Authorities claim that most motor failures result from one or more of the following abuses: (1) overloading; for example, calling on a ¼-hp motor to do the work of a ½-hp size; (2) improper wiring and low voltage; (3) overfusing, which results in heating; (4) using the wrong type of motor for the job; (5) improper lubrication; (6) improper installation; (7) too tight belts, and (8) allowing excess dust, water, oil, or fumes to enter the motor.

A suggested remedy for most motor troubles can be found in Table 18.

HOW CAN YOU PROTECT YOUR MOTORS FROM OVERLOADING?

Several things can cause overloading even though you have the proper size and type of motor. For example, a foreign object can get into a machine and cause it to become locked. When this happens, the current continues to flow through the locked rotor and, within seconds, enough heat is generated to burn the insulation off the windings unless the motor circuit is properly protected. Therefore, it is essential to provide the proper kind of fuses and other protection so as to avoid burn outs from overloads. Motor protectors are of two general types, built-in and manual reset.

Motors With Built-in Protection

Figure 225 shows a ⅓-hp motor which came from the factory equipped with a built-in protector. This device breaks the circuit when excess heat causes a metallic strip to expand. This expansion has the same effect as a switch. When the metal strip cools sufficiently, it comes back into position and thus reconnects the circuit. Built-in protectors may be manual or automatic.

Manual Switch Protection

A manual reset control is the type of protector suitable for shop machines and other electric devices that are handled. Why is this so? It would be dangerous to have an automatic reset control for a bench saw or jointer because the automatic reset could reestablish the current while you are working on the machine and thus cause a serious injury. Therefore the automatic control is adapted to refrigeration, water pumping, and other continuous processes, not shop machinery. The protection device (sometimes called a relay) may be connected to the motor itself or it may be mounted in a separate unit.

Electric motors of ½ to 3 hp in size may come from the dealer without built-in protection. If so, install the proper-size manual-start switch or magnetic switch.

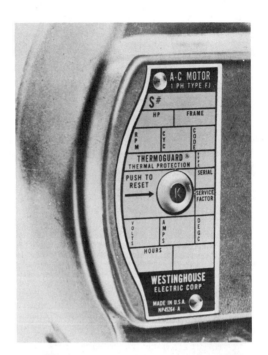

FIGURE 225. *Top right:* Built-in protector. When overload causes the motor to cut out, wait two minutes, then reset the switch by pressing the button.

FIGURE 226. *Right:* Manual reset switch. When overloaded motor cuts out, it will not start again until the start button is pushed.

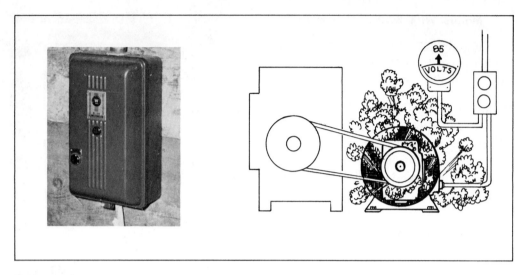

FIGURE 227. A magnetic switch (left) affords protection for low voltage as well as overloads. The diagram (right) illustrates the result of low voltage when the motor is unprotected.

Manual-Start Switch. This type of switch is suitable for use on ½- to 3-hp motors. It operates on the principle of excess heat breaking the circuit. The circuit is broken when a metal strip in the protector expands a certain amount and thus breaks the circuit. Another type of switch operates by spring tension and still another breaks the circuit when a ribbon of solder melts. To reestablish the circuit, reset the switch by hand.

Magnetic Switch With Built-in Protection. This type of protector is used for 5-hp motors and up. The cost is excessive for small motors. The protection principle is the same as in other types previously discussed. An additional feature of the magnetic switch, however, is that a magnetic relay protects the motor against both heating and low voltage. If the voltage drops below a safe operating level, the magnetic switch opens the circuit. The Code specifies that motor controls must be visible from the motor and not be over 50 feet away.

Protection by Time-Delay Fuses

On page 109 you will find an illustration of some time-delay fuses. This type of fuse allows a motor to draw up to six or eight times its normal running current while it is starting. If the motor continues to draw this much current for more than 20 to 30 seconds however, the fuse will blow. The common trade names of this type fuse are Fusetrons and Fustats. The amperage rating must not exceed the motor current rating over 25%, and no other appliance may be operated on the same circuit, otherwise the protection is lost.

Protection by Special Starter

For 5-, 7½-, or 10-hp motors it may be necessary to use a *current-limiting* starter, sometimes called a resistor starter. This device controls the amount of current while the motor is getting up to full speed. A current adjustment arm can be moved so as to increase the current from starting position up to full speed.

The value of such control is to prevent excess dimming of lights while a large motor gets started, without damage to motor.

HOW SHOULD AN ELECTRIC MOTOR BE INSTALLED?

Many good motors are ruined every year because of improper installation. It is essential that several steps be properly followed in the installation of a motor.

Install the Motor in the Correct Operating Position

It is essential that sleeve-bearing motors be mounted in a horizontal position unless they are manufactured especially for vertical installation. A motor that is tilted as much as 15 degrees is considered as being in a vertical position. Ball-bearing motors may be mounted in a vertical position.

The Motor Should Be Mounted Securely

It is important to have your motor mounted securely. A concrete base will provide a substantial mount and result in a quiet running motor. Figure 230 shows an inexpensive method of mounting a farm shop sander. This machine will operate without vibration. Materials needed for building this project include a discarded oil drum, ½ yard of concrete, a ball-bearing shaft, and sanding drum.

For heavier motors which require a secure and adjustable anchor, you need

FIGURE 228. *Top:* One easy way to provide motor protection is to install a weather-proof socket in the hot line(s) of the motor circuit and insert the proper size Fusetron.

FIGURE 229. *Bottom:* The motors in *A*, *B*, and *C* have sleeve bearings and therefore must be mounted in a horizontal position. The oil cups must be positioned up by rotating the end bell of the motor. The ball-bearing motor in *D* may be operated in a vertical position.

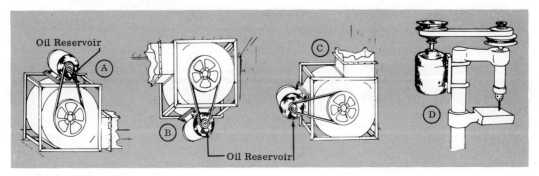

FIGURE 230. *Top:* Author's son using drum sander constructed by author. Ball-bearing shaft anchored to concrete in upper half of oil drum. Motor is 1/3-hp split phase type; separate toggle switch and extension cord. Cost, about $50, including motor, and 1/2 yard concrete.

FIGURE 231. *Bottom:* Slide rails for mounting heavy-duty motors provide about 2½ inches of horizontal adjustment. Anchor the rails to a solid foundation.

an adjustable metal frame similar to that in Figure 231. This type may come with a heavy-duty motor or it may be purchased separately. The adjustable frame is convenient for tightening or loosening belts. If you are handy with the farm welder, you can construct one using angle iron for the main rails.

The motor frame should be fastened to a substantial base, perhaps concrete or heavy floor boards.

Due to vibration, motor mounts, pulleys, and other parts often tend to work loose. Therefore, it is necessary to tighten all anchor bolts and pulley screws occasionally. Loose motor parts will result in excessive wear.

When to Use a Rigid Shaft

A rigid shaft makes it possible to operate several machines with one motor. It is important that the shaft be located out of the way, or, better still, have it enclosed. An exposed shaft is dangerous to those working around it. To operate successfully, a rigid shaft should be installed level with the floor. It should be equipped with good bearings which can be reached for ease of lubrication. Transmission of power from the shaft to a machine may be by V-pulley and V-belts or by gears. Make certain that the pulleys and belts or the

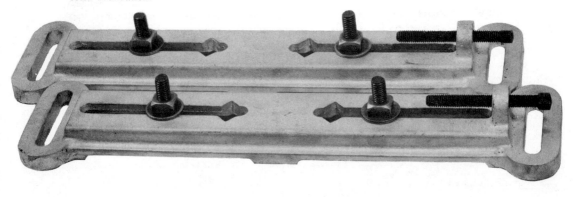

gears are properly aligned and not too tight.

How to Make a Small Motor Portable

A ½- to ¾-hp motor can be mounted on a portable frame, as illustrated in Figure 232, and used to operate several different machines on the farm. This arrangement is especially adapted for use in the farm shop. Study the diagram in Figure 232 to see how you can construct one of these projects.

How to Mount a Large Motor on a Portable Cart

Another convenient arrangement for portable farm power is a large motor on a cart. This type of project makes it possible to operate a feed grinder, a barn cleaner, a silo unloader, and other machines with a single large motor.

FIGURE 232. Exploded view of a portable motor. One motor can thus be used to operate several machines. Install one four-step cone pulley on the motor and one on the machine.

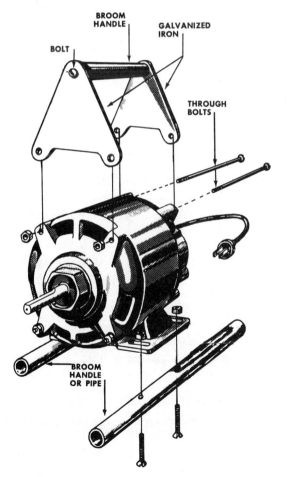

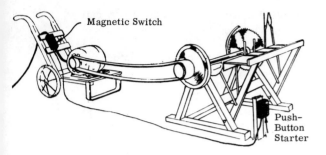

FIGURE 233. A 5-hp motor mounted on a cart made from junk materials. Note the wiring for the motor and the magnetic protector switch.

Whatever type of mount you use, the motor should be properly aligned with the machine it is to operate. The motor shaft should be parallel with the machine shaft so that no "twisting" action is present. The motor pulley should be in line with the machine pulley, since misalignment causes excessive wear and friction and also reduces efficiency.

The most desirable position of the motor is level with the machine shaft, but this may not be possible. Your motor will operate above or below the machine-shaft level but with less efficiency.

HOW SHOULD YOU CARE FOR YOUR PULLEYS AND BELTS?

Good transmission of power and long life of motors and belts will depend on a few common-sense practices. It will pay you to follow these.

How to Anchor a Pulley to a Shaft

Light-load pulleys may have a single setscrew for anchoring to the shaft. It is sometimes necessary to drill and tap another hole in the pulley collar to prevent slippage. The second setscrew should be installed one quarter of the way around the pulley collar, not on the side opposite the original setscrew.

Pulleys that are designed to do heavy work usually have a keyway corresponding to a keyway on the shaft of the motor or the machine to be operated. A setscrew holds the key in place and anchors the pulley so that it cannot slip. In buying a new pulley that requires keying to the shaft, it is necessary to state the dimensions of the keyway—$3/16$ x $1/4$ inches, for example. The diameter of the pulley bore must be stated also.

The larger, heavy-duty, two- and three-groove pulleys usually have an inner cone which is universal for a given line of pulleys. The outer part of these pulleys may be interchanged for any desired size and can then be mounted on the universal cone.

How to Care for Pulleys and Belts

Rubber-base belts should never be oiled, as oil causes rubber to rot. Special compounds are available for use on flat belts to reduce slippage.

Care should be taken not to adjust a belt too tightly. Correct tension for a V-belt is determined as follows:

1. Grasp both sides of the belt and force together with the hand.
2. The belt should have enough slack to allow for closing in by one fourth of the distance between them.

Too much tension will cause the bearings in your motor and machine to wear out, and the belt may be ruined also.

In installing a belt, see that the pulleys are lined up to make the belt run true and not bind in any way. Always loosen your motor sufficiently to allow for installing the belts without undue stretching. Never use a stick or crow bar to force a belt onto pulleys.

The best operating position of a motor is to have its shaft horizontal to the machine shaft. Other operating positions, however, are possible and the loss in efficiency is not excessive even when mounted directly overhead.

Pulleys should be removed from the motor shaft by using a puller. Pounding on the rim of a pulley may ruin it. Do not scar the pulley opening or the end of the shaft, since this will interfere with removal of the pulley. If you are attempting to remove a pulley from a battered shaft, you may have to file off the metal burr.

WHAT ROUTINE MAINTENANCE JOBS SHOULD YOU DO?

1. Replace worn brushes as soon as you notice sparks between the brushes and the armature. The armature will require truing up in a lathe if it becomes out-of-round or acquires high spots.

2. A motor should not be installed where the surrounding air gets above 160° F (up to 180° F on new models).

3. A motor shaft should turn freely. A slight misalignment of the shaft with relation to the bearings will cause heating and ruin the motor if it is operated in this condition. This can be checked by removing the belt from the drive pulley and spinning the shaft by hand.

4. Avoid excessive belt tension, as this causes motor bearings to wear too fast.

5. Never blow out a motor with compressed air as this forces dust into the windings and bearings. To clean a motor, first disassemble it, then brush out all dust and other foreign matter.

FIGURE 234. The key in the motor shaft must fit keyway in the pulley arbor of a rigid flange.

FIGURE 235. (A) These belts are too loose as indicated by flapping. (B) Correct and incorrect alignment of belts. See that motor and machine shafts are parallel.

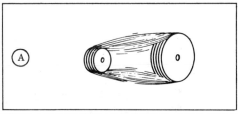

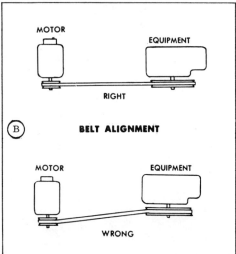

Bearings should be washed in solvent or kerosene.

HOW SHOULD YOU LUBRICATE YOUR ELECTRIC MOTORS?

The lubrication of electric motors varies according to the type of bearings the motor has. As stated before, improper lubrication is one of the main causes of motor failures. Always use the kind of lubricant recommended by your service manual and follow instructions in applying it.

How to Lubricate a Sleeve-Bearing Motor

Since sleeve-bearing motors ordinarily come from the factory with the oil wells dry, the bearings must be oiled before the motor is operated. Figure 236 shows how wool yarn is used to assure continuous lubrication of the bearings in motors up to 2 or 3 hp.

FIGURE 236. For large motors, yarn packing in the sleeve-bearing housing maintains an even distribution of oil.

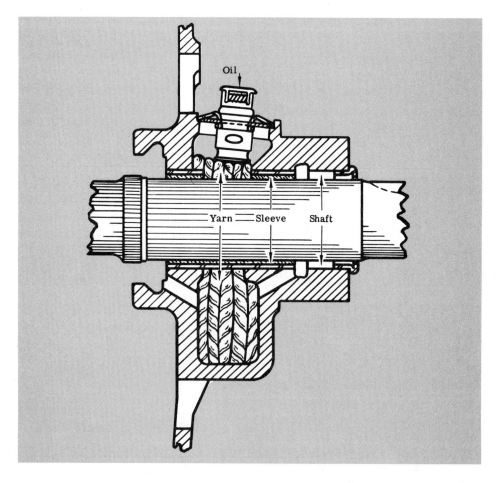

HOW TO CARE FOR AN ELECTRIC MOTOR

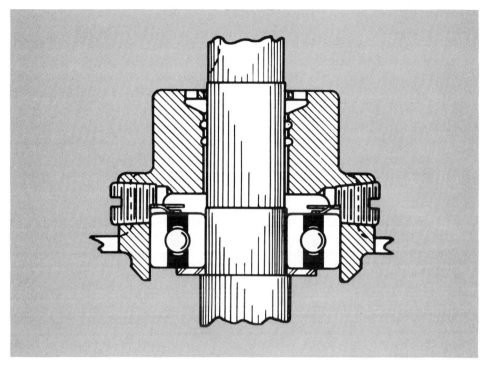

FIGURE 237. Cross-sectional view of ball bearings, shaft, and bearing race. Note the space for ball-bearing grease.

A good grade of SAE-10 cylinder oil is satisfactory for lubricating sleeve bearings. Heavier oil will not give proper lubrication. For average farm use, fill oil wells twice a year.

(NOTE: Sleeve-bearing motors can be mounted on side walls or ceiling. However, the end plates must be removed and rotated so that the oil wells will always be up, otherwise the oil will drain out.)

How to Lubricate a Ball-Bearing Motor

The type of grease used for lubricating a ball-bearing motor will affect the life of the motor and the quality of service it will give. Special ball-bearing grease recommended by the manufacturer should always be used. Ordinary cup grease or pressure grease is not satisfactory. You must have a lubricant that will not melt and drip out of the bearings.

The following steps are essential in getting the long life you expect from your ball-bearing motor: *

1. A new motor should be inspected when installed, and lubricated if necessary.

2. Ball-bearing motors should be lubricated approximately once a year.

---

* Wagner Electric Corp., *To Grease or Not to Grease*, St. Louis, Missouri (no date).

Certain ball-bearing motors should be lubricated at shorter intervals if the original instruction has indicated that additional attention will be required.

3. Ball-bearing motors for general farm use should be lubricated once a year as follows:

- (a) Wipe off all dirt and give the exterior of the motor a thorough cleaning.
- (b) Remove the pipe plug from both lubrication and drain openings.
- (c) Run a rod or wire partly into the lubrication opening to determine if the grease is dry and hard.
- (d) If the grease is hard, run the motor while adding a little oil to soften up the grease. Operate the motor until the grease starts to run out of the drain opening.
- (e) Stop the motor and add grease until the old black grease has passed out of the drain opening and new grease starts to come out. The grease should be added to the housing by means of a standard grease gun.
- (f) Operate the motor and allow the bearing to force out excess grease.
- (g) Stop the motor and run a wire into the lubrication and drain openings to force out a quantity of the grease. This allows for grease expansion after the motor warms up.
- (h) Replace both pipe plugs.

TABLE 18* TROUBLE-SHOOTING CHART FOR ELECTRIC MOTORS

Probable Cause	Test and Remedy
I. MOTOR WILL NOT START	
1. Blown fuse or tripped protective device	Test motor bearings to see if they are in good condition and well lubricated. Machine and motor should turn freely without binding. Test voltage at motor terminals against line voltage. See if protection devices are open or fuse blown. Replace motor fuse with Buss Fusetron (time-delay) fuse of the proper amperage (see nameplate of motor). After replacing fuse or resetting relay, allow motor to operate for awhile to see if it goes off again. If so, check further before using the motor.
2. No voltage or low voltage	Test voltage at motor terminals with switch closed. It should not be more than 10 percent lower than the voltage called for on the nameplate.

* The Wagner Electric Corp., *Servicing Wagner Single-Phase Motors*, St. Louis, Missouri (no date). (Slightly condensed and re-arranged.) (NOTE: Codes for motor types, Repulsion-Induction—RA; Split-Phase—RB; Capacitor-Start—RK.)

TABLE 18 (Continued) TROUBLE-SHOOTING CHART FOR ELECTRIC MOTORS

Probable Cause	Test and Remedy
1. MOTOR WILL NOT START (continued)	
3. Improper current, wrong voltage, or wrong frequency	Single-phase current will not operate three-phase motors and three-phase current will not operate single-phase motors without a converter. Some power companies do not recommend converters. When purchasing an electric motor, check current, voltage, and frequency against your power supply.
4. Improper line connections	Check your motor connections with the diagram which is sent with the motor. Wiring connections for 115 volts are different than for 230 volts. Motor connections must match the voltage on your line.
5. Open circuited field (types RB & RK)	Motor hums when switch is on. Examine for broken wires, loose connections, faulty switch, or open protector.
6. Open circuited field or armature (type RA)	Excessive sparking appears on starting. Motor may refuse to start altogether at certain positions of the rotor, or motor may hum when switch is on. See if protector is tripped, check for loose wires, or see if there are burned segments in the commutator. Check commutator for foreign metallic substance which might short-circuit it.
7. Condenser short-circuited	See item 1 above.
8. Worn or sticking brushes	Poor contact of brushes with commutator will result in slow or weak starting. Brushes may be worn or may stick in holders. Brush springs may be weak, or commutator may be dirty. Clean commutator with piece of fine sandpaper, not emery cloth. Do not oil or grease commutator.
9. Improper brush setting	Brush holder or rocker arm should be opposite index and locked in position. If new armature has been installed, this may vary a little.
10. Excessive load	If items 1 through 9 are all right, motor should start with load on it. If it fails to do so, test by starting without load. If it will start idle and refuses to start loaded, motor is overloaded. Have an electric motor shop to test for starting torque.

TABLE 18 (Continued) TROUBLE-SHOOTING CHART FOR ELECTRIC MOTORS

Probable Cause	Test and Remedy
II. MOTOR RUNS HOT	
1. Bearing trouble	See section on "Excessive Bearing Wear."
2. Stator coils are short-circuited	Short-circuited coil will be much hotter than others.
3. Rotor rubbing against stator	Check bearings for excessive wear and see if foreign substance is present between stator and rotor.
4. Excessive loads	Check pulley ratios to determine whether machine is turning too fast (faster speed requires more horsepower). Check current with ammeter. Amperes should not exceed amount stated on nameplate.
5. Low voltage	Check voltage at motor terminals. It should not exceed 10% of stated voltage on nameplate.
6. High voltage	Same test as item 5.
7. Incorrect connections to motor leads	Check motor connections with the diagram sent with the motor. Be sure that connections match voltage on your lines.
III. NOISY MOTOR	
1. Unbalanced rotor	Run motor without load, holding the palm of your hand on the case. If shaft is bent or rotor is unbalanced, you can feel the vibration. This is a repair job for a motor shop.
2. Worn bearings	Worn bearings will result in "side play" in the shaft. Replace with new bearings (have this done at a motor shop) and check for cause of bearing wear. (See also section on "Excessive Bearing Wear.")
3. Switch rattles	Install new switch hub and felt washer.
4. Rough commutator or improperly seated brushes	Noise occurs only during starting but should be corrected to avoid further trouble.
5. Excessive end play	End play should be kept at or near zero. Bearings should not bind, however, in making this adjustment. Use washers furnished by the factory and be sure rotor turns freely.
6. Improper alignment of motor and driven machine	Line up pulleys and correct all other misalignments.

TABLE 18 (Continued) TROUBLE-SHOOTING CHART FOR ELECTRIC MOTORS

Probable Cause	Test and Remedy
III. NOISY MOTOR (continued)	
7. Loosely mounted motor	Tighten up all connections. Check motor base to see if it fits the floor or table on which it is mounted. A loose motor will vibrate.
8. Loose motor accessories	Tighten up capacitor box, switch box, pulleys, and other parts. Excessive noise and vibration will result from loose accessories.
9. Amplified motor noise	Use rubber mounts to reduce motor roar.
IV. BURNED-OUT MOTOR	
1. Frozen bearings	Check section on "Excessive Bearing Wear."
2. Prolonged and excessive overload	Before replacing the burned-out motor, locate and correct the cause of overloading. The driven machine may be at fault as a result of worn bearings, lack of lubrication, bent wheel shafts, and so on. Figure the present work being done by the machine as compared with its original capacity. Check pulley ratios; an oversize pulley on the motor would cause overloading of the motor.
V. EXCESSIVE BEARING WEAR	
1. Belt tension too great; misalignment of motor and machine; improper meshing of gears	Flat or V-belts should have only sufficient tension to provide power without slippage. At midpoint between the motor and machine pulleys, you should be able to compress the two sides of the belt about one fourth the distance between them. A flat belt will require slightly greater tension. Slippage of belts will cause squeaking noise and make the pulley heat. If machine operation requires that the belt be too tight, replace with heavier belt or use two-groove or three-groove pulleys. Check for correct distance from motor to driven machine. (See also section on "How to Select Motor Drive.")
2. Too much, too little, or wrong kind of lubricant	Follow manufacturer's instructions for proper lubrication. Use oil and/or grease recommended for the motor being lubricated. (See page 236.)

TABLE 18 (Continued) TROUBLE-SHOOTING CHART FOR ELECTRIC MOTORS

Probable Cause	Test and Remedy
V. EXCESSIVE BEARING WEAR (continued)	
3. Dirty bearings	Disassemble motor and wash bearings with solvent. Dry with clean cloth and reassemble after lubricating with proper oil or grease. Excessive dirt or dust in bearings indicates that a dust-proof motor enclosure may be needed.
VI. EXCESSIVE BRUSH WEAR	
1. Dirty commutator	Clean with piece of fine sandpaper.
2. Brushes make poor contact with commutator	Brushes may have worn off too short to reach commutator. Replace short brushes with a new set. See that brushes move freely in their slots; springs should hold brushes firmly against commutator without excessive pressure.
3. Excessive load	If brush wear is due to excessive load, you can tell this by timing the starting cycle. Brushes should lift within five seconds.
4. Brushes fail to lift and stay off during running	Check conditions under "Motor Runs Hot."
5. High mica	Glazed appearance of surface of commutator. Take motor to shop and have a light cut taken off surface of commutator.
6. Rough commutator	See preceding item.
VII. MOTOR OPERATES WITHOUT RELEASING BRUSHES	
1. Dirty commutator	Clean with fine sandpaper.
2. Governor mechanism or brushes sticking; brushes worn too short for contact	Replace worn brushes with new set. Check governor mechanism to see whether it works freely by hand. Faulty governor must be replaced with new one. See that brushes work freely in slots.
3. Frequency of current not correct for motor	If motor varies more than 10% from speed on nameplate while running idle, this indicates that motor does not match the frequency cycle of your current. Ordinarily, rural electric power lines carry single-phase, 60-cycle current. In some sections, three-phase current may be available also. Motor must match the current you have available.

TABLE 18 (Continued) TROUBLE-SHOOTING CHART FOR ELECTRIC MOTORS

Probable Cause	Test and Remedy
VII. MOTOR OPERATES WITHOUT RELEASING BRUSHES (continued)	
4. Low voltage	See that voltage is within 10% of nameplate voltage with the switch closed.
5. Line connections not correct and not making good contact	Tighten up all terminal connections, solder any loose wires which appear to be corroding (after cleaning), and check wiring diagram furnished with the motor.
6. Incorrect brush setting	See that rocker arm setting corresponds with index mark.
7. Incorrect adjustment of governor spring	Governor should throw off brushes at about 75% of speed on nameplate. If it does this at less than 65% or over 85% of nameplate speed, the governor must be repaired or replaced.
8. Excessive load	Motor may start a heavy load but fail to attain sufficient speed to throw off brushes. Check motor bearings for tightness and correct lubrication. Or, motor may throw off brushes for a time but they come back on commutator. Reduce load on motor or replace with larger motor.
9. Shorted stator	See "Motor Runs Hot."

SUMMARY

Electric motors may burn out or be ruined in other ways if not properly protected and correctly installed.

The first essential in caring for a motor is to see that it is properly fused or has some other type of protection. For a water pump or refrigerator motor, an automatic reset protector is best. But for a farm shop machine, a manual reset protector should always be used—never an automatic type.

A separate circuit for an electric motor can be fused with a Fusetron and thus provide protection against overloads. To restore service, it is necessary to put in a new fuse each time.

Ball-bearing motors may be installed in a vertical position, but sleeve-bearing motors must be operated level or very nearly so. Fifteen degrees out-of-level is considered vertical because the lubricating oil will drain out.

Special installations include rigid shafts and portable mounts. A large motor mounted on a cart provides power for several different jobs. The same is true with a rigid shaft. A small portable motor is handy around the farm shop for operating several different machines.

V-belts are easily ruined by operating when improperly adjusted. A belt that is too tight will wear rapidly and may ruin the motor bearings as well. If

TABLE 19* WIRE AND FUSE SIZE FOR ELECTRIC MOTOR CIRCUITS, SINGLE-PHASE, TWO-PHASE, AND THREE-PHASE, 60-CYCLE CURRENT

Motor H.P.	Volts	Wire Size for Motor Circuit (AWG or B&S Rubber Covered)							Fuse Size for Motor Circuit
		14	12	10	8	6	4	2	
		Maximum Length of Span in Feet							
FOR SINGLE PHASE									
¼	115	80	130	210	280	520			15
¼	230	360	520	840	1,160				10
⅓	115	70	105	215	245	435			20
⅓	230	280	420	860	980				15
½	115	60	95	150	235	375	600		25
½	230	240	380	600	940	1,400			15
¾	115	40	65	105	165	260	420	670	30
¾	230	160	260	420	660	1,040			20
1	115	35	60	95	150	240	375	610	35
1	230	140	240	380	600	960	1,500		20
1½	115			65	105	165	260	415	45
1½	230			260	420	660	1,040		25
2	115			50	85	130	210	335	60
2	230			200	340	520	840	1,340	30
3	115			35	55	85	140	225	90
3	230			140	220	340	560	900	45
5	115			20	35	55	90	145	150
5	220			80	140	220	360	580	80
FOR TWO OR THREE PHASE									
½	220	150	225	300					15
½	440	600	900	1,200					15
¾	220	150	225	300	450				15
¾	440	600	900	1,200	1,800				15
1	220	150	225	300	450				15
1	440	600	900	1,200	1,800				15
1½	220			300	450				15
1½	440			1,200	1,800				15
2	220			250	400				20
2	440			1,000	1,600				15
3	220			150	250				30
3	440			600	1,000				15
5	220			100	175	300	500		45
5	440			400	700	1,200	2,000		25

* The Deming Pump Co., *Deming Pump Catalog C-57* (Salem, Ohio, 1957), p. 71. (NOTE: Data based on National Electrical Manufacturers' Association rating for a maximum of 2 percent voltage drop with allowance of 25 percent for heavy starting loads. Table slightly re-arranged.)

too loose, a belt will tend to pull apart, tear, and frazzle at the edges and will overheat.

In lubricating motors, the manufacturer's instructions should be carefully followed. It is essential that the proper lubricant be used.

Most motor troubles can be detected before they become serious. The trouble-shooting chart in this chapter describes clues to possible future difficulties. In general, excessive noise, heat, vibration, or the presence of smoke is an indication of trouble.

QUESTIONS

1. What is meant by protection against overloading? How does it affect electric motors?
2. Why is it necessary to use a manual-start protector on a table saw instead of an automatic restart protector?
3. Why will a sleeve-bearing motor be ruined if operated at 15 degrees or more out of horizontal (level) position shaft-wise?
4. Why does a motor repairman never pound on a pulley or motor shaft?
5. Why is it necessary to use special grease in lubricating a ball-bearing motor?
6. What would cause a motor to become too hot to touch with your hand?

ADDITIONAL READINGS

Allis-Chalmers Manufacturing Co., *A Guide to Care of Electric Motors.* Milwaukee, Wisc., 1964.
American Association for Agricultural Engineering and Vocational Agriculture, *Farm Electric Motors, Selection, Protection, Drives.* Agricultural Engineering Center, Athens, Ga., 1964.
Brown, R.H., *Farm Electrification.* McGraw-Hill, New York, N.Y., 1956.
Edison Electric Institute, *Electrical and Basic Controls used in Agricultural Production.* New York, N.Y. (no date).
Fairbanks Morse and Company, *Electrical Machinery Catechism.* Freeport, Ill., (no date).
Fuchs, J. David and Stephen W. Garstang, *Electrical Motor Controls and Circuits.* Indianapolis, Ind., 1963.
Industry Committee on Interior Wiring Design, *Farmstead Wiring Handbook.* New York, N.Y., 1964.
Lytel, Allan, *ABC's of Electric Motors and Generators.* Howard W. Sams and Co., Indianapolis, Ind., 1966.
National Fire Protection Association, *National Electrical Code, 1968 Edition.* Boston, Mass., 1968.
Potomac Edison Co., *Introduction to Electric Motors, Selection, Operation and Care.* Hagerstown, Md., 1964.

Promersberger, William J.; Frank E. Bishop; Donald W. Priebe, *Modern Farm Power, Second Edition*. Prentice-Hall, Inc., Englewood Cliffs, N.J., 1971.

Texas Education Agency, *Electrical Controls and Basic Control Circuits Used in Agricultural Production*. Vocational Agricultural Education Division, Austin, Texas, 1967.

University of Illinois, Vocational Agriculture Service, *Electric Motors for Farm Use*. College of Agriculture, Urbana, Ill., (no date).

SUGGESTED PROJECTS FOR PROBLEM-UNIT FOUR

Several illustrations used in Chapters 9 and 10 will make excellent projects:
  1. Provide Fusetron protection for motor circuits. (See Figure 194.)
  2. Construct permanent mount for sander. (See Figure 230.)
  3. Make ½-hp motor into portable style. (See Figure 232.)
  4. Construct cart and mount heavy-duty motor on it. (See Figure 233.)
  5. Reverse direction of capacitor or start-and-run motor. Interchange leads 1 and 2 as shown in Figure 238; motor will run in opposite direction. See motor manuals for reversing direction of other types of motors.

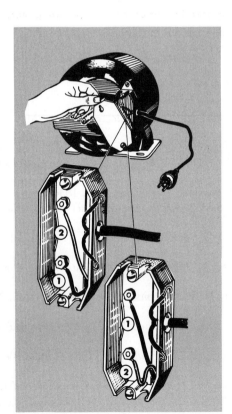

FIGURE 238. By interchanging leads number 1 and 2, as illustrated, the direction of split-phase and capacitor motors can be reversed.

GLOSSARY FOR PROBLEM-UNIT FOUR:

Rotor The rotating part of a motor or generator.

Stator The stationary part of a motor or generator. In most small motors, the stator contains the field windings. In larger motors and generators, the rotor may contain the field poles.

Pole That part of the motor magnetic circuit around which the field windings are wound. The purpose is to confine the magnetic flux to given locations within the motor since iron will support many times as much magnetism as the same volume of air. The pole in the alternating-current motor is made of laminated iron.

Commutator A device which changes the direction of the flow of electricity through the rotating part. It consists of narrow copper bars insulated from each other by layers of mica. The ends of the rotor windings are soldered to the commutator bars or segments.

Brushes Small carbon blocks which make contact with the commutator of an electric motor. These are used to carry electricity from the stationary to the rotating parts of the motor.

Condenser An electric device in which an electric charge may be stored temporarily. It consists of two or more electric conductors or plates separated by a thin insulating material.

Time-Delay Fuse A fuse which has the ability to carry overload currents of short duration without melting. The heavier the overload, the less is the time required for the fuse to blow. In motor circuits, where the starting currents are high, a time-delay fuse or other special device is necessary to permit the motor to be started. When properly selected and installed so as to carry only the current of a single motor, such fuses can also be used to provide motor running (overload) protection. Like the common fuse, the time-delay fuse is also made in plug and cartridge types.

Torque The measure of the tendency of a force to rotate the body upon which it acts. (EXAMPLE: A pull on the spoke of a wheel.) A pull of 1 pound at 1 foot from the center of rotation equals 1 foot-pound of torque.

Phase Converter A device which makes it possible to operate three-phase motors on single-phase current.

PROBLEM-UNIT V

FARMSTEAD LIGHTING AND WATER SYSTEMS

FIGURE 239. The water pump at the right is necessary in the operation of this modern farm.

Good light and plenty of water are necessary for successful farming and comfortable living. Almost every phase of farming and home life depends, to some extent, upon these two essentials.

As farms become larger and more specialized, the need for water and lighting is likely to increase. So, in planning your water and lighting systems, take the future into account. Many farmers and other homeowners install electric equipment two or three times before they get adequate systems.

In this problem-unit, you will find suggestions on how to select and care for water systems and lighting equipment. The lighting section presents many scenes of proper lighting arrangements including the kinds of equipment needed for special purposes. The chapter on water systems includes basic principles to consider in selecting the proper types and sizes of pumps and pump motors for the home and farmstead.

Some principles of irrigation and suggestions on how to select irrigation equipment are presented in the latter section of Chapter 11. When all the factors are favorable, irrigation holds great promise for good profits. Many things, however, can cause complete loss of profits in irrigation.

11 HOW TO SELECT AND CARE FOR ELECTRIC WATER PUMPS AND RELATED EQUIPMENT

Running water has not always been so commonplace in the American home. It was almost a luxury in rural areas before electric power became plentiful. Most farms with running water in earlier days used engines or windmills for power.

Rural electrification has changed this situation. A census publication shows that about 90 percent of all non-dilapidated farm homes have access to running water. This figure includes water in the home and on farm grounds.*

An adequate supply of water for the home, although essential, is no more important than water for farming and agri-business. In fact, it would be almost impossible to operate a modern animal-production unit without running water. Most farming operations are at least partially dependent upon an adequate water supply.

In selecting and caring for farmstead water systems or irrigation equipment, it is necessary to understand and be able to do the following things: (1) select the location and type of well; (2) determine the size of pump needed; (3) select the type of pump needed; (4) determine the size of water pipes required; (5) determine the size of the motor required; (6) identify and correct pump troubles; (7) select pumps and accessories for irrigation; and (8) select and care for irrigation equipment.

WHAT SHOULD YOU CONSIDER IN CHOOSING THE LOCATION AND TYPE OF WELL FOR THE FARMSTEAD?

By selecting a safe location for your well, you can help to insure the health

* United States Bureau of the Census, *Statistical Abstract of the United States*, 89th ed., Washington, D.C., p. 709.

249

of your family. Outbreaks of typhoid fever and other diseases have been traced to contaminated water supplies. The cause of this may be the improper placement of a well or use of untreated surface water.

The diagram in Figure 240 shows three wells that are safe because they are located on a slope above the level of the farm buildings where there is no danger of seepage. On the other hand, the three wells to the right of and below the buildings are all unsafe. Why? Because of seepage from a septic tank, a barnyard, and an open pool. The water from any one of these wells would be unsafe to use.

Type of Well

The diagram in Figure 241 shows five types of wells. Dug wells are suitable for shallow depths in soils that do not cave in. A dug well must have a wall of some kind. This can be stone, concrete, wood, or other durable material.

A driven well is the easiest and least expensive to put down; that is, in soils where it is practical to use this type. Notice the sharp "drive point" at the bottom of the casing. This point is necessary in driving the casing down to the water supply. A special cap is installed on top of the well casing, then driving force is applied to the cap. A tractor driving unit may be used for this, or a commercial well driller will have proper driving equipment. Where casing is to be driven, the soil must be of a permeable nature and must be free of large stones.

The other illustrations show tubular and drilled wells—the types most often used in deep-well construction and in

FIGURE 240. Safe and unsafe locations for water wells.

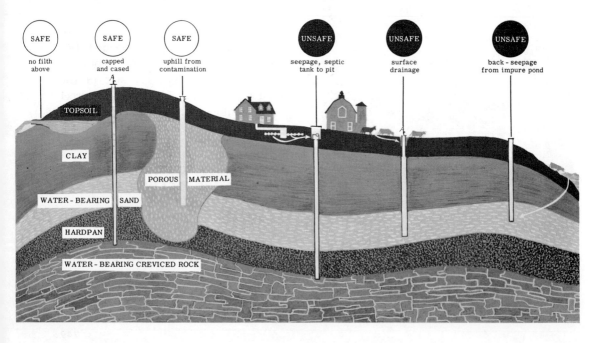

hard soils where driving is not practical. The diameter of drilled wells varies from 2 inches up. A 2-inch casing, however, is too small to provide an adequate supply of water for the average farm. It will pay you to make the well large enough during the first drilling—at least 4 inches.

How to Protect Your Well from Pollution

Your water supply can become contaminated even though its location is ideal. The following precautions should be observed.

1. Take the necessary steps to direct surface drainage *away* from the well.

2. See that the septic tank, the cesspool, or the disposal field is located at least 150 feet from the well.

3. Have a sample of water tested at least once a year.

WHAT SIZE PUMP SHOULD YOU HAVE?

The capacity of farmstead water pumps is rated in *gallons per hour* or *gallons per minute*. This is stated as *gph* or *gpm* throughout this chapter. A

FIGURE 241. Five types of wells for the farmstead.

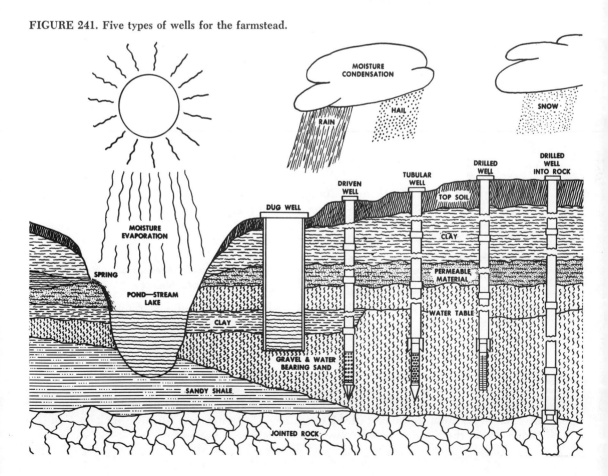

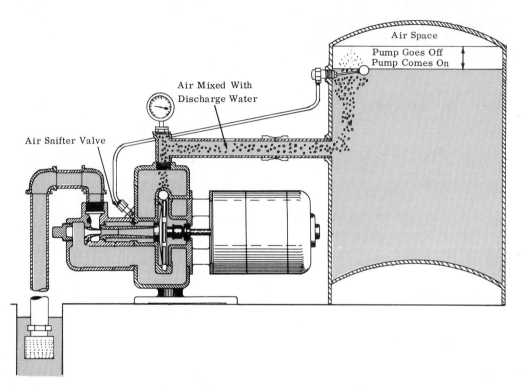

FIGURE 242. Diagram of a pressure tank showing an 8-gallon operating space in a 42-gallon tank.

250gph pump costs less than a 720gph size. No doubt this is why some farmers buy pumps that are too small.

One farmer got into the following difficulties when he bought a pump that was not large enough:

1. During peak-use periods there was a shortage of water, and this prevented the needed expansion of the dairy.

2. After two years the farmer finally decided to install a larger pump, whereupon he found that he needed a larger well.

3. In addition to the cost of the new equipment, the farmer had to pay for removing the old pump as well as installing the new one.

Didn't this farmer have an expensive water system?

How to Determine the Number of Faucets Needed

In determining pump size, disregard the amount of water in a pressure tank because the few gallons it holds are not of any real help to a pump. To illustrate, only 8 gallons of water can be drained from a 42-gallon tank before the pump takes over. Refer to the diagram of a pump and pressure tank in Figure 242 and notice how small the space for water is. After the pump starts, you can get only the amount of water that the pump delivers and no more.

Since the pump must meet the requirements of the farm and farm home during peak periods, the most accurate way to figure pump size is to determine the amount of water required during peak-use periods. You will find it convenient to use the following formula for determining the size of pump needed by a particular farm.

1. Identify the peak period use.
2. Count the number of faucets that will be running continuously (1½ minutes or longer).
3. Figure at the rate of 3gpm for each ½-inch faucet counted.
4. Convert to gph (gpm × 60) and buy this size.

PROBLEM: Assume that you have ½-inch faucets in the following locations: 50-cow dairy barn, milk house, holding area for cows, pasture or holding area for young stock, farm home, vegetable garden and lawn, tenant house, and farm shop. List the number of faucets as shown in Table 20.

The situation in Table 20 is based on a peak-use period occurring early in the morning during the milking chore. Water is used at both kitchens, and in the milking barn; however, the only faucets to be counted as *continuous* are one for a bathtub in the farm home, two for filling stock watering tanks, and one for garden or lawn watering.

The result is:

Four ½-inch faucets × 3gpm = 12 gpm
12gpm × 60 (minutes) = 720gph

ANSWER: A 720gph pump is needed to furnish water during the peak period. By watering the garden or lawn in the evening, a three-faucet setup would be adequate. This would take a 540gph pump.

Watch Out for Fire. Protection of farm buildings against fire requires a pump of at least 500gph capacity. Moreover, your savings on insurance rates with an adequate pump may pay the additional cost of a larger size. Remember that friction loss in long runs of pipe results in low pressure and less water, unless the pipe size is adequate. For further information, see the section on selection of pipe size.

Well Capacity Must Equal or Exceed Pump Capacity

Before buying a pump, make certain that your well has an output equal to or greater than the capacity of the pump you plan to buy. If your well has a limited capacity—say 400gph—the 720gph pump could not be used since it would pump the well dry in a little over 30 minutes. Emptying a well completely stirs up the water and may cause trouble in the pump.

If your well does not have sufficient output to supply the various needs at the proper time, you may need to install a large reservoir or overhead tank. A smaller pump could then be used to fill the reservoir over a longer period of the day. One precaution to remember is that if a pumping period longer than 3 to 4 hours in each 24 hours is required, it may be necessary to have a special type of motor; that is, one that will operate continuously.

Larger Motors Require Larger Pressure Tanks. If the pumping load requires a motor larger than ½ hp, a larger pressure tank should be used. This will allow

TABLE 20 TOTAL NUMBER OF FAUCETS AND NUMBER THAT WILL COUNT AS CONTINUOUS DURING PEAK-USE PERIOD FOR FARM AND TWO RESIDENCES

Location of Faucet	Total Number of Faucets		Number To Count *	Number Continuous
	Hot	Cold		
Kitchen sink, farm home	1	1	1	0
Bath tub or shower, farm home	1	1	1	1
Lavatory in bathroom, farm home	1	1	1	0
Tenant house	1	1	1	0
Dairy barn, milking room	1	1	1	0
Milk house	1	1	1	0
Holding or lounging area (tank)		1	1	1
Young stock area (tank)		1	1	1
Farm shop		1	1	0
Vegetable garden and lawn		1	1	1
Totals	6	10	10	4

* Each pair of faucets, one hot and one cold, counts as *one* faucet only.

the necessary time between pumping intervals for the larger motor to cool. It is recommended that an 82-gallon tank be used for motors from ½ to 1 hp. One hundred and twenty-gallon, or larger, tanks should be used for motors larger than 1 hp. Pressure tanks range in size from 12 to 525 gallons capacity.

WHAT KIND OF PUMP SHOULD YOU CHOOSE?

For a well that is less than 25 feet deep, you can use a *shallow-well* pump, and it will cost less than a *deep-well* pump of equal capacity. For well depths greater than 25 feet, however, it is necessary to use deep-well pumps. Both types are discussed in detail a little later.

How to Determine Whether a Deep-Well Pump Will Be Required

The normal (standing water) level in a well is not the true depth load for a pump. Other factors include (1) the amount of drop or "drawdown" that occurs during pumping, (2) the elevation of the pump above ground level, and (3) the amount of power lost to friction in the suction pipes.

Therefore, in deciding whether or not a deep-well pump will be required, the following factors must be considered: Normal or standing water level plus drawdown plus pump elevation plus friction loss. If the total is greater than 25 feet, choose a deep-well pump; if less than 25 feet, select a shallow-well pump. The water level and pump elevation can be measured, but the amount of friction loss must be figured for each well, and this will vary according to the size of the suction pipe and rate of flow.

How to Figure Friction Loss. In this problem, you will be interested only in the friction loss on the suction side of the pump load. This includes the water lines and fittings leading from the pump into the well. Both horizontal and vertical runs of water pipes are included in computing friction losses of the suction side. Horizontal runs are not counted in well depth, of course, only the friction loss in these runs.

Tests have been conducted to determine the exact amount of friction loss incurred in different sizes of pipes at different rates of flow. These data are presented in Table 21. First, locate the size of pipe you will use; assume ½-inch. Second, locate the rate of flow; assume 300gph, or 5gpm. Third, read along this line until you find the figure directly under the pipe size. The answer is 41.0 feet.

This simply means that in 100 feet of ½-inch pipe (at 5gpm) friction would have the effect of adding 41 feet of pipe. In short, friction would "steal" about 40 percent of the horsepower required in pumping 5gpm through ½-inch pipe. Friction loss would be less in ¾-inch pipe at the same rate (5gpm).

(NOTE: Friction loss is measured in pounds pressure but is easily converted

FIGURE 243. *(A)* Diagram of factors involved in figuring pump size. *(B)* Shallow and deep wells.

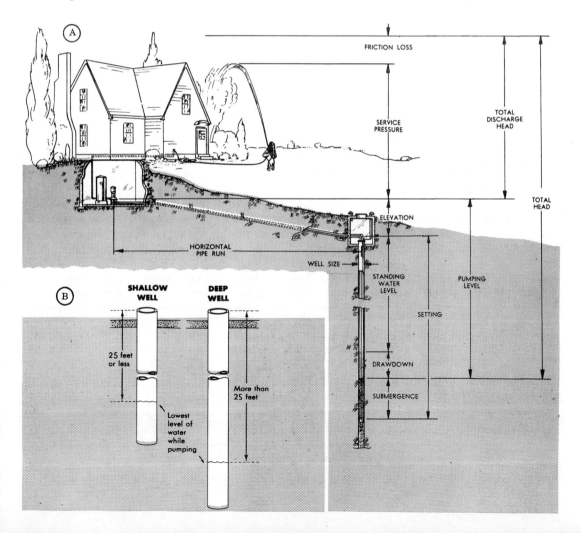

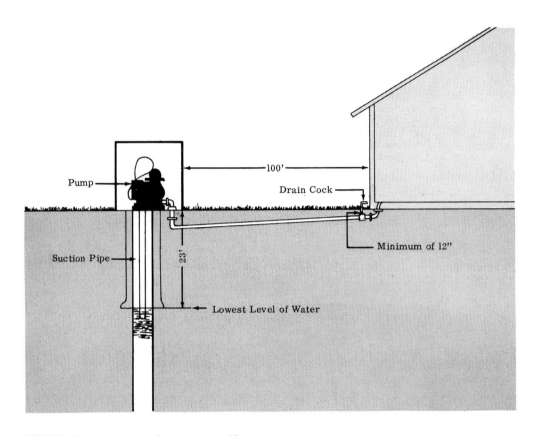

FIGURE 244. Diagram of a pump problem.

TABLE 21 FRICTION HEAD LOSS IN FEET PER 100 FEET OF PIPE FOR VARIOUS RATES OF FLOW *

Gallons per Hour	Minute	½-inch pipe	¾-inch pipe	1-inch pipe	1¼-inch pipe	1½-inch pipe	2-inch pipe	2½-inch pipe	3-inch pipe
240	4	27.0	7.0	2.1	.6				
300	5	41.0	10.5	3.25	.84	.39			
360	6		17.17	4.55	1.20	.56	.20		
480	8		25.0	7.8	2.03	.95	.33	.11	
600	10		38.0	11.7	3.05	1.43	.50	.17	
720	12			16.4	4.3	2.01	.70	.24	
900	15			25.0	6.5	3.0	1.09	.36	.15
1200	20			42.0	11.10	5.2	1.82	.61	.25
1500	25			64.0	16.6	7.8	2.73	.92	.38

* Based on William and Hazen Formula for ordinary iron water pipe.

to feet by the formula: pounds pressure × 2.3 = feet. This gets all the suction load into feet and the total is expressed as *feet of head*. The data in Table 21 have already been converted into feet. To convert feet of head to pounds pressure, divide feet by 2.3.)

PROBLEM: Keeping the four preceding points in mind, work out the problem as shown in the diagram in Figure 244 and decide whether or not you could use a shallow-well pump equipped with 1-inch suction pipe. Use the 720gph pump (or 12gpm) from a previous example.

SOLUTION 1:
1. Measure the standing
   water level   20.00 feet
2. Determine the drawdown
   during pumping   3.00 feet
3. Measure the pump
   elevation   0.00 feet
4. Figure the friction loss
   in 23 feet of 1-inch
   pipe (See Table 21)   3.77 feet

   Total suction head   26.77 feet

CONCLUSION 1: Since the total suction head of 26.77 exceeds the 25-foot maximum suction lift of a shallow-well pump, a shallow-well type could *not* be used with 1-inch suction pipe.

SOLUTION 2: Substitute 1¼-inch suction pipe and refigure the problem as follows:

1. Total pumping level, as
   determined in solution 1   23   feet
2. Friction loss in 23 feet of
   1¼-inch pipe (See
   Table 21)   .99 feet

   Total suction head   23.99 feet

CONCLUSION 2: By using 1¼-inch suction pipe, a shallow-well pump would work successfully since the total suction load is less than 25 feet.

(NOTE: Figures in Table 21 are for 100 feet of pipe. Therefore the friction in 23 feet of suction pipe in the preceding example is found by taking .23 of the appropriate figure in Table 21.)

CAUTION: If your farm is located at a high altitude, the amount of suction

TABLE 22 FRICTION LOSS FOR VARIOUS SIZE FITTINGS * (Stated As "Feet" of Loss)

Size of Fitting	Elbows		Tees	Valves		
	90°	45°		Gate	Globe	Angle
½	2.0	1.2	3	.4	15	8
¾	2.4	1.5	4	.5	20	12
1	2.9	1.7	5	.6	25	15
1¼	3.7	2.3	6	.8	35	18
1½	4.9	3.0	7	1.0	45	22
2	7.0	4.0	10	1.3	55	28
2½	8.0	5.0	12	1.6	65	34
3	10.0	6.0	15	2.0	80	40

* Add these values to the total lift and other loads on the pump.

TABLE 23 SUCTION LIFT OF PUMPS AT DIFFERENT ALTITUDES

Altitude	Maximum Suction in Feet
Sea level	25
¼ mile (1,320 feet) above sea level	23
½ mile (2,640 feet) above sea level	22
¾ mile (3,960 feet) above sea level	21
1 mile (5,280 feet) above sea level	20
1¼ miles (6,600 feet) above sea level	19
1½ miles (7,920 feet) above sea level	18
2 miles (10,560 feet) above sea level	16

lift of shallow-well pumps will be reduced. See Table 23 for further information.

Things to Consider in Choosing a Shallow-Well Pump

There are two types of shallow-well pumps commonly used on the farm. Each has certain advantages (as well as disadvantages) depending upon the situation. These types are the centrifugal pump and the piston pump.

When to Choose a Centrifugal Pump. The centrifugal pump is the simplest in design of any type. It contains only one moving part called the *impeller*. Figure 245-A shows a diagram of this type of pump, with the main parts labeled. This impeller has slots in the disk which rotates at high speed inside a chamber.

FIGURE 245. *(A)* Diagram of a centrifugal pump. *(B)* Cutaway section of a vertical-type centrifugal pump.

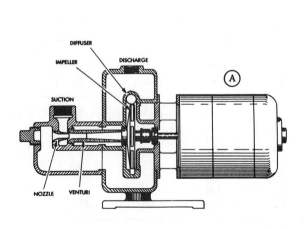

Notice that the impeller is mounted on a shaft that is connected to the motor shaft. As the impeller turns, it creates a suction that pulls water into the impeller chamber. At the same time, water is forced from the opening at the top.

The centrifugal pump is quite efficient up to a total lift of about 18 feet. Above that, the efficiency drops. Then you get less pressure and less output of water.

This type of pump is quiet running and costs little to maintain. When repairs are necesary, the pump is easy to dismantle and put back together again.

Another popular style of centrifugal pump is shown in Figure 245-B. The motor and pump in this style are mounted in a vertical position.

When to Choose a Shallow-Well Piston Pump. If your water table is between 18 and 25 feet deep, a piston pump may give better service than a centrifugal type. The piston pump shown in Figure 246-A produces positive action and will maintain high pressure in long lines and against high working heads. The pump shown here is a single piston, double-acting type.

Unlike the centrifugal pump, the piston pump puts out a constant volume of water regardless of lift; that is, as long as the pistons make the same number of strokes. The pressure drops in the centrifugal pump as well depth increases.

A "duplex" piston pump is shown in Figure 246-B. This style has two pistons which are both double acting. You can get up to 125 pounds pressure with this piston pump. It is advocated for supplying long runs of pipe.

The main disadvantage of the piston pump is that there are many moving

FIGURE 246. *Top:* (A) Cutaway section of a single-piston pump. *Bottom:* (B) Cutaway section of a duplex- (2-) piston pump.

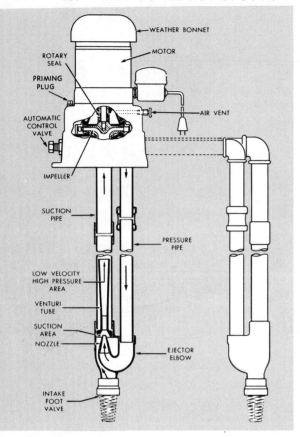

FIGURE 247. Diagram of a deep-well jet-pump installation. Notice the alternate offset at right for use where the pump cannot be located directly over the well.

parts; therefore, repair jobs are required more frequently than in centrifugal or jet pumps. Piston pumps are also noisier than centrifugal types.

Things to Consider in Selecting a Deep-Well Pump

If the vertical lift of water in your well is 25 feet or more, you will need a deep-well pump. There are five major types of deep-well pumps on the market and three of these are widely used for farmstead water systems. The three are (1) the deep-well jet or ejector pump; (2) the plunger or cylinder-type pump; and (3) the submersible pump. The other two deep-well types, which are less used for farmstead water systems, are the rotary pump and the turbine.

There are several variations in these types. For example, the jet pump can be bought in a horizontal or vertical style; the same holds true with the cylinder pump. Some models of the jet pump are made so that they can be converted from shallow to deep-well operation. Another variation in pump types is the incorporation of several "stages" (impellers). This pump is called the multi-stage jet. It will develop higher pressures than will the single-stage jet.

When to Choose a Deep-Well Jet Pump. The centrifugal jet, as shown in Figure 247, is practically the same as a shallow-well jet with the exception that the pumping mechanism is equipped with an ejector. This is located down in the well instead of being in the pump. Pumps of this type are satisfactory for depths up to 80 feet. They will operate at depths greater than this but with less efficiency. The same advantages apply to deep-well jets as to shallow-well jets; that is, quiet and economical operation. Where high pressures are required, several sets of impellers (stages) are used. Notice the alternate offset suction pipes at right, showing that the pump can be located somewhere away from the well.

When to Choose a Deep-Well Plunger Pump. The plunger-type pump, as illustrated in Figure 248, will work successfully at depths to 850 feet. A plunger (sucker rod) operates a piston, which is down in the well under water. Each stroke of the piston forces water

ELECTRIC WATER PUMPS 261

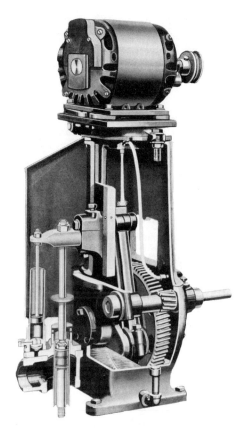

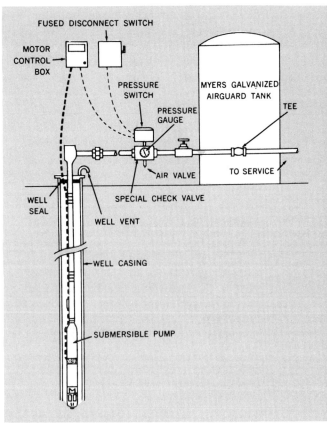

FIGURE 248. *Left:* Cutaway section of a plunger-type deep-well pump.

FIGURE 249. *Right:* A typical submersible well (pump) installation.

up through the casing and into the tank. Again, this type of pump is noisier than jets and may require periodic repairs. The working head must be set directly over the well. The plunger pump develops good pressure so long as the cylinder and piston are in good condition.

When to Choose a Submersible Pump. The submersible pump, which is illustrated in diagram form in Figure 249, is becoming popular for well depths to 400 feet. The pump and motor are located down in the well under water.

Being located down in the water, a submersible pump "pushes" the water out of the well. This type has no sucker rod or other moving parts between the pump and the tank. The pumping mechanism consists of several stages of turbines for building up pressure. The electric cable to the motor is insulated with waterproof plastic. The pump and motor parts are permanently lubricated and sealed to give long and trouble-free service. The well casing for a submersible pump must be four inches in diameter or larger. The special-built motor for this type of pump is shaped to fit into this small diameter.

A word of warning is in order here. *The submersible pump will not hold up in sandy conditions.* If your well contains loose, fine sand, a submersible pump will not be satisfactory.

The advantages of the submersible-type pump are (1) long, trouble-free service, (2) little or no noise, (3) no priming, and (4) the tank and controls may be placed at any convenient location—at a distance or nearby, whichever is most practicable.

Two other types of pumps are available, though these are not as widely used as the preceding five types. These are the rotary and the turbine.

When to Choose a Rotary Pump. The rotary pump is simple in construction. It consists of two or more close-fitting gears mounted inside a housing. The gear teeth "trap" a certain amount of water as they turn, forcing it out of the opening with considerable pressure. The rotary pump delivers water at a constant volume without pulsation, but the gears will wear out quickly if there is any sand in the water. This type of pump is most often used for moving oils and other heavy liquids.

When to Choose a Turbine Pump. The turbine pump consists of a bronze disk mounted on a shaft and fitted inside a case with very little clearance. As the disk or impeller turns, it forces water through the opening that leads to the water tank. For deep wells and large volumes of water, several impellers (stages) may be used to build up pressure and increase volume.

If you need help with your water problems, call on a representative of any well-known pump manufacturer. Ordinarily, you must know three things about your water problem before these agents can help you pick the right pump. You must determine (1) the number of "continuous" faucets for the farm and home, (2) the total head of your system, and (3) the total length and elevation of water pipes from the tank to the supply points. You should also know the approximate capacity of your well.

WHAT SIZE WATER PIPE WILL YOU NEED?

When you worked a problem to determine the amount of water needed for a farm and two residences, you allowed 3gpm for each ½-inch faucet. For a ¾-inch faucet you should allow 5gpm. These figures are based on a pressure range of 20 to 40 pounds for a farmstead water system.

In long runs of pipe—say 400-500 feet beyond the tank—you will have the problem of low pressure due to friction loss. Since it is not practical to operate a farm pump at pressures higher than 40 or 50 pounds, the only way to maintain pressure in long lines is to use larger pipe.

How large should a long run of water pipe be? This will depend on the amount of water delivered and on the length of run. Figure 250 contains data that will enable you to select the correct-size pipe for any farm water problem up to 650gph and up to 1,000 feet of pipe. Study the following example and see how easy it is to use the chart in Figure 250.

PROBLEM: Determine the pipe size needed to supply the two ½-inch faucets which you counted as "continuous"

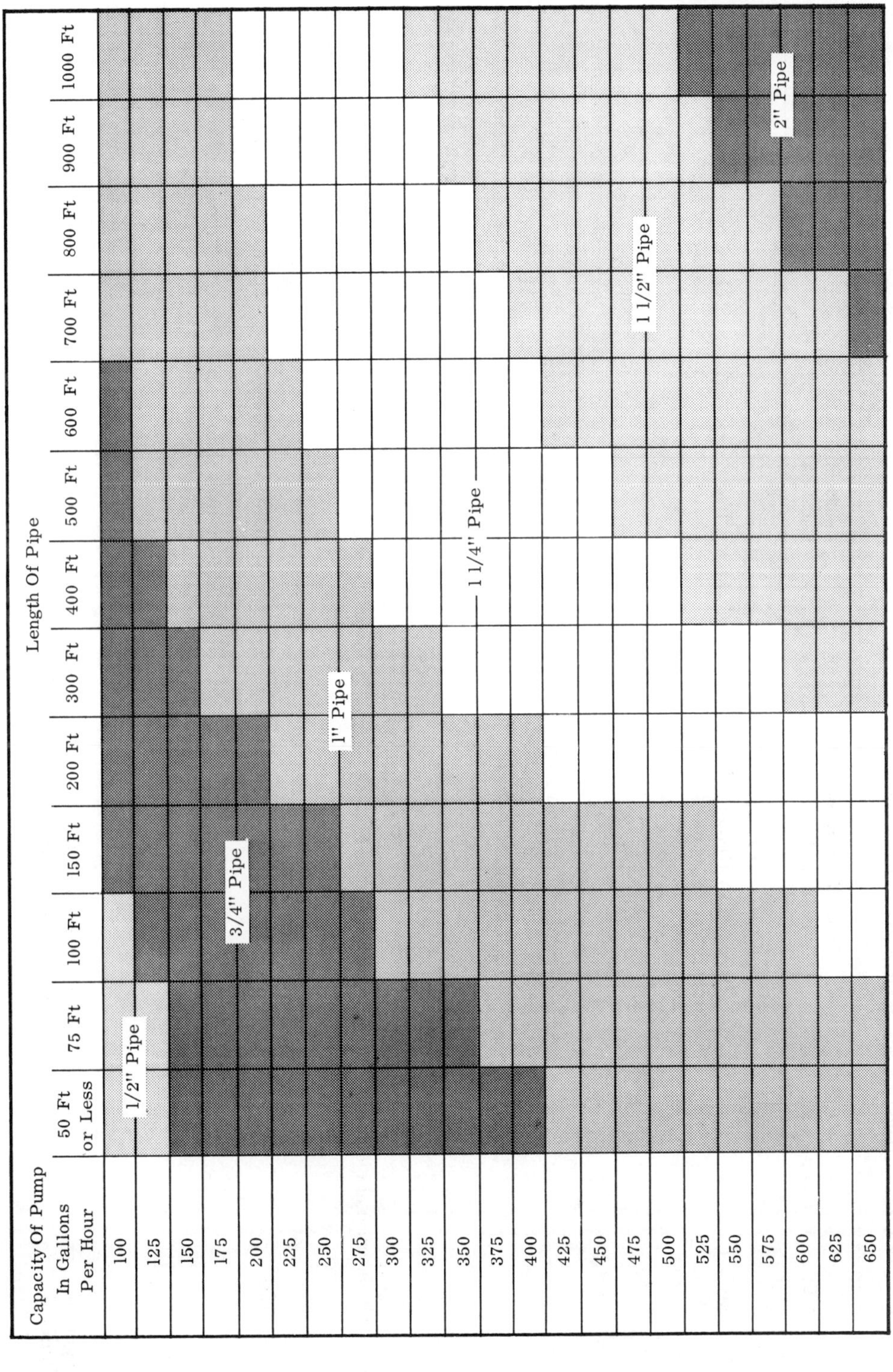

FIGURE 250. Pipe size selector chart.

TABLE 24 CHARACTERISTICS OF MAJOR TYPES OF PUMPS

Type of Pump	Type of Service for Which Best Suited	Practical Location	Other Considerations
	FOR SHALLOW WELLS		
Centrifugal Jet	Lifts water to 18 feet Rather low pressure	Over the well or off the well	Quiet operation Low first cost No moving parts in well May require priming Few repairs Can be converted for deep well service
Piston	Lifts water from 18 to 25 feet Produces high discharge pressure for long water pipes Constant amount of water pumped	Over the well or off the well	No moving parts in well May be noisy Can be converted for deep well service Requires more service than jet pump
	FOR DEEP WELLS		
Deep-well centrifugal ejector or jet	Lifts water up to 80 feet	Over the well or off the well	One moving part Adaptable to wells having 2- or 3-inch casing Quiet operation Self priming pump
Submersible	Lifts water from 60 to 400 feet	Pump and motor are mounted in the well under water Tank and controls may be installed wherever most convenient	No oiling or greasing Quiet operation Easy to install Motor is cooled by well water Long life Original cost may be more than other types
Plunger or Cylinder	Lifts water to 850 feet Delivery is uniform	Over the well	Noisy May vibrate

in an earlier problem. Allow 3gpm each (or 6gpm total) and figure the length of run to the two stock watering tanks at 400 feet.

SOLUTION:

1. Convert 6gpm to gph as follows, 6gpm × 60 minutes = 360gph.

2. Refer to Figure 250. Reading downward under heading "Capacity of Pump in Gallons Per Hour," find 360, which is the amount of water needed to operate two ½-inch faucets. The nearest figures to this are 350 or 375.

3. Reading along either of these lines to the right, locate the block directly under the column "400 ft."

4. This falls in the 1¼-inch zone.

CONCLUSION: To keep two ½-inch faucets going full force at the stock watering tanks, 400 feet from the pressure tank, 1¼-inch pipe is required. Notice that 1¼-inch pipe is also satisfactory for operating the two faucets up to 700 or 800 feet from the pressure tank.

You can use the data in Figure 250 to determine pipe size for any average farm water problem.

WHAT SIZE AND TYPE OF MOTOR WILL YOU NEED?

Usually, a water pump comes from the factory equipped with the correct size and type of motor. Sometimes, however, the pumping load on a motor increases after it is installed. For example, the water table in your well may drop, or you may have to move your tank to a higher elevation. These or other changes in your water system may require a larger and perhaps different type of motor. Therefore, you may need to know how to figure the size and type of motor needed for a water system.

The motor required for a water system is determined by (1) the capacity of the pump and (2) the total working head (or load) that is on the motor. Once you know these two factors, you can easily apply the horsepower formula.

How to Determine Pump Capacity

Earlier in this chapter you observed an example in which a farm with two residences required four "continuous" faucets during the peak-use period. A 720gph pump was selected to keep these four ½-inch faucets running. The 720-gph size is used in the problem that follows later.

How to Determine the Total Working Head of a Water System

The *total head* refers to the total load that a pump must work against, and therefore represents the load that the pump motor must pull. The total head is divided into two parts: (1) the *suction head,* or the total load on the suction side of the pump; and (2) the *discharge head,* or the total load on the discharge side of the pump.

Suction Head. The method of computing the suction head is presented on page 257. By referring to that topic, you will see that the total suction head consists of the depth to the normal standing water level plus the amount of drawdown during pumping plus the elevation of the pump above ground level plus the friction loss in the suction lines and fittings. Total suction head is expressed as *feet of head.*

Discharge Head. All the load on the discharge side of the pump is referred to as discharge head or pressure head.

It consists of the tank pressure plus the elevation of the tank above the well plus the friction loss in the discharge pipes and fittings. Tank pressure is converted to feet before being added. See the example that follows for more details.

How to Apply the Formula for Water Horsepower

From previous experience you should know that work is force or weight moving through distance, and horsepower is the rate at which work is done. One horsepower equals 33,000 foot-pounds of work in 1 minute or any equivalent of this. Therefore, one simple formula for computing horsepower is:

$$hp = \frac{ft\ lb\ per\ min}{33,000}$$

The corrected formula for water horsepower is:

$$hp = \frac{gpm \times lbs\ per\ gal \times total\ head}{33,000 \times pump\ efficiency}$$

This formula is exactly the same as the former with the exception that pump efficiency has been included. This may be taken as 50 percent for farmstead pumps. Since water weighs 8.3 pounds per gallon, you can further reduce the formula to the following form:

$$hp = \frac{gpm \times 8.3 \times total\ head\ (feet)}{33,000 \times 50\%}$$

and this will reduce to approximately the following:

$$hp = \frac{gpm \times total\ head}{2,000}$$

This latter formula is used in the following example.

PROBLEM: 1. Your pump capacity is 720gph, or 12gpm. The suction and discharge pipe is 1¼-inch. Your tank operates on a range of 20 to 40 pounds pressure.

2. The drawdown water level in your well is 42 feet and the elevation of your tank is 3 feet above ground level. Allow for three 90 degree elbows in the two lines. What size motor will you need?

SOLUTION:
1. Determine pump capacity   12 gpm
2. Determine working head
   (a) Vertical lift, drawdown
       level   42 feet
   (b) Additional elevation of
       tank   3 feet
                                                     ————
       Total amount of lift   45 feet
   (c) Friction loss in 45 feet
       of 1¼-inch suction pipe
       at 12gpm* =
       .45 × 4.3 (see Table 21) 1.9 feet
   (d) Friction loss in three
       90 degree elbows** =
       3 elbows × 3.7
       (see Table 22)   11.1 feet
   (e) Convert tank pressure
       to feet, or
       40 pounds × 2.3***   92 feet
                                                      ————
       Total working head   150 feet

Substituting these values in the formula, you can solve for hp:

$$hp = \frac{12gpm \times 150\ feet\ head}{2,000}$$
$$= 9/10\ or\ .9\ hp$$

* Table 21 shows that friction loss in 100 feet of 1¼-inch pipe at 12gpm is 4.3 feet.
** Table 22 shows that friction loss in one 90 degree elbow (1¼-inch) is 3.7 feet.
*** To convert pounds pressure to feet, multiply by 2.3.

CONCLUSION: Your water system would require a .9-hp motor, but you would have to buy a 1-hp size.

By applying the method demonstrated in the preceding example, you can determine the size of the motor needed for any given size pump and depth of well.

Type of Motor Needed. For detailed information on types of motors, refer to Chapter 9. Generally, pumps requiring up to a ¾-hp motor come from the factory equipped with a capacitor-start motor. The next range in size, 1 to 5 hp, usually is of the capacitor-start or the repulsion-induction type for pumps. The 5- to 7½-hp size may be of the newer repulsion-capacitor type. This new type of motor is well adapted to the farm since it draws less starting current than other types of single-phase motors.

If three-phase service is available, in all likelihood you would choose a three-phase, general-purpose motor for any load above ¾ hp, or use three-phase motor with converter.

WHAT KIND OF PUMP HOUSE SHOULD YOU HAVE?

The farmstead water system deserves a good house for protection against weather and pollution. The footings and floor slab of a pump house should prevent surface water from seeping into the well, and proper construction should protect the system from frost. Blueprints and plans for pump

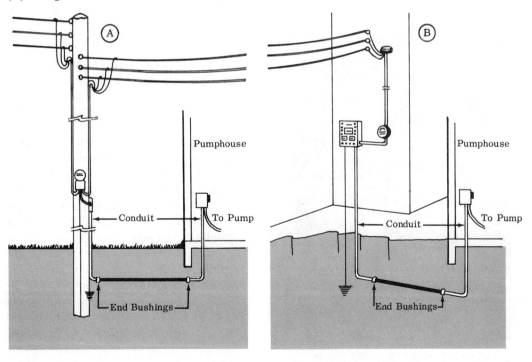

FIGURE 251. Underground wiring for a pump motor: *(A)* wiring from a pole meter; *(B)* wiring from a meter located on a farm building.

houses to fit weather conditions in different sections of the country can be obtained from the nearest land-grant college.

HOW SHOULD THE PUMP MOTOR BE WIRED?

Information on all phases of wiring can be found in Problem-Unit Three. For detailed instructions on selecting and installing wiring materials, refer to the appropriate section.

The trend in wiring pump houses is toward an *underground feeder circuit* that is wired ahead of the main disconnect switch. As previously discussed, this arrangement provides the best fire protection possible, and the wiring is safe from storm damage.

FIGURE 252. A farmer greasing a pump in accordance with the operator's manual.

WHAT CARE SHOULD BE GIVEN TO THE FARMSTEAD WATER SYSTEM?

The best insurance for long, trouble-free service from your water system is to follow the operator's manual in every detail. Figure 252 shows a farmer greasing his pump according to instructions in the operator's manual. It pays to follow directions as this farmer is doing.

Trouble Shooting

You may save yourself a lot of inconvenience and expense by knowing how to take care of ordinary pump and motor troubles. The pump troubles listed in Table 25 are the ones most often encountered in keeping up a farmstead water system. With a little experience and study, you should be able to handle most of your pump troubles yourself.

Again, use your operator's manual, if it covers the particular trouble, in preference to this chart.

WHAT ELECTRIC EQUIPMENT IS NEEDED FOR IRRIGATION?

In recent years farmers and experiment stations have reported many exciting results from irrigation. Some crop yields have been increased as much as 500 percent; in years of extreme drought, crops have been saved outright; strawberries, as well as other tender crops, have been saved from killing frost by sprinkler irrigation.

Despite these results, however, irrigation is not a guarantee of large farm profits. If irrigation is to pay, it must be used in conjunction with appropriate crops and correct farming practices, and the soil must be adaptable to irrigation.

TABLE 25* TROUBLE-SHOOTING CHART

Trouble	Cause
Motor runs, no water delivered	1. Pump needs priming 2. Air volume control lines leaking 3. Check valve stuck 4. Water level in well too low 5. Water suction pipes plugged 6. Working head too great for pump 7. Pump or motor speed too slow 8. Wrong direction of rotation 9. Air pocket in suction line
Motor runs, not enough water delivered	1. Leaky suction pipes 2. Water level too low 3. Low line voltage, motor not running up to full speed 4. Strainer or suction pipes partially blocked 5. Pumping mechanism worn 6. Suction pipe or valve too small 7. Working head too great for pump or motor
Low pressure in the system	1. Speed too slow 2. Air in water 3. Pumping mechanism worn 4. Foreign matter under valves
Pump starts too often	1. Tank water-logged; drain and refill with air 2. Tank, valves, or pipes leaking 3. Pressure switch out of adjustment
Pump is noisy	1. Too much working head, pump overloaded 2. Bent or misaligned shaft 3. Parts binding or too loose 4. Worn pump or motor bearing 5. Driving unit not properly aligned with pump
Too much power required	1. Speed too high 2. Pressure set too high 3. Misaligned or bent shaft 4. Pump parts binding 5. Pump and driving unit not properly aligned
Pump leaks	1. Packing worn or improperly lubricated 2. Packing improperly inserted 3. Wrong kind of packing 4. Shaft scored

* This chart applies to several common types of pumps; therefore, some items listed will not fit every pump.

FIGURE 253. "Irrigation on Wheels" keeps field of snap beans at top-level production in Illinois.

The location of the water supply is another very important factor in irrigation.

Electric Power Supply

In localities where only single-phase service is available, the largest motor permitted will likely be 5, 7½, or 10 hp. Where this is so, irrigation by electric power is limited to small areas.

Where three-phase service is available and large electric motors can be used, they are considered one of the best sources of power for irrigation.

Things Involved in Planning an Irrigation System

The planning of an irrigation system is a job for a specialist. However, you should be prepared to discuss several important things about your situation when the specialist arrives. The following are among the most important points.

Water Supply and Crops to Be Grown. If the source of your water is a well, you can figure about 10gpm capacity per acre; that is, a 500gpm well will furnish water for about 50 acres of crops.

If your water source is a lake or pond, figure 2 to 3 acre-feet per acre of crops to be irrigated. (NOTE: 1 foot of water over 1 acre equals 1 acre-foot.) Also, allow for evaporation of approximately 1½ acre-feet per acre of water during a growing season. For example, a 30-acre lake, averaging 4 feet in depth, would be adequate for 30 to 40 acres of crops.

If there is any question about excess minerals in your water, have it tested before going ahead. Much acreage has been ruined by excess salts in the water.

The total amount of water required, as well as the rate of application, will vary for different crops. For example, a heavy crop of corn uses at the rate of .2 inches of water per day during the peak season of growth; strawberries use about .15 inches per day; and alfalfa uses about .3 inches. These amounts will vary somewhat, depending upon local temperatures. You can apply water faster to alfalfa and other close-growing crops than to clean cultivated crops.

A selection of crops that mature at different times allows for greater use of an irrigation system. Remember, however, that vegetables, fruits, and similar intensive-type crops offer the greatest returns. For example, results of irrigation of tomatoes, beans, and sweet potatoes in Mississippi resulted in increases of yields by 100 to 300 percent.*

---

* Mississippi Agricultural Experiment Station, *Irrigation for Truck Crops*, Circular 163 (1951), and *Irrigation for Vegetable Crops*, Circular 182 (1953), State College, Mississippi State University (Revised 1965).

FIGURE 254. Underground irrigation pipe ready for lowering into a ditch.

FIGURE 255. *Top:* One of the problems in supplemental (overhead) irrigation is moving the sprinkler equipment. Mounting "mains" on wheels reduces labor requirement. Corn in this scene should produce at least 200 bushels per acre.

FIGURE 256. *Bottom:* Three-phase service makes it possible to use a 50-hp motor (or larger if needed).

Topography and Types of Soil. The lay of your land will be a large factor in determining whether or not you can use the furrow or flooding method. On other than flat land or gentle slopes, irrigation is limited to the overhead or sprinkler method.

The type of soil to be irrigated will determine the rate of application of water. A loamy soil will absorb water at a much faster rate than will a tight clay. Therefore, a clay soil requires a longer setting of the equipment than does sandy loam. Sandy soils, on the other hand, have a lower water-holding capacity and therefore require more frequent applications.

Labor Available. Considerable labor is required in operating all irrigation systems. The moving and setting up of sprinkler equipment, for example, is

Dig down to main root system. Take a handful of soil and "ball" it in your hands. If it looks like this, it's time to start irrigating.

Coarse-Sandy Soil

Hand Test For Moisture

Tends to ball under pressure—but won't hold together when bounced in hand.

Here's how the soil looks when it's down to 50 per cent or less of moisture-holding capacity. Here you have waited too long.

FIGURE 257. When to start irrigating.

When squeezed, soil appears dry, won't form a ball under pressure.

quite time consuming. This should be remembered in planning your farm irrigation system.

In sprinkler irrigation the trend is toward smaller diameter pipe of shorter lengths to enable one man to move the setting. Also, you can have your designer plan a sprinkler system that can be moved between 6 and 7 a.m. and again between 6 and 7 p.m., thereby causing a minimum of interference with the regular work day.

Amount of Power Required. The amount of power required to pump a given amount of water against a given head in a given period of time is a simple horsepower computation. In fact, the same hp formula used in figuring motor size for the farmstead system can be used: hp = gpm × total head ÷ 2,000.

(NOTE: This formula is based on a pump, motor, and system that is rated at 50 percent efficiency. Some irrigation pumps and motors may have higher efficiency. Check this with your dealer.)

PROBLEM: Assume that you wish to put 1½ inches of water on 30 acres of corn. Your well is 150 feet deep and you will operate your pump at 40 pounds pressure for eight hours a day. You wish to complete this job in 5 days (40 hours).

ELECTRIC WATER PUMPS   273

Medium Texture

Forms ball—is somewhat plastic—sticks together slightly with pressure.

Fine Clay

Forms ball, ribbons out between thumb and forefinger, has slick feeling.

Soil is somewhat crumbly but it will hold together in ball from pressure.

This soil is somewhat pliable, and will form a ball under pressure.

What size motor will be required?

SOLUTION: 1. Determine the gpm rate required. Apply the rule that 453-gpm pumping for 1 hour = 1 inch on 1 acre. Then the rate required to pump 1½ acre-inches per hour would be 453 × 1.5 = 679.5gpm.

Therefore, the rate required to pump 1½ inches on 30 acres in 1 hour would be 679.5 × 30 = 20,385gpm. Since this rate is for 1 hour, you can determine the rate required to do this job in 40 hours by dividing by 40; thus: 20,385 ÷ 40 = 509.6gpm.

Rounding this off, you see that a rate of 510gpm would be necessary to put 1½ inches of water on 30 acres in 40 hours.

2. Total head:
 (a) 40 pounds pressure
      × 2.3 =   92.0 feet
 (b) 20% friction loss in
      laterals = 8 pounds,
      then 8 × 2.3 =   18.4 feet
 (c) 20% friction loss in
      mains =   18.4 feet
 (d) Total elevation
      (well depth + field
      elevation) =   150.0 feet
                            ──────────
                  Total head 278.8 feet

Round this figure off to 279 feet.

(NOTE: One pound pressure equals 2.3 feet of head. Friction loss of 20 percent (on 40 pounds pressure) is the maximum amount that is usually designed into the system.)

3. Apply formula: hp = gpm × total head ÷ 2,000.

Substituting values found in 1 and 2, you get 510 × 279 ÷ 2,000 = 71.1hp.

ANSWER: You could use a 75-hp engine. Take note that a 60-hp, three-phase electric motor would do as much work as the larger engine. Moreover, you could get by with a 20-hp electric motor by increasing the 8-hour day to 24 hours.

Cost of Irrigation

Studies in various parts of the country show that the total investment in irrigation varies from a low of $65.00 per acre, where a natural source of water is available, to $200 per acre for deep-well systems. The average investment is around $100 to $110 per acre.

The overall cost of irrigation ranges from $4.50 to $7.50 per acre-inch of water applied. Therefore, the cost of 1½ inches of water on the 30 acres of corn in the preceding problem ranges from about $200 to $300, or a maximum of $10 per acre per application of 1½ inches.

A rule of thumb on well cost is to figure $1.25 per inch of diameter per foot of depth. On this basis a 6-inch well should cost around $7.50 per foot of depth. Add $20 to $40 per foot of screen needed at the bottom of the well, depending on size. In the irrigated sections of the country, well size goes up to 24 inches or so.

Aluminum pipe for a sprinkler system on wheels, costs about $1.25 a foot for 4-inch size in 20-foot lengths. Longer lengths are less costly per foot but are more difficult to handle.

Operation

The most general rule about starting to irrigate is to start your pump when the soil water capacity is down to 60 or 75 percent. For best results 50 percent is considered too late. After irrigation has been started, the top 12 inches of soil should be kept in a good moist condition until the crop is matured. Figure 257 shows a practical hand test for determining when to irrigate.*

Engines will last from 3 to 10 years depending on the type and upon the care received. Electric motors should last 10 to 30 years with proper lubrication and correct wiring.

SUMMARY

The first essential in getting a pure water supply is to select a well site that is safe from pollution. If a well is located on a slope below a barnyard, cesspool, or other unclean spot, contaminated water may find its way underground into the well. Typhoid fever and other diseases may be spread in this manner.

Well depths under 25 feet are considered *shallow* wells; depths over 25 feet are considered *deep* wells. This is the basis of determining whether a shallow- or a deep-well pump is required.

---

* "When and How Much to Irrigate," *Successful Farming* (May, 1957), pp. 146–148.
Note: Test still valid in 1970.

Pump size is determined by counting the number of continuous faucets that operate during the peak-use period for the farmstead. A continuous faucet is one that runs continuously for 1½ minutes or more. For each ½-inch continuous faucet, 3gpm is allowed; for each ¾-inch faucet, 5gpm. Example, a 12gpm peak rate requires a 720gph pump (60 min $\times$ 12gpm = 720gph).

Two types of shallow-well pumps are common to the farm: (1) the centrifugal pump and (2) the piston pump. The centrifugal pump is suitable for depths to 18 feet. The piston pump is better adapted to well depths of 18 to 25 feet and gives higher, more uniform pressure at these depths than the centrifugal pump. Quieter operation and fewer repairs are the main advantages of the centrifugal pump.

The three most common types of deep-well pumps are (1) the jet or ejector, (2) the plunger, and (3) the submersible. The deep-well jet pump is suitable to well depths to 80 feet. By using several stages of jets, greater pressure can be developed. The jet type can be located some distance from the well. The plunger pump is suitable for well depths to 850 feet. The *working head* of this type of pump must be located directly over the well. The submersible pump is so called because the pump and motor are located in the well under water. This type pump is suitable for wells of 400 feet in depth, but the water must be *free of sand* since the pump turbines will quickly wear out when sand gets into them. The submersible type of pump is one of the most dependable of those on the market where pumping conditions are favorable for it.

Due to friction loss, long runs of pipe result in a drop in water pressure. Where it is necessary to use long runs, larger pipe must be used in order to maintain the proper volume of water. The proper-size pipe can be chosen from data in Figure 250.

The size of a motor required for a given pump and situation can be determined by the formula hp = gpm $\times$ total head $\div$ 2,000. Total head refers to the total load on the pump, both suction and pressure.

In areas where frost may damage water pipes, an insulated, heated pump house is necessary. The house should also have means of preventing surface water from seeping in around the footings.

Most pump troubles can be identified by referring to the trouble-shooting chart on page 269. A maintenance program, based on the operator's manual is the best insurance against pump troubles.

Irrigation may result in as much as 400 to 500 percent increase in crop yields. On the other hand, it may not pay at all. The first requirement for irrigation is to have an adequate water supply. A widely used rule of thumb is to figure 10gpm well capacity for each acre of crops. Allow 3 to 4 acre-feet in lake capacity for each acre of crops. This figure includes an allowance for loss by evaporation of 1 to 1½ feet of water per acre.

The type of soil, type of crop to be grown, lay of the land, labor supply, and amount of power are important factors that enter into the design of a system. Because of the large number of factors to consider, it is best to have an expert design the irrigation system. A good many farmers have been ruined financially by investing in an irrigation system that did not fit their situations.

The cost of irrigation ranges from about $4.50 to $7.50 per acre-inch of water applied, and total investment ranges from $65 to $200 per acre. The lesser amount here applies to farms having a natural water supply (lake, pond, or stream) and other favorable factors. The top figure applies to drilled wells, which often run to $7.50 per foot of depth (6-inch well).

QUESTIONS

1 Why is a deep-well pump required for depths over 25 feet?
2 Why must friction loss be added to well depth in computing the total depth load?
3 Why is larger pipe required for longer runs of water lines?
4 Why will sand ruin a submersible pump?
5 How does a pump and motor react to a water-logged tank?
6 How can you determine the number of acres a well will irrigate?
7 When should irrigation be started for a given crop?
8 Why is it necessary to have an irrigation specialist design the irrigation system?

ADDITIONAL READINGS

Brown, R.H., *Farm Electrification.* McGraw-Hill Book Co., New York, N.Y., 1956.
Member Services Division, National Rural Electric Cooperative Association, *Planning Water Systems That Work.* Washington, D.C., 1968.
National Association Domestic and Farm Pump Manufacturers, *Manual of Water Supply and Equipment.* Chicago, Ill., 1953.
Schaenzer, J.P., *Rural Electrification,* Fifth Edition, Bruce Publishing Co., Milwaukee, Wisc., 1955.
Southern Association of Agricultural Engineering and Vocational Agriculture, *Planning Farm Water Systems.* University of Georgia, Athens, Ga., 1955.
Wright, F.B., *Rural Water Supply and Sanitation,* Second Edition. John Wiley and Sons, New York, N.Y., 1956.

12 HOW TO PROVIDE GOOD LIGHT FOR HOMES, GROUNDS, AND FARM SERVICE BUILDINGS

A replica of Edison's first incandescent light bulb, which he perfected in 1879, is shown in Figure 258. Ever since that memorable date, the American farmer and his family have used electric light more than any other form of electricity. Indeed, for many years after the beginning of rural electrification, lighting was regarded as the major use of "high-line" electric power.

Only since the 1940's has electric power been used much to do farm work. Despite the fact that it has largely replaced human labor on many farms, a large number of farmers continue to use it mainly for lighting.

A recent check by Edison Electric Institute showed that 98 percent of all non-dilapidated farm residences in the country were lighted with electricity.* This does not mean that all were well lighted. Only a small percentage of them, in fact, could have been classified as having *good light*.

Plenty of good light seems to be so essential to the production and efficiency of the farm and to the health and safety of farm people that the lack of proper lighting appears to be one of the major farm problems in this country today.

WHAT IS GOOD LIGHT?

Light is considered good light only when there is the *right amount of the right kind in the right place*. The need for light at different places on the farm and in the residence varies in both kind and amount.

---

* Edison Electric Institute, *Questions and Answers About the Electric Utility Industry*, 1968 ed. (New York), p. 14.

FIGURE 258. *Top:* Replica of Edison's first successful incandescent lamp.

FIGURE 259. *Bottom:* The Illinois farmstead shown here is well lighted by a mercury vapor lamp—now very popular.

Amount of Light

Amount of light refers to the *brightness* of light at a given spot. To illustrate, a 100w bulb will produce more light than a 50w bulb, provided they are the same type and are located at an equal distance from the lighted spot.

The overall problem of how to provide the right amount of light for a given need involves three smaller problems: (1) determining the amount of light needed for different areas on the farm and in the farm home, (2) choosing the proper color schemes for interior walls and ceilings, (3) placing lamps and fixtures to get the proper amount of light from a given size bulb or tube.

Amount of Light Needed for Different Locations and Different Jobs. The standard measure of light is the *foot-candle*—the *amount of light falling on a surface 1 foot away from a standard candle*. The only accurate method of measuring light is to use a *light meter* as shown in Figure 260.

The inexpensive meter in Figure 260 is used to check light in residences, schools, and other similar places. Perhaps you can borrow one to use at home. Ask your local agriculture teacher, club agent, or power representative about this.

The foot-candle meter is easy to read. You can measure and record the amount of light you have in the different localities at your farm and farm home in a few minutes. These figures will tell you whether or not the amount at each spot is adequate. Simply check your readings against those listed in Table 26. If your readings show too little light at any given spot, the size of bulbs or tubes should be increased. This may

TABLE 26 AMOUNT OF LIGHT NEEDED FOR HOMES, GROUNDS, AND SERVICE BUILDINGS

Lighting Task	Foot-candles
GENERAL OR "FILL IN" LIGHT	
Kitchen	10
All other rooms and areas in the farm home	5–10
Farm grounds	2–5
Interior of barns and poultry houses	3–5
Farm shop and pump house	5
LOCAL OR CONCENTRATED LIGHT	
Fine sewing on dark fabrics	150
Fine sewing on light fabrics	100
Medium sewing and machine sewing	40
Prolonged reading, studying, writing, shaving, handicraft	40
Close work in kitchen	30–40
Casual reading, of good type on white paper, facial make-up, coarse sewing, reading music, table tennis, card playing, laundry work	20
Milk house	30–50
Milking room, at milking station	20
Egg handling	10–15
Vegetable grading	30–40
Farm shop	
Close work	20–40
General work	15–20
Tobacco grading	100
Feed handling	10
Pump house, pump repairing	20
Animal grooming	10–15
Seed cleaning	20
Poultry slaughter	40–50

TABLE 27 PERCENT OF LIGHT REFLECTED BY FIFTEEN BASIC COLORS

Color	Approximate Percent Reflectance
White	85
Ivory	75
Lemon peel	72
Baby blue	71
Pink	65
Yellow	65
Tan	60
Medium gray	60
Light blue	60
Light green	55
Apricot	50
Cardinal red	20
Dark brown	10
Olive green	10
Dark blue	5

require some new circuits as well as new fixtures. Sometimes, a change to a lighter color scheme will solve the lighting problem.

How to Choose the Proper Color Schemes for Interiors. If walls and ceilings are dark blue, only 5 percent of the light from a lamp will be reflected. This is why it is hard to get adequate light in dark-colored interiors. Therefore, if your light-meter readings are low, you should check the color schemes of your interiors. In addition, you should check the sizes and the types of light bulbs being used.

The reflective powers of 15 different basic colors are listed in Table 27. Mixing two or three of these colors will result in a reflection that is an average for the colors mixed.

If you plan to buy new lighting fixtures, the cost can be reduced by using light colors on your interior walls and ceilings. This will allow you to use smaller, less costly fixtures.

Rules of Thumb on Interior Colors. Pick color schemes for your ceilings that will reflect 60 to 90 percent of the light; for the side walls, 35 to 60 percent; for the floors, 15 to 35 percent.

Take meter readings again after the painting is completed. They should be higher if your interiors were dark colored before. An extra-high ceiling will also reduce the amount of light reflected by a given size bulb.

How to Place a Lamp to Get Proper Light From It. A light fixture that is mounted too high will not cast as much light on the work area as it should. An example of this is a light fixture mounted in the top of a high-ceiling haymow. For this type of interior, a reflector lamp or a reflector-type bulb should be used.

An ordinary light bulb (inside frosted) in an open fixture without a reflector will waste light when mounted in an open area. This fact is illustrated by the smoke-box pattern in Figure 261-A. Notice the amount of light falling on the ceiling. This light would be lost if the fixture is located outdoors; it would be lost on the inside of buildings if the ceiling is high or has very little reflective power. Compare the patterns of light produced by reflector-type bulbs as shown in Figures 261-B, C, and D. From two and one-half to four times more light is directed onto the work area by a reflector-type bulb. White porcelain reflectors used with ordinary bulbs have a similar reflective power.

These tests show that some kind of reflector or reflector bulb is necessary for large buildings with high ceilings and for outdoor lighting. Otherwise light (and electricity) will be wasted.

FIGURE 260. A meter that measures footcandles of light.

On the other hand, a yard light can be too low. The result is an excess amount of light directly under the lamp and not enough spread. A yard light that is too low will also produce a blinding effect.

The diagrams in Figures 262-A and B illustrate the point here. A low yard light causes a long shadow to be cast. Be sure to place your outside lights at least 15 feet high. This height will require at least a 200w bulb.

Sizes of Bulbs, Tubes, and Circlines for the Farm and Farm Home. For ordinary farmstead use, incandescent bulbs are made in sizes ranging from 10w to 300w. Fluorescent tubes and Circlines come in sizes from 15w to 60w.

For a lighting arrangement that calls for 250w or 300w, it is a common practice to use a fixture with three to seven small bulbs instead of one large bulb. Better light diffusion can be produced this way because of the greater bulb area.

FIGURE 261. Smoke-box light patterns of four different types of light bulbs.

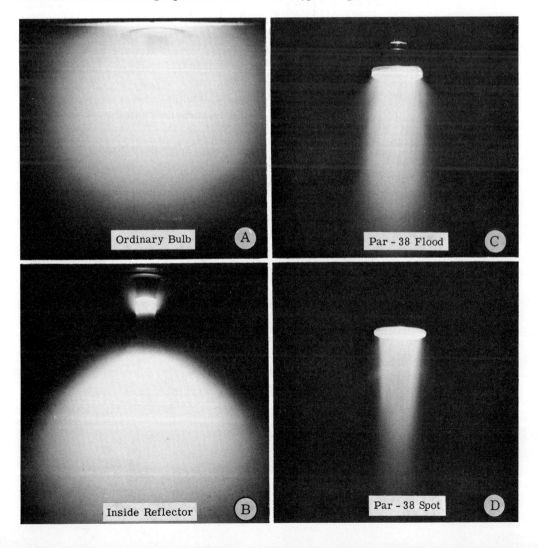

Important Things to Know About Kind of Light

Kind of light has reference to quality and light source, which may be incandescent or fluorescent.

How to Provide Good Quality Light. In lighting, *quality* refers to an even distribution of light so that there is no glare or shadow present. A contrast between bright and dark spots in a room can cause eyestrain.

The human eye has a lens to control the amount of light that may enter it. When you look at a bright spot these lenses will contract. Then, if you quickly look at a dark area, they will expand again. "Spotty" light in a room where you read or work may damage your vision. This can result from eyestrain brought on by too much contracting and expanding. Your vision is too valuable to take chances with poor quality lighting.

Getting quality light is closely related to the selection and placement of types and sizes of fixtures and lamps. Generally, the larger the diffuser or bowl, the better the quality of light. A large diffuser distributes light over a wider area, yet does not reduce the amount of light produced. The glass bowl of a lamp fixture serves as a type of diffuser too. The white finish on a light bulb also acts as a diffuser.

The manner of directing or reflecting light into a room or on a particular spot has a strong influence on quality. According to the method of directing the light, lamps and fixtures are classified into five groups. These are illustrated and described in Figure 263.

For reading, sewing, preparing food, and other close work in your farm home, choose *indirect lighting*. *Semi-indirect* lighting can be used for a few special needs in the residence.

For outside lights and for most farm work, other types of lighting can be used and will be less expensive.

Important Things to Know About Light Sources. The two light sources common to the farm and farm home are *incandescent* and *fluorescent*. Both types have advantages as well as disadvantages. No doubt you will find a need for both in your farmstead lighting system.

FIGURE 262. *(A)* The lamp is too low—note the large shadow. *(B)* The lamp is mounted at the correct height (30 feet). Compare the shadow with the one in *A*.

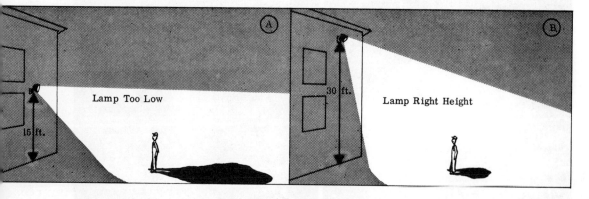

LIGHT FOR HOMES, GROUNDS, AND FARMS

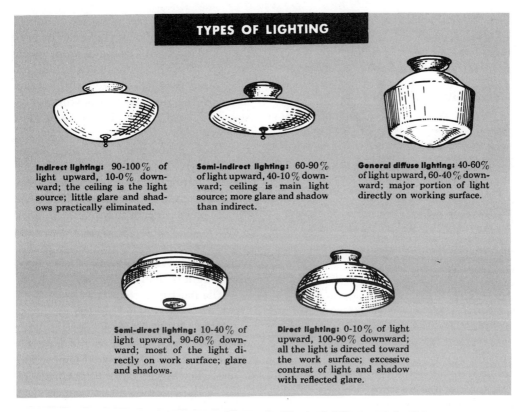

FIGURE 263. Five types of lighting according to direction and diffusion of the light.

Moreover, you may find a need for heat lamps and other special sources of light.

Incandescent Bulbs. The light bulb invented by Edison (Figure 258) was of the incandescent type. This is still the most widely used on the farm or in the farm home. However, there has been a trend in recent years toward the use of more fluorescent lighting.

Notice the details of the bulb shown by diagram in Figure 264. The globe is filled with certain inactive gases. A current flows through the metal base and on through the metal filament. In the presence of the inactive gases, the electric current causes the filament to become white hot and to give off light.

An ordinary household light bulb is built to last from 700 to 1,000 hours. The wrong voltage may reduce this life span however.

Fluorescent Tubes and Circlines. The soft, daylight appearance of fluorescent light is almost ideal for the milk room and for reading, writing, sewing, or other close work. This type of light is produced by a different principle than is the incandescent bulb.

Upon examining the diagram in Figure 265, you will see that there is

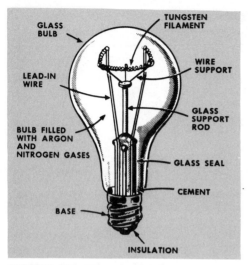

FIGURE 264. Diagram showing parts of an incandescent light bulb. Light is produced when current causes the tungsten filament to become white hot.

By varying the phosphor coating, different shades of white light can be produced. Tubes can be bought in seven or more different shades of "white" light. The most popular shade for the farm home is the de luxe warm white. The standard cool white is recommended for the laundry.

How Fluorescent Lighting Compares with Incandescent. Fluorescent lighting equipment is somewhat more expensive than incandescent. However, the life of fluorescent tubes and Circlines is three to five times greater than incandescents; fluorescent tubes, watt for watt, produce about three times more usable light than incandescents.

The most important features of both types are compared in Table 28.

Other Light Sources. Several light sources for special uses on the farm include heat lamps, germicidal lamps, and black-light insect traps. Illustrations of some of these can be found near the end of this chapter.

HOW CAN YOU PROVIDE GOOD LIGHT IN YOUR HOME?

You will be concerned with two systems of lighting for your home: (1) good *general* light for each room or other area, and (2) good *local* light for all close work.

no filament or wire through the tube. The space inside the tube becomes filled with mercury vapor when an electric current begins to flow. The current "arcs" across the space inside the tube through the aid of these mercury particles. The inside walls of the tube are coated with a substance called phosphor. The bombardment of the current on the mercury particles produces ultraviolet light. This, in turn, acts on the phosphor coating to produce fluorescent light.

FIGURE 265. Diagram of a fluorescent tube.

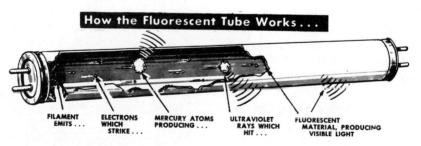

LIGHT FOR HOMES, GROUNDS, AND FARMS

TABLE 28 COMPARISON OF INCANDESCENT AND FLUORESCENT LIGHT SOURCES

Fluorescent Does This	Incandescent Does This
Produces a line of light	Produces a spot of light
Casts little or no shadow	Casts shadows and produces glare if open to view
Different wattages not interchangeable	Different wattages are interchangeable in standard sockets
Slightly more expensive	Most economical in first cost
Lasts about 2,500 hours	May last from 700 to 1,000 hours
Older models are slow to start, but newer ones start instantly	Starts instantly
Gives off very little heat	Gets hot while in use
Special fixture containing ballast is necessary	Requires no ballast
Produces 2½ to 3 times more light than incandescent	

How to Provide Good General Light

For your living room, bedrooms, and hallways, you will need from 5 to 10 foot-candles of general light; for your kitchen, 10 foot-candles. This is necessary for fill-in lighting so that there will not be bright and dark spots in a room.

Type and Placement of Fixtures for General Lighting. Incandescent ceiling fixtures are the kind that are most often used for general lighting. A recessed type of fixture, shown in Figure 266, is being used more than surface-mounted or suspended fixtures and is considered "standard" now. The fixtures shown in this scene come prewired for easy and correct installation. Each fixture in the illustration carries two 60w or 75w bulbs.

Without general light in this living room, there would be dark areas around the bright spots of each portable lamp.

Close-to-Ceiling Fixtures. The ceiling fixture in the bedroom in Figure 267 is a more common type. It carries five 40w bulbs and has a 17-inch shield or diffuser.

Cornice Lights. Notice also the two cornice lights over the windows. These furnish general light, but, in addition, they supplement the local light at the dressing table.

FIGURE 266. The background (general) light from two ceiling fixtures keeps the table lamps from making bright spots in the room.

FIGURE 267. A five-lamp ceiling fixture and two valance lights provide adequate general lighting for a bedroom.

Fluorescent Special Fixtures. A modern trend in kitchen lighting is illustrated in Figure 268. The fluorescent dinette fixture shown carries two 40w tubes. In the same scene (not visible) a recessed fluorescent fixture is over the sink. Although the latter is local light, it

FIGURE 268. A large fluorescent center fixture in the kitchen gives good general light and also supplements local light.

supplements the general light in the kitchen.

Combination Fixtures. The scene in Figure 269 illustrates the use of a combination of fixtures and lamps for general and local light. Notice the long runs of cornice lights around the two walls. A three-unit suspended fixture is located over the breakfast table. Portable lamps complete the lighting system for this large room. The light here is good light for viewing television.

Methods of Wall Lighting. The diagram in Figure 270 shows details of three methods of wall lighting. Fluorescent tubes and brackets, which can be bought at any good appliance store, are installed on the wall. A board is mounted in front of the tube to break the direct light rays. The result is a very fine quality light.

Porch and Step Lights. The scene in Figure 271 shows how general light for the front porch and steps is provided by the recessed fixture installed directly overhead. If this fixture were located on the porch wall, the light would have a "blinding" effect on any one approaching the building. Notice the attractive fixtures recessed into the steps. Perhaps you could do this wiring job as a project.

A word of caution about fixtures. A high priced fixture may not produce the best light. In highly decorated lamps, you pay for the decorations and finish. *Be sure you get your money's worth in good light first.* Decoration and finish should come second, if at all.

How to Provide Good Local Light for Close Work

The amount of light required for close work ranges up to 150 foot-candles

Cornice Valance Wall - Bracket

FIGURE 269. *Top left:* A lighting arrangement for den, playroom, and breakfast room includes valance lights, a drop fixture, and reading lamps.

FIGURE 270. *Top right:* Three arrangements of wall lighting using fluorescent tubes and brackets. Build these yourself.

FIGURE 271. *Bottom:* Built-in step lights and a recessed porch fixture.

for sewing dark-colored fabrics. However, you will be concerned not only with providing a sufficient amount of light for the many special needs in your home; you will also want to provide good quality light.

Things to Look for in Choosing Lamps or Fixtures for Local Lighting. The four most common types of portable lamps are shown in Figure 272. All of the types shown can be bought in either table or floor models. All four styles produce indirect lighting. Check on the following points in buying or arranging lamps.

1. A *large diffuser* is perhaps the most important single feature to look for in a portable lamp. If you cannot afford a lamp with a diffuser, the white indirect bulb in Figure 272-C should be chosen for close work, in preference to ordinary inside frosted bulbs. Though all four lamps shown will give indirect lighting, some cost more than others. You will be wise to study a good catalog of lighting fixtures before buying.

2. Another important feature of a lamp is a *three-lite* arrangement. Sizes usually range as follows: 30-70-100w, 50-100-150w, and 100-200-300w. The latter size has a large base called a *mogul*

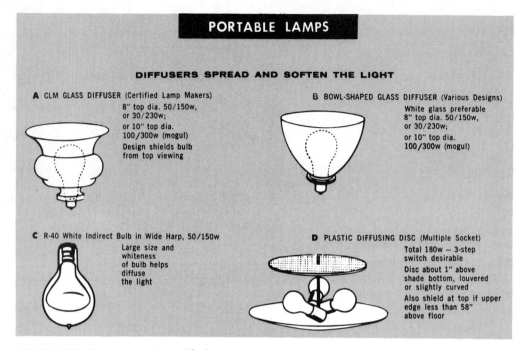

FIGURE 272. Four types of portable lamps.

base. The advantage of a three-lite lamp is that three levels of light can be had from one lamp. *Mogul* can be changed to regular by an adapter.

3. Choose lamp shades that are wide enough at the bottom to throw light across the work area and wide enough at the top to allow good upward projection.

FIGURE 273. Correct placement of table lamps. The bottom of the lamp shade should be level with eye height.

4. The lower edge of the lamp shade should be level with the eye. Observe that the table at left in Figure 273 must be higher than the one at right in order to have the bottom of the lamp shade at eye height. This is due to the shorter base of the lamp on the left.

5. The material of which a lamp shade is made should be light in color, or the inside surface should be white. Dark colored shades absorb and waste light. Lamp shades should be dense enough to prevent light from shining through them.

6. The depth of a lamp shade should be sufficient to conceal the bulb from the bottom or top view, whether you are sitting or standing.

7. Convert your one-lite table lamps to three-lite by installing a converter

switch which can be bought at most appliance stores for $1.50; a white indirect bulb will cost another $1.25.

How to Provide Good Light for Reading. The correct postion of a three-lite, 100-200-300w, senior floor lamp is shown in Figure 274. Study the placement dimensions of the lamp in relation to the reader. The bottom edge of the shade is 47 inches from the floor. Observe that the table lamp in Figure 275-A is lower than the senior floor lamp (Figure 274). This is due to the lower wattage and smaller shade of the table lamp, which is a three-lite, 50-100-150w size.

Figure 275 shows (A) table lamp and (B) pull-down lamp correctly placed for reading. Note the dimensions shown in the diagrams.

There are many other possible arrangements for good reading. The ones shown here, however, are the most common ones.

FIGURE 274. *Top:* Senior floor lamp correctly placed for reading.

FIGURE 275. *Bottom:* (A) Table lamp and (B) pull-down lamp correctly placed for reading. Note the dimensions.

 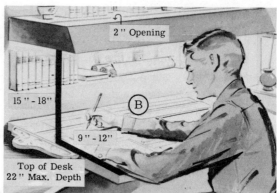

FIGURE 276. *(A)* Two pin-to-wall lamps provide shadowless light for studying. *(B)* A fluorescent study lamp that can be built at home.

How to Provide Good Light for Studying and Writing. If you are to make the best possible progress in school work, you must have good light for studying at home. Figures 276-A and B show two arrangements, either of which will provide the 40 foot-candles needed for this important task.

In Figure 276-A, two pin-to-wall lamps are mounted 30 inches apart, 15 inches above the desk top, and 17 inches from the front of the desk. Each lamp is equipped with a 6-inch plastic bowl and carries a 100w inside frosted bulb. Lamp shades are light-colored and have a wide bottom. Two lamps placed as shown in Figure 276-A will eliminate almost all shadows.

The fixture shown in Figure 276-B carries a 33-inch, 25w, deluxe warm white tube. You can purchase the tube and the channel at any good appliance store for about $8 each. The height of the desk bracket after mounting is 15 to 18 inches above the desk top.

How to Provide Good Light for Sewing and Reading Music. According to Table 26, hand sewing requires more light than any other task done in the home. The senior swing-arm floor lamp in Figure 277-B will provide the 100 to 150 foot-candles needed for this difficult visual activity. This lamp carries a three-lite, 100-200-300w bulb and a 10-inch white bowl diffuser. An even better arrangement would be the same style lamp equipped with a 12-inch diameter, 32w, de luxe warm white Circline tube. The bottom of the lamp shade is 47 inches above the floor.

In homes where one or more members of the family play the piano or read other music, the same senior swing-arm floor lamp can be used to provide the 40 foot-candles of light called for in Table 26. Figure 277-B shows the correct placement dimensions for the lamp at a piano. Note that the bottom of the lamp shade is 47 inches from the floor.

How to Provide Good Light for Shaving. Table 26 shows that 40 foot-candles are needed from three sides for easy shaving. The ideal setup for this is shown in Figure 278-A. Two 24-inch,

LIGHT FOR HOMES, GROUNDS, AND FARMS

FIGURE 277. *(A)* A senior swing-arm floor lamp furnishes ideal light for sewing. *(B)* The same lamp used for reading music. Note the proper placement for both activities.

20w, shielded fluorescent lamps are centered 60 inches above the floor, one on each side of the mirror; these lamps are spaced 30 inches apart. One 18-inch, 15w, fluorescent fixture is mounted over the mirror, about 15 inches above head height and 12 to 18 inches out from the wall. De luxe warm white tubes are best for this bathroom arrangement.

An incandescent fixture combination can be used instead of the fluorescents. The globe diffusers for these fixtures should be of opal or ceramic-enameled glassware. At least one 60w bulb should be carried in each wall and ceiling unit.

How to Provide Good Light for Grooming. In the bedroom, 20 footcandles of special light are needed at a dresser. Figure 278-B shows a good arrangement of dresser lamps for make-up. The two lamps shown are centered 36 inches apart, with the shades 22 inches above the dressing table. This is the correct height for a standing position.

Dresser lamps do not have a diffuser but rather have an ivory or translucent shade (9-inch diameter), which produces a strong light from two sides. The bulb may be either a three-lite, 30-70-100w, or a 100w white bulb.

How to Provide Good Light for Work in the Kitchen. The family's health demands that plenty of good local light be provided in the kitchen where food is prepared and utensils are washed. Table 26 shows that from 30 to 40 footcandles should be provided. The light source should be de luxe warm white fluorescent tubes or Circlines, if possible.

Notice the excellent lighting setup at the sink in Figure 279-A. This fluorescent fixture carries two 33-inch, 25w

FIGURE 278. *(A)* The proper set-up for shaving requires light from three sides. *(B)* Two dresser lamps properly placed for grooming. Note the placement of the fixtures in both scenes.

tubes mounted behind a faceboard. You can purchase standard tubes and channel at any good lighting appliance store and build the wood bracket as a project.

A lighting arrangement for the food preparation counter in the kitchen is shown in Figure 279-B. In this setup, a fluorescent tube and channel is mounted underneath a cabinet less than 52 inches high. If the cabinet is more than 52 inches high, the fixture should be mounted on the wall underneath. Similar fixtures and arrangements can be placed over the range.

How to Provide Good Light for the Laundry. Approximately 20 foot-candles

FIGURE 279. *(A)* Fluorescent tubes and bracket mounted over a kitchen sink and behind a face board. *(B)* A fluorescent tube and bracket mounted underneath a cabinet over a food preparation area. Build these yourself.

LIGHT FOR HOMES, GROUNDS, AND FARMS

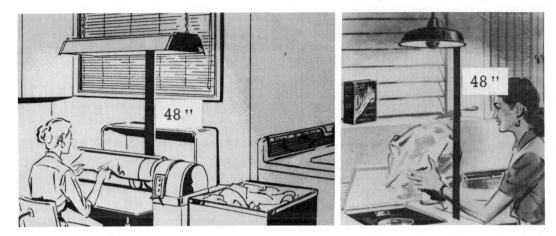

FIGURE 280. Either the fluorescent fixture at left or the reflector lamp at right is suitable for the laundry. Note the dimensions for placement.

are needed for ironing, washing, and other laundry work. Figure 280 (left) shows an ideal setup for ironing. A rectangular fluorescent fixture is mounted 48 inches above the work. It carries two 33-inch, 25w, standard cool white tubes.

Another good arrangement is shown in Figure 280 (right). This arrangement uses a standard dome reflector equipped with a 150w silver-bowl bulb. In both scenes, the reflector is mounted 48 inches above the work. Recessed fixtures are especially adapted to use in the laundry, where head room is often scarce.

HOW CAN YOU PROVIDE GOOD LIGHT FOR YOUR HOME AND FARM GROUNDS?

A dark farmstead is dangerous. An investment of a few dollars in a yard light may pay for itself many times over by preventing accidents and increasing chore time. A good rule of thumb is to have at least 2 or 3 foot-candles of light on all major chore paths on the farm-stead. Barnyards and building entrances should be lighted separately if the farmstead buildings are scattered. It may be possible to get adequate light with one yardpole equipped with a battery of PAR 38 lamps as shown in Figure 281. The PAR 38 bulb is a weatherproof type that has sealed-beam features similar to an automobile headlight.

FIGURE 281. One yardpole and a battery of P A R 38 flood bulbs may provide enough light for the entire farmstead.

Lighting Equipment Needed for the Farm Grounds

The barnyard in Figure 282 is lighted by two reflector lamps mounted at opposite corners of the barn. Reflectors used outdoors must be of the weatherproof type as shown in Figure 282 (upper right), and must always be equipped with porcelain sockets. Metal sockets will corrode if used outdoors for any length of time.

HOW CAN YOU PROVIDE GOOD LIGHT FOR YOUR FARM SERVICE BUILDINGS?

Good light in your farm service buildings is essential to farm production, convenience, and sanitation. The figures in Table 26 call for as much as 50 footcandles (in the milk house) of the right kind of light. Special lighting equipment is required to provide this much light.

Special Equipment Needed for Lighting Farm Buildings

The high walls and dark-colored interiors usually found in farm buildings do not reflect light very well. Therefore, ordinary household lighting equipment does not provide good light. Generally, good light for farm service buildings can be provided by special equipment as shown in Figure 283. These are described in the following:

Reflectors. Figure 283-A shows a standard dome reflector which is widely

FIGURE 282. Two reflector lamps placed near the corners of a barn produce light for the barnyard at both ends and at the side of the barn. At upper right is a weatherproof reflector for outdoor use.

used for inside lighting. Interior reflectors also come in the shallow dome style.

Reflector Bulbs. Reflector bulbs come in several types. The flood lamp shown in Figure 283-B is for either outside or inside use. A reflector-type bulb has a silvered inner surface which directs the light onto the desired area. Reflector bulbs are not as efficient as reflectors, but the result is satisfactory for most inside uses. The smoke-box light patterns in Figure 261-B illustrate the effectiveness of reflector bulbs.

Bulb Protectors. The fruit jar fixture shown in Figure 283-C is for protection against breakage and excessive dust. The jar can be wiped clean every day or two as needed and will prevent dust from getting into the socket.

Fluorescent Fixtures. The fluorescent fixture shown in Figure 283-D carries two 40w tubes. The 40 to 50 foot-candles of light needed for close work in the shop or in the milk room can be provided by a fixture of this type. In fact, there is a trend toward the use of fluorescent tubes in many types of farm work.

Special Lamps. A battery of heat lamps is shown in use at a washing vat in Figure 284. Heat lamps are moderate in cost, are easy to install, and have a

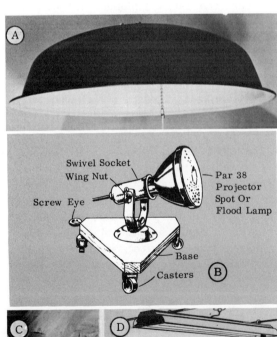

FIGURE 283. *Left:* Special equipment needed for the inside lighting of farm buildings.
FIGURE 284. *Top:* A battery of hard-glass heat lamps keeps the operator warm while working in a milkhouse.

wide range of uses. For information on cost and operation of heat lamps, refer to page 319.

SUMMARY

Good light consists of the right kind and the right amount at the right place. Good vision and healthy eyes are dependent upon good light. The foot-candle is the standard measure of amount of light. The right amount for the farmstead ranges from 2 or 3 foot-candles for general lighting on the farm grounds to 150 foot-candles of local light for sewing dark fabrics. The amount of light for a given need can be varied by increasing the wattage of light bulbs, by increasing the number of bulbs, by selecting the proper color combinations for interiors, and by placement of the lamps or fixtures.

Good quality light can be had by choosing the correct type of bulbs or tubes and by using the best type of lighting for a given need. Either fluorescent or incandescent fixtures can be purchased in five types of lighting according to the direction of light. These are (1) indirect, (2) semi-indirect, (3) general diffuse, (4) semi-direct, and (5) direct.

For the farm home, indirect or semi-indirect fixtures and lamps should be used. A large diffuser or globe usually provides better quality light than a small one. If indirect fixtures or lamps are not available, a good substitute is to use white indirect light bulbs. Another possible improvement of farm home lighting is to convert ordinary table lamps to three-lite styles. This can be done by installing a special three-lite converter switch and a three-lite bulb.

The right places for light are (1) proper locations for general light in the farm home, on the farm grounds, and in the farm service buildings; (2) local light for special tasks in the farm home and in the farm service buildings.

General light for the farm home is usually supplied by ceiling fixtures and wall brackets, either fluorescent or incandescent types. Among these are dozens of different styles, sizes, and finishes, which can be selected to suit the individual taste and match the furnishings of the home.

Fluorescent fixtures, although superior for most lighting needs, are more expensive than incandescent bulbs. But fluorescent tubes last longer and use less electricity per watt of output than do incandescents. The life of incandescent bulbs is considerably shorter than fluorescents. For use in the kitchen and milk house and for grading fruits and vegetables, fluorescent lighting is well worth the extra cost.

Local light for the farm home is usually provided by one or more portable or wall lamps. Portable styles include a wide variety of both table and floor models. A well-placed table lamp, of good design and proper size, can be used with as good effect as a more expensive floor lamp. Correct placement and proper design is more important in getting good light for close work than is having costly floor models that may be highly decorated. The color and placement of lamp shades also influence the kind of light provided.

Lighting requirements for farm grounds are not great, but a little light is essential in the interests of safety and convenience in doing after-dark chores. One yard light may take care of the farm grounds except at barnyards and entrances to buildings, providing the farmstead is not scattered. This one yard light may be a battery of PAR 38 lamps. Additional lighting may be needed around certain buildings. For this, reflector units or spot bulbs may be installed at the gable ends of buildings, or more poles can be used.

Generally, good lighting inside farm buildings requires reflector units, reflector bulbs, or spot bulbs. Ordinary light bulbs are not satisfactory in farm buildings because of the large, open interiors. The walls and ceilings, moreover, do not reflect much light in most farm buildings.

Local light for farm buildings and special tasks is usually provided by reflector units mounted at proper places, by spot and reflector bulbs.

Special lamps used on the farm include germicidal, infrared heat lamps, insect killers, and burglar repellers.

QUESTIONS

1 What is meant by the right amount of light? Is it possible to have too much light for a given task?
2 Why is fluorescent light considered better for reading than incandescent light?
3 How can improper lighting cause permanent damage to the eyes?
4 What happens to the 95 percent of light that is not reflected in a room painted dark blue?
5 How can lighting be used to increase farm income?
6 Why are reflectors or reflector-type bulbs required in lighting farm buildings that have high ceilings or large, open spaces?

ADDITIONAL READINGS

Agricultural Engineering Research Division, Agricultural Research Service, *Planning Your Home Lighting*. U.S. Government Printing Office, Washington, D.C., 1968.
General Electric Corporation, *Residential Fixture Lighting Guide*. Cleveland, Ohio, 1969.
General Electric Corporation, *Residential Structural Lighting*. Cleveland, Ohio, 1969.
General Electric Corporation, *See Your Home in a New Light*. Cleveland, Ohio (no date).
General Electric Corporation, *The Light Book: How to be at Home with Lighting*. Cleveland, Ohio (no date).
Illuminating Engineering Society, *Lighting Education*. New York, N.Y., 1959.
Sylvania Electric Products, Inc., *Better Lighting for Your Farm*. Salem, Mass. (no date).
Sylvania Electric Products, Inc., *Primer of Lamps and Lighting*. Central Advertising, Buffalo, N.Y., 1965.

SUGGESTED PROJECTS FOR PROBLEM-UNIT FIVE

1. **Concrete Watering Tank.** By writing to the nearest Portland Cement Association office you can obtain blueprints for the concrete watering tank shown in Figure 285. Ask your local agriculture teacher or club agent for the address. To prevent freeze-ups, the tank should be equipped with an electric water warmer.

2. **Convert Post Lamp to Automatic On-Off Style.** Figure 286-A shows simplicity of converting a yard lamp to on-at dusk, off-at dawn style. Screw Vigilite adapter, Figure 286-B, into socket; replace bulb.

FIGURE 285. Concrete watering tank.

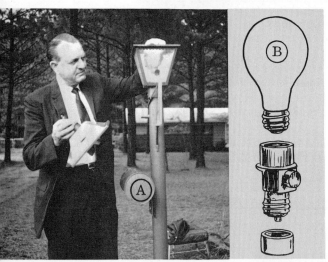

FIGURE 286. In *(A)* college professor demonstrates the "easy way" to do a lighting project. Inset *(B)* shows Vigilite attachment that screws into lamp socket—turns light on at dusk, off at dawn.

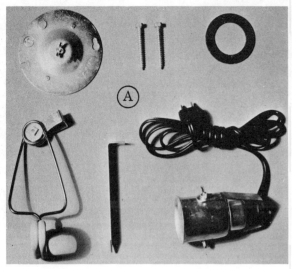
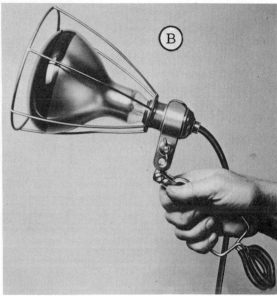

FIGURE 287. Parts for a portable heat lamp.

3. Portable Heat Lamp. In Figure 287-A, note the layout of parts for constructing a heat lamp project. The assembled project is shown in Figure 287-B. A flood bulb can be substituted for the heat lamp to give emergency light. The clip at lower left holds the lamp in place.

4. Portable Emergency Light. Obtain a swivel socket, a PAR 38 bulb, 50 feet of rubber cord, and a plug-in at a hardware store. Other parts can be made from junk material. Have on hand a 6-foot length of 1-inch iron pipe, one floor flange, and material for the base (either 2×4 inch lumber or iron pipe). Construct as shown in Figure 288. Note the swivel base at lower left.

FIGURE 288. Homemade portable emergency lamp.

PROBLEM-UNIT VI

HOW TO SELECT AND CARE FOR OTHER ELECTRIC FARMING AND AGRI-BUSINESS EQUIPMENT

Previous sections of this book have stressed opportunities for using electricity to improve the farm. The remaining chapters deal with the selection and care of electric equipment needed in taking advantage of some of these opportunities. Since electricity is used in more than four hundred ways on the farm, however, it has been necessary to include only those uses that are rather common to the country as a whole. If you need information on equipment not included in this book, you should check with professional agricultural workers and reliable equipment dealers in your locality.

Five sections on electric equipment are included in these latter chapters: (1) equipment for dairy farming, (2) equipment for poultry farming, (3) equipment for feed handling, (4) equipment for crop drying, and (5) power tools for the farm shop.

The problem of equipping a modern farm raises several questions on economics that should be studied before going into the selection of a given machine. A brief discussion follows.

WHAT MANAGEMENT FACTORS SHOULD BE CONSIDERED IN EQUIPPING THE HOME, FARM, AND AGRI-BUSINESS?

A book on electricity cannot go very deeply into economics. However, it is a fact that the wrong decision on equipping a farm—whether the equipment be electric or otherwise—can result in financial disaster. This is equally true of buying equipment that can not be justified economically or not buying equipment that is badly needed on the farm.

By giving careful study to your situation before equipping your home or business, you can avoid these mistakes. You should find it helpful to refer to Chapter 1 and review the section on management problems involved in equipping the home, farm, or business. Give careful attention to the following points: (1) the need for better equipment to increase production; (2) the need for equipment that will reduce human labor requirements in the home or agri-business; (3) the need to keep up with market trends, especially those that may change future demands for certain products; (4) the need for improved storage facilities, especially those that improve the quality of products and reduce the risk of loss in storage; and (5) the need to keep your business up to par in overall efficiency.

Consider All Factors Involved in Cost

As has been pointed out before, an electric machine may pay for itself in several forms: through increased production, better market prices, reduced risk, and so on. Some economists say that a good rule of thumb is that a machine or an electric system is a sound investment if it will pay for itself in *three* to *five* years. The point to remember here is that total cost involves more than one factor. Make certain that you consider all of these items before investing in a machine. The following outline of fixed and operational costs should give accurate enough figures.

Fixed Costs. The fixed costs for electric equipment include (1) annual depreciation at one-twelfth the original purchase price (deduct salvage value before dividing by 12); (2) annual cost of housing at 2 percent of the original purchase price; (3) annual interest at 8 to 9 percent on the average investment; (4) annual charge for taxes and insurance at 2 percent of the original purchase price; (5) installation (wiring, etc.) at one-tenth the original cost of wiring and other installation expenses.

Cost of Operation. The major costs of operation of electric equipment include the cost of electricity and the cost of repairs and maintenance of equipment. The method of estimating the annual cost of electricity has been discussed in previous sections. In brief, the steps required are (1) obtain the amperage rating of motors, (2) convert this to wattage, (3) determine the approximate number of hours used during a year, (4) convert to kwh, and (5) apply the average price of electricity to find the annual cost. For more complete information on estimating costs of electricity, refer to Chapter 2.

The cost of repairs and maintenance on electric equipment generally runs about 5 percent annually, figured on the original purchase price of a machine.

Things to Consider After Deciding to Purchase a New Appliance

Other very important considerations in purchasing a new appliance are:

1. The dealer should be located within 100 miles of your farm in order to provide emergency services.

2. The appliance should be a standard brand for which you know you can obtain parts without difficulty.

3. The dealer should have a reputation for reliability, fair play, and prompt service.

13 HOW TO SELECT AND CARE FOR ELECTRIC EQUIPMENT FOR DAIRY, POULTRY, AND LIVESTOCK AGRI-BUSINESS

Each year a larger share of the labor required in dairy-poultry-livestock farming is being taken over by electric equipment. Indeed, many of the appliances being used in modern farming are not only electric but are automatic as well. This trend is almost certain to continue, and the resulting competition is likely to force all farmers to electrify every operation on the farm, wherever it is at all economical to do so.

WHAT MAJOR ELECTRIC APPLIANCES SHOULD YOU HAVE FOR YOUR DAIRY?

Since 1930, the electrification of the dairy has brought about a 70 percent reduction in labor requirements per cow. A fully electrified dairy that saves so much labor probably is equipped with a combine (pipe-line) milker, bulk-tank cooler, water heater, gutter cleaner (in stanchion barns), silo unloader and other mechanical feeding devices, as well as automatic watering equipment. Other appliances probably include ventilators, a milk house heater, clippers, and cow trainers (in stanchion barns).

Things to Consider in Selecting a Milker

In addition to the points already considered, the selection of a milker should be based on herd size, barn arrangement, and amount of time available for milking.

Figure 289 shows a modern arrangement of a *combine* milker equipped with an in-place cleaning system. The milk flows through the stainless steel (or glass) pipes directly into a bulk-tank cooler. This milker consists of a vacuum

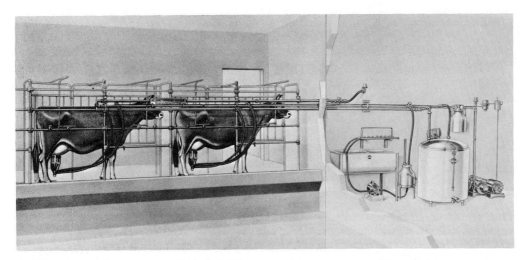

FIGURE 289. Cutaway view of a milking parlor and milk room equipped with combine (pipe line) milker, in-place cleaning system, and bulk-milk cooler. Note the ¾-hp motor and in-place cleaning pump underneath the wash vat; a 2-hp milker unit is at far right.

pump, a ½-hp electric motor, a ¾-hp electric motor, suction lines and pulsators, teat cups, stainless steel milk lines, and in-place cleaning equipment.

The setup shown in Figure 291 is a conventional milker. It differs from the combine type in not having milk lines and in-place cleaning equipment. With this type of milker, it is necessary to empty the milk into cans or bulk tanks by hand. This may require walking back and forth to the milk room, and therefore requires a longer milking period. The type of vacuum pump and motor for a two-unit conventional milker is about the same as that used with the combine milker except that it is smaller.

Figure 292 shows an inexpensive milker, often called the *cow-to-can* type. With this equipment you should be able to milk from 8 to 10 cows per hour with one unit or 16 to 20 cows per hour with two units. (NOTE: A unit is the pail and

FIGURE 290. Close-up view of an in-place cleaning unit. A ¾-hp motor and pump circulate cleaning fluids through the milking equipment.

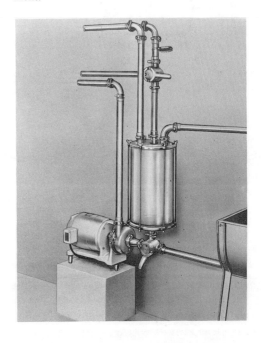

304 USING ELECTRICITY

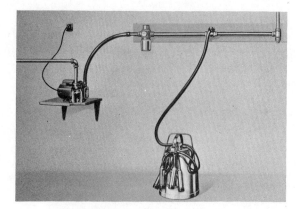

FIGURE 291. *Top:* The basic parts of a conventional pail-type milker showing only one unit. The ½-hp motor and vacuum pump here will operate two milker units.

FIGURE 292. *Bottom:* (A) A "Cow-to-Can" milker in a two-place milking parlor is suitable for herds up to 15 head. (B) A one-place arrangement is suitable for 8 to 10 head. (C) Filter arrangement.

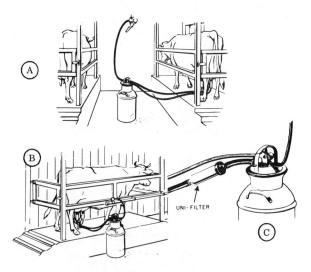

milking equipment used to milk one cow; thus, with two units, you could milk two at a time.)

The cow-to-can milker differs from the conventional milker in Figure 291 in that the milker pails are eliminated; that is, the milk goes directly into the can. The milk lines contain filters that strain the milk as it flows to the cans. If no expansion is planned, the cow-to-can milker is the least expensive system you can own for small herds up to 25 cows.

How to Care for Milker and Motor. Pulsators and drains should be removed and wiped dry with a clean cloth once a week. The motor and all milker parts should be lubricated according to instructions in the operator's manual. Motor belts should be tight and properly aligned at all times. It is a good practice to keep extra belts on hand. If your lights dim when the cooler or milker comes on, check your voltage and wire size.

Summary of Comparison of Milkers

A "modern," combine milker will handle about 50 cows per hour (one operator) in a four-place, walk-through milking parlor; or 36 cows per hour in a three-place, tandem-style milking parlor (three-units); or 25 cows per hour in a stall barn. The purchase price and installation costs range from $2,000 to $5,000. The combine milker is most efficient when used in a milking parlor.

A conventional, "modern" milker will handle from 20 to 25 cows per hour, whereas a single-unit outfit will handle about 15 cows per hour. The cost of a two-unit milker installed is around $700.

A one-unit cow-to-can milker, costing about $90 installed, will handle about 10 to 12 cows per hour.

Important Things to Consider in Selecting a Milk Cooler

Among the advantages claimed for the bulk-tank cooler over the wet type are easier handling of the dairy chores,

ELECTRIC EQUIPMENT FOR AGRI-BUSINESS 305

 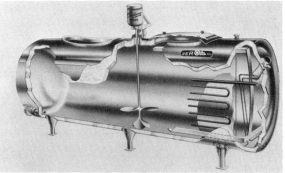

FIGURE 293. *Left:* Owner of a Mississippi all-electric dairy services vacuum pump regularly and in accordance with the operator's manual. The pump operates a 4-place milking room and pipe-line milker.

FIGURE 294. *Right:* Cutaway view of a direct-expansion bulk cooler. Note the refrigerating coils at right. Note also the propeller type agitator which makes the tank self-cleaning. The vacuum feature allows the milk lines to empty directly into the tank without vacuum releases.

less labor required, no milk cans to handle, and greater sanitation along with lower bacteria counts.

The main disadvantage of the bulk-tank system is the higher initial cost of the tank; also, the larger sizes require extra wiring. A 10-can, wet-type cooler can be purchased for $800 or so, whereas a 400-gallon bulk tank costs from $2,500 to $3,000. A well-built 600-gallon bulk tank costs around $3,600 and should last from 20 to 30 years.

It is claimed by manufacturers and dairy specialists that savings on hauling, elimination of cans, and so on, will normally pay for a 400-gallon tank in about 4 years when production averages 800 to 900 pounds or more daily. The tank should be large enough to hold two days' production, since hauling is usually done every other day. Of course you can buy bulk tanks in smaller sizes and cheaper models. A "cheap" tank should last from 10 to 15 years in comparison to 20 to 30 years for a better model.

Types of Bulk Tanks. According to the method of cooling used, the two types of bulk tanks are classified as direct-expansion and ice-bank. In the direct-expansion tank, the refrigerant is in direct contact with the walls of the tank. The ice-bank type, as the name indicates, makes use of a bank of ice that forms around the refrigeration coils. Cold water from this ice bank is pumped against the walls of the tank.

Dairy specialists say that the overall cost of the direct-expansion tank is about the same as the ice-bank type when prorated over a 12 to 15 year period. The ice-bank tank costs less to purchase but uses more electricity than does the direct-expansion tank. These two factors balance over a period of time.

Either type of tank is designed to cool the milk to 50 degrees F. or below in one hour.

Another type of tank is referred to as a vacuum tank (Figure 294) because

TABLE 29 COMPARISON OF SELECTED FEATURES OF DIRECT-EXPANSION AND ICE-BANK TANKS (300-GALLON SIZE)

Feature	Direct-Expansion	Ice-Bank
Original cost	Somewhat higher	Somewhat lower
Compressor-unit motor	2 or 3 hp *	1 or 1½ hp
Agitator motor	¼, ½ hp *	⅛ hp
Condenser-fan motor	⅙ hp	⅙ hp
Water-pump motor	None	¼ hp
Hours of operation per day	5 to 6	15 to 20
Amount of electricity to cool 100 lbs. milk (every other day pickup)	.8 to 1.1 kwh	1.2 to 1.6 kwh

* Some manufacturers use up to 1-hp motors on the agitator of the direct-expansion tank. They claim faster cooling with better circulation of milk inside the tank.

FIGURE 295. *(A)* A vacuum release is necessary where a combine milker and an open (non-vacuum) bulk tank is used. *(B)* A dairyman pours milk from a milker pail into a bulk-tank cooler by hand.

it operates under vacuum. Milk lines lead from the cows directly into the tank. No vacuum releaser is necessary. A releaser is required to release milk from pipe lines into an open (non-vacuum) tank.

For obtaining production records of each cow, you need a meter or weigh jar installed in the milk lines of a pipeline system.

Wet-Type Coolers. A 10-can immersion cooler costs around $800. The compressor for this size cooler is equipped with a 1- to 2-hp motor. About 5 kwh of electric energy are required to cool ten 10-gallon cans of milk to 50 degrees F. The motor operates from 2 to 4 hours per day, depending upon the outside temperature and the motor size.

Another wet-type cooler operates by having cold water sprayed onto the milk cans. The principal advantage of this type is that the cans may be placed into the cooler from the side, thus eliminating most of the lifting. The main difference between the spray-type and immersion coolers is that a small (1/6- to 1/4-hp) motor is required to operate the pump. The consumption of electricity is about the same for both types, but the first cost of the spray type is somewhat greater.

What Care Should You Give Your Cooler?

As has been stressed throughout this book, the first and most important rule on caring for an appliance is to *follow the operator's manual*. Your main concerns will be to keep the tank clean and to make certain that the cooler is properly installed so that the compressor gets plenty of circulation. Also make certain that the wiring is adequate.

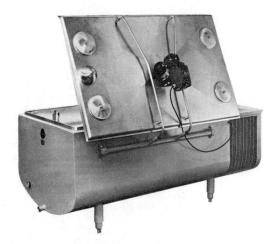

FIGURE 296. A direct-expansion bulk cooler with counterbalanced lid makes opening and closing easy.

Things to Consider in Selecting a Water Heater

Most water heaters used in dairies are of the pressure type. The non-pressure type is less expensive and may be plugged into a convenience outlet without additional wiring. Wattage usually ranges from 300w to 1,500w. About 15 gallons of hot water per day is the maximum amount you can get from the nonpressure heater. Water must be poured into this type of tank since it is not connected to the water system. Of course, you can run a water line to the tank and install a faucet to empty directly into it. A non-pressure water heater is suitable only where a few head of cows are kept.

In the pressure-type water heater, the water pipes are connected to the tank and fresh water enters as hot water is drained.

This type of heater ranges from 30 to 150 gallons in size. An 80- to 100-gallon heater is considered adequate for

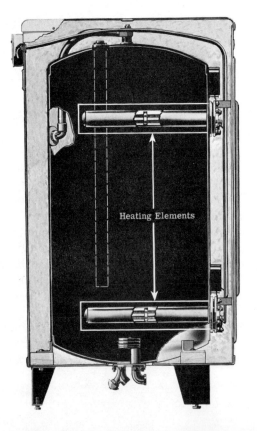

FIGURE 297. *Top:* A wet-type cooler (B) requires that the cans be lifted over the side. The electric hoist relieves the backache of the dairyman. The compressor unit (A) should be installed where it can get adequate ventilation.

FIGURE 298. *Left:* Double element, 80-gallon water heater rated at 4,000w.

a 40- to 50-cow dairy equipped with a pipe-line milker and in-place cleaning.

Figure 298 shows a cutaway section of a double-element water heater of 2,500w to 4,000w. This heater is capable of rapid heating or "recovery."

A single-element heater is less expensive but recovers hot water more slowly. Wattage may range as low as 1,500w for a 20-gallon tank or 2,400w for a 60-gallon size.

A 60-gallon, double-element water heater costs in the neighborhood of $130 installed. This heater uses about 1 kwh of electricity in heating 4 gallons of water to 160 degrees F.

If you live in a section of the country where you can get off-peak rates for heating water, your electric bill will be much less. Also, some water heaters are equipped with a house-heating attachment that may be used on the off-peak cycle. This allows the milk house to be heated in the early morning hours while the off-peak rate is still in effect. A well insulated house will stay warm for several hours, usually through the milking chore. Figure 299 shows a combination water/house heater.

How to Care for a Water Heater. A modern water heater is fully automatic and requires very little attention. Leaky faucets are expensive and should be repaired. If your water is hard, you should use a water softener to keep lime deposits from forming in the tank.

Things to Look for in a Gutter Cleaner

During the six to eight months of the year (in the northern sections of the country) that a herd of 50 cows are kept inside, you handle 200 to 250 tons of manure. If you do this job by hand (hand scoop and conveyor) you handle this tonnage twice.

Types of Gutter Cleaners. One of the first types of barn cleaners to be used extensively was a scoop fastened to a cable that wound up on a drum located at the outer end of the run. Since the scoop is made to fit the gutter, it cleans the gutter as it picks up a load of manure on the trip out. A motor (or manual

FIGURE 299. *(A)* Combination water and milkhouse heater. *(B)* House heating element.

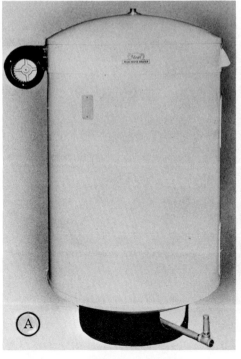

FIGURE 300. *Top:* Scoop-type gutter cleaner in operation.

FIGURE 301. *Bottom:* Shuttle-stroke gutter cleaner: *(A)* inside installation; *(B)* elevator unit to the dumping pit.

operation) pulls the scoop back to the starting point again. Also, an overhead track and conveyor has been used extensively in cleaning dairy barns.

One of the newer "shuttle stroke," paddle-type, barn cleaners is shown in Figure 301. On the forward stroke, the paddle stands out at a right angle to the gutter, thereby pushing the manure forward the length of the stroke. On the backward stroke the paddle folds back against the side of the gutter in a dragging position. The effect of this shuttle action is to move the manure in the gutter a certain distance at each stroke; within a few minutes it reaches the dumping pit. It is better to arrange your equipment to unload the manure directly onto a spreader.

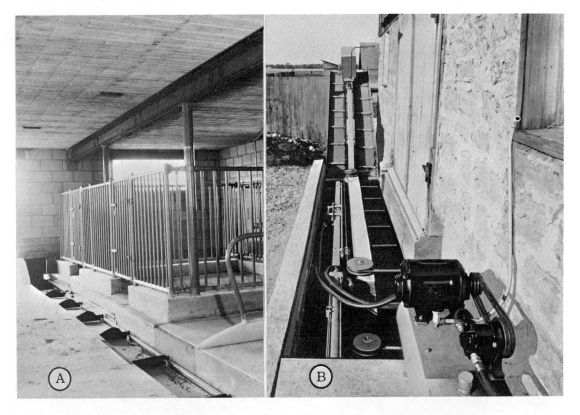

The installation shown is a hydraulic unit consisting of continuous bars, chains, and paddles that deliver the manure to an elevator. The corner construction of the chain allows for turns of 90 degrees or more. The elevator is independent of the gutter mechanism and is equipped with a 1½-hp, single-phase motor.

A 2-, 3-, or 5-hp, single-phase motor is used to operate the gutter unit, according to the size of the barn.

The cost of an installation of the type shown in Figure 301 varies, depending upon the size of the barn. Also, you may save considerable money if your barnyard has sufficient slope so that unloading can be done without the elevator. Total investment ranges from $900 to $2,000, the latter sum including the elevator.

Records show that ½ to 1 kwh per cow per month is used in operating a barn cleaner.

Care of Barn Cleaners. If your barn cleaner is properly installed, it should operate for many years with very little attention. A link in the chain may have to be replaced occasionally, and parts must be lubricated according to the instructions in your manual. Improper installation leads to rapid wear of parts and expensive breakdowns.

Clean and lubricate motors as instructed by the operator's manual. Be sure that your motors are properly wired and fused.

Miscellaneous Equipment

See your local appliance dealers and your electric power representative for information on ventilation systems, cow trainers, clippers, automatic fountains, and similar equipment.

FIGURE 302. *Top:* Endless-chain gutter cleaner installation.
FIGURE 303. *Bottom:* The corner installation in *A* shows how to turn 90° corners with an endless-chain gutter cleaner. The chain link (*B*) is easy to install.

WHAT EQUIPMENT IS NEEDED FOR FEEDING LIVESTOCK OR DAIRY CATTLE?

Several large manufacturers are now marketing mechanical feed-handling systems. Some of them operate automatically and do the complete chore.

Several combinations of feed-handling units are used in these systems. The basic parts of such systems include a combination of the following: (1) holding bins or tanks, which are often located overhead so as to operate by gravity; (2) elevators or screw conveyors for moving feed from a wagon to a holding bin or from one bin to another; (3) a feed grinder or a crimper; (4) a mixer; (5) feed distributors; and (6) controls for the system.

How to Arrange a Feed-Handling System Using a Crimper Mill

In the feed-handling system shown in Figure 305, several types of grain, from overhead bins, are metered into the crimper mill, which is operated by a 5-hp, single-phase motor. The crimped feed passes into a mixer at the left end of the unit. The mixed feedstuff is then elevated back to overhead holding bins. The elevator shaft can be seen in the background near a ceiling light fixture. The feed can then be handled in several ways. For example, it can be dumped by gravity into a self-unloading truck, or it can be distributed directly into feed bunkers by gravity, by conveyors, or by other types of feed distributors.

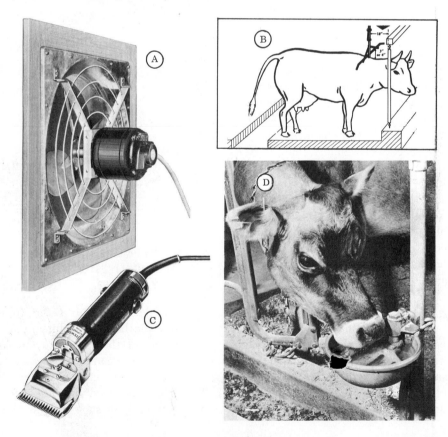

FIGURE 304. Other electric equipment for the dairy: *(A)* wall-type fan; *(B)* cow trainer; *(C)* animal clippers; *(D)* automatic drinking fountain.

ELECTRIC EQUIPMENT FOR AGRI-BUSINESS

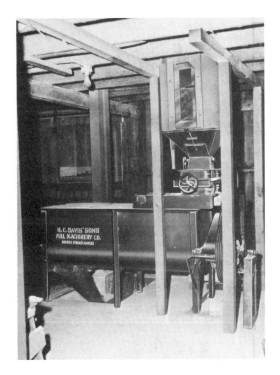

FIGURE 305. These parts of a feed-handling installation include a metering unit, a crimper mill, a mixer, and an elevator.

How to Arrange a Feed-Handling System Using Hammer Mill and Dial Controls

One manufacturer is marketing a unit often referred to as a "dial feed mill." At the upper left in Figure 307 is shown the mill unit and the dials for setting the proportion of each ingredient to be included in the ration. A 2-hp, single-phase motor powers the mill. Grains and supplements are fed by gravity from overhead holding bins into this mill. The mill is of the hammer type, as shown at upper right. A close-up view of dials and control augers is shown at the lower part of Figure 307. Notice the timer switch, which is the extreme right hand dial. The mill grinds and mixes 2,000 pounds of feed per hour and shuts itself off at the proper time. This unit is popular with poultry producers since it works well with automatic feed dispensers and reduces man labor for the feeding chore to almost nothing.

FIGURE 306. *(A)* The feed rollers in a typical crimper mill, with the housing removed. *(B)* The shaft and agitator unit of the system in Fig. 305. A 5-hp motor operates the mill, mixer, and elevator.

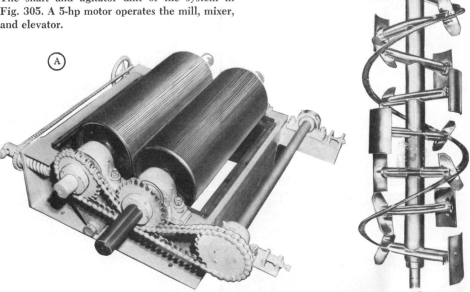

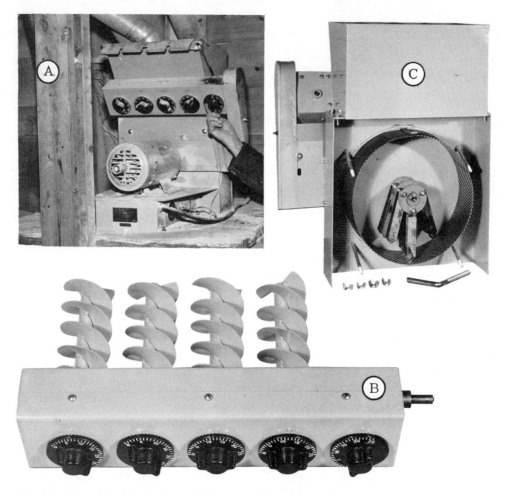

FIGURE 307. *(A)* "Dial" mixmill ready for operation: the operator sets the time switch for 1 hour of operation. *(B)* Mixer unit of the same installation. *(C)* Inside view of a hammer mill.

Equipment for Handling Hay

Chopped hay is easily handled by a self-unloading wagon as shown in Figure 308. In this scene, chopped green hay is going into a silo. Hay can be blown into the mow or unloaded onto an elevator and conveyed into the storage.

In the large mow in Figure 309, baled hay is easily handled by a home-made conveyor of the endless-chain type. A ¾-hp motor takes the backache out of hay handling for this farmer. The cost of this unit was under $300.

Other Equipment for Handling Grain

In Figure 310 you see a portable, screw-type elevator. This is one of the most useful elevators you can have. In this scene, the farmer is transferring grain from one truck to another. He can use the elevator to unload grain from the large truck into an overhead bin. Many other jobs are also possible with this equipment.

Equipment for Handling Silage

A silo unloader eliminates the climbing and most of the hard labor in handling silage. The two general types are the surface unloader, shown in Figure 311, and the bottom unloader as shown in Figure 312.

The first cost of silo unloaders ranges from $1,000 to $1,500 for the surface type and from $1,500 to $2,000 for the bottom type. Prorated over a period

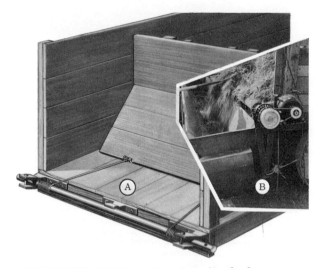

FIGURE 308. (A) Rear view of a self-unloading wagon. The motor and gear unit are mounted on the shaft as needed. (B) The motor and gear unit is unloading chopped hay.

FIGURE 309. *Left:* Homemade endless-chain elevator-conveyor moving baled hay into the barn.
FIGURE 310. *Right:* Portable screw-type elevator transferring grain from a pickup truck to a large wagon.

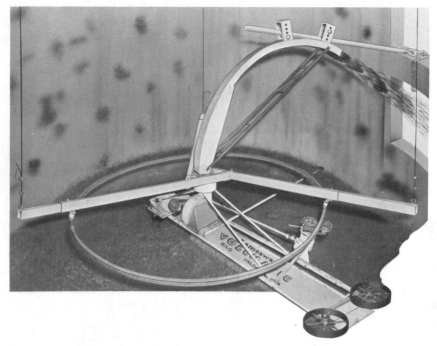

FIGURE 311. *Top:* Surface-type silage unloader: a 3- to 5-hp motor is required, depending upon the type of silage to be handled.

FIGURE 312. *Bottom:* Attendant examines mechanism for unloading silage from the bottom of silo. Auger at lower center distributes silage; controlled by timer switch.

of 12 years useful life, studies show that the per-ton cost for surface unloading ranges from 60 cents to $1.50. The range in cost is due to variation in the kind and total amount of silage handled during a season. In one or more studies, the cost of electricity accounted for 5 to 9 cents a ton. Note: Per-ton cost has been reduced by 1970.

The total cost for bottom-type unloading, as reported in the same study, ranged from 50 cents to $1.76 per ton. Naturally, the larger operations (500 tons per season) gave the lower costs. The 100-ton operations resulted in the highest cost per ton.

FIGURE 313. *(A)* Complete silage feeding system, using a surface-type unloader and a screw-type conveyor. *(B)* Close-up of the screw-type conveyor in operation.

In both surface and bottom unloaders, a 3- to 5-hp motor is required. If tough grass silage is handled, the 5-hp size is needed, but corn or sorghum silage can be handled by a 3-hp size. A silo unloader seems to be justified for a 40-cow herd. It is questionable, however, to use a silo unloader for herd sizes of less than thirty.

The silage conveyor shown in Figure 314 operates on the shuttle-stroke principle similar to the barn cleaner described on page 310. This arrangement can be used in feeding systems for both dairy and beef cattle, of course, and will handle chopped green hay or silage equally well. Feed bunkers are constructed of concrete and are fitted with side rails. In order to provide adequate drainage, the bunkers should have at least 3 inches fall per 100 feet. For further information on this particular type of equipment, see the discussion on barn cleaners.

FIGURE 314. *Top:* Shuttle-stroke silage conveyor: *(A)* cows eating from a homemade bunker; *(B)* paddles in closed position during the back stroke; *(C)* paddles in open position on the forward stroke—same as the shuttle-stroke gutter cleaner.

FIGURE 315. *Bottom:* A hover-type brooder, 76 inches across, takes care of 350 turkey poults; this size brooder would handle 500 to 600 chicks.

WHAT APPLIANCES SHOULD YOU HAVE FOR YOUR POULTRY FARM?

It is possible for you to produce broilers with less than one minute of man labor per bird—if your operation is properly equipped, and it is large enough. An individual farmer can handle up to 36,000 laying hens without hired help.

Things to Consider in Selecting Poultry Equipment

The major items of electric equipment used in poultry houses are electric brooders, electric feeders, automatic waterers, litter cleaners, egg-handling machines, egg-storage coolers, and ventilators. Some specialized equipment is also required for on-the-farm slaughtering.

Methods of Brooding.* Until recent years, hover brooding with oil heat was the most widely used method of raising chicks; while this method is still widely used, electricity is becoming one of the most popular ways of heating hover brooders. Other methods of electric brooding are being used more and more each year.

Not only is electric heat economical, but it is less likely to start fires than other fuels and does not use up oxygen that is needed by the chicks.

1. Hover Brooding. Provide 7 to 9 square inches of hover space for each chick or 12 to 14 inches for each turkey poult. For example, a round hover 76 inches in diameter would cover about 4,500 square inches of floor space, enough for 500 to 600 chicks or 350 turkey poults.

Wattage must be sufficient to meet the temperature demands, which are different from one section of the country to another. If your brooder house is properly constructed, a 1,000w heater for a 76-inch hover with side curtains should take care of any normal temperature situation. In the South, a 750w heating element is adequate for this size hover.

Your investment in hover brooding equipment will average about $75 for each 800-chick outfit or $90 for each 1,200-chick outfit. The average cost of wiring will increase this to $100 and $125 respectively.

Care and Operation of Hover Brooders. Turn your brooders on and observe their operation for 24 hours before putting chicks in the brooder house. During the first week, maintain the temperature at 90 to 95 degrees F; thereafter drop 5 degrees per week until the temperature is down to 72 degrees.

If thermostats fail to cut off, check the voltage and wiring of your brooder circuit. Check your house ventilators also.

2. Infrared Lamp Brooding. Infrared lamps are becoming popular for brooding chicks as well as other young farm animals. Bulbs come in three standard sizes—125w, 250w, and 375w. The life of an infrared bulb is rated at 5,000 hours. When designing a homemade brooder, or in selecting a ready-made one, see that the lamps are spaced about 20 inches apart. Also, have a total of 4 watts of heating capacity per chick in

---

* Information on brooding has been drawn largely from the personal notes of Mr. R. C. Jaska, Department of Agricultural Engineering, The Agricultural and Mechanical College of Texas, College Station, Texas.

FIGURE 316. Each four-lamp infrared brooding lamp, as shown, takes care of 100 to 150 chicks. The cardboard guards will be removed after the first few days and automatic feeders will be substituted for hand feeding.

FIGURE 317. A workman laying heating cable for underheat brooding. The cable will be covered with a layer of concrete.

regions where the house temperature may drop to 20 degrees F. or lower. (See Table 30 for additional information.)

The regular type of infrared bulbs cost from $2 to $2.50 each, while the hard-glass type costs $4.50. At least one manufacturer is putting out a bulb for seventy-five cents. The total investment for brooding 500 chicks by infrared heat will average $50; for 1,000 chicks, $70.

Care and Use of Infrared Lamps. Locate 250w lamps 18 inches above the floor, 125w lamps 16 inches above the floor. Thermostats for infrared brooding must be set to operate at room temperature. Infrared heat cannot be measured directly by a thermometer; therefore, the thermostat should be mounted at some point on the wall, not under the lamps.

3. Underheat Brooding. The most recent trend in brooding is to bury heater elements or soil-heating cable in concrete or blacktop (asphalt) floors. If a power failure occurs, the heated floor will provide warmth for about seven hours after the failure began. See Figure 317.

At 2 cents per kwh, the cost of electricity for underheat brooding averages about .3 cents per chick, or one-fourth the cost of infrared brooding. Your total investment in this setup would be about $75 for 1,000 chicks. For each chick, allow 1.6w of heater cable. Bury this in sand and plaster over with concrete or blacktop material.

Two lengths of soil-heating cable at 230 volts will produce the 1,750w needed for 1,200 chicks. This method of brooding is becoming popular because of the ease of controlling heat and the low investment in wire and equipment. Also, power failures do not mean immediate danger because heat is retained in the floor for six to eight hours.

TABLE 30 THERMOSTAT CONTROL TEMPERATURES AND NUMBER, SIZE, AND LOCATION OF HEAT LAMPS FOR BROODING CHICKS *

Minimum Room Temperature Degrees F	Lamps		
	No.	Size in Watts	Height in Inches
50	6	125	16
40	8	125	16
30	6	250	18
30	12	125	16
20	8	250	18

* First week, 500 chicks.

Things to Consider in Choosing Electric Feeding and Watering Equipment

The feeding installation shown in Figure 318 operates by means of a reciprocating motion of the feed troughs. This shaking back and forth in ¾-inch strokes causes the feed to move from the 500-pound hopper along the entire length of the feed trough. Feed travels at the rate of 10 to 12 feet a minute. In this

FIGURE 318. *(A)* This automatic feed hopper holds 500 pounds of poultry feed. *(B)* A shaker-type drive unit distributes the feed automatically as needed; the motor is a ¼-hp capacitor type.

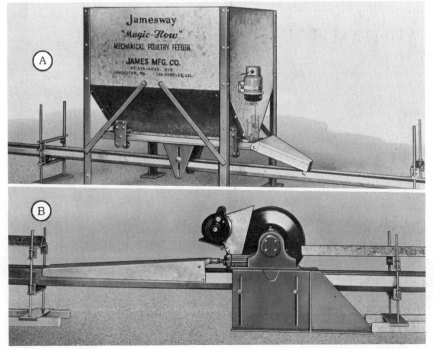

FIGURE 319. A cut-off device automatically stops the shaker unit when the feeders are full.

One ¼-hp capacitor motor will keep two lines of feeders equally full. This system can be made fully automatic and continuous by providing a bulk-feed supply overhead. The device shown in Figure 319 cuts off the motor when the feed troughs are full. This feeding system, as well as many other mechanical makes and styles, can be adapted to broiler, egg, and turkey production. During the first few days, baby chicks are fed by hand methods. But after the first week you can put your automatic feeder to work. Feed troughs for turkeys must be somewhat larger, but the basic system described here works equally well.

The feeding system shown in Figure 318 will cost from $500 to $800 installed, depending upon the size of the system. Only 4 to 6 kwh of electricity are used in each 24-hour period. The length of feeder lines and the amount of feed handled determines this. Mechanical feeders have been built by farmers and farm boys throughout the country. Perhaps you too can build your own.

setup, no moving parts operate inside the troughs. Other makes, as well as many of the homemade feeders found on farms, use some type of endless chain to drag the feed along the length of the trough.

FIGURE 320. Automatic waterers relieve the poultryman of this chore and also increase egg production: (A) trough-type waterer; (B) pail-type waterer.

Waterers. Tests in various parts of the country have shown that automatic watering systems for laying hens pay large returns. For example, a 300-hen flock watered by automatic waterers produced 5,100 more eggs in one year than a similar flock watered by hand. Figure 320 shows two types that are widely used. No doubt you will prefer a waterer that can be installed in the water line, thus completely eliminating the handling of water.

The cost of an 8-foot automatic waterer, as shown in Figure 320-A, is $30. A good electrification project is a homemade watering trough for your poultry enterprise.

Operation and Care of Feeders and Waterers. There is little skill but considerable care required in handling poultry feeders and waterers that have been properly installed. Keep motors and moving parts lubricated in accordance with the operator's manual and keep the water

FIGURE 321. *(A)* This poultryman is plugging in the thermostat which controls the ventilator fan in his modern laying house. *(B)* Note the multiple roosts, which have the effect of increasing floor area.

container clean. Protect your waterers against frost by using heating cable.

Poultry House Ventilation

Electric ventilator fans and other modern equipment have made it possible to put more laying hens into a given size house. For example a 40- by 80-foot house equipped with cages or multiple roosts will accommodate 2,000 hens. This averages just a little over 1½ square feet per hen in comparison with four square feet formerly required.

FIGURE 322. Illinois poultryman and wife "process" more than a million eggs per year. Cleaning, grading, and packing are handled in single operation by electric equipment.

Forced-air ventilation helps to regulate temperature and humidity as well as reduce odors. Also, good ventilation will aid in preventing diseases that can occur as a result of damp conditions.

Conditions from one part of the country to another vary so much that it is not practicable to give detailed instructions for ventilation in this book. Check your ventilation problems with the local farm serviceman.

Egg-Handling Equipment for Modern Poultry Farming

If you now have or plan to have a large flock of layers, no doubt you will be interested in some of the newer egg-handling machines. Among these are egg cleaners, egg graders, candlers, and refrigerated storage units.

Egg Cleaner-Candler. The farmer shown in Figure 322 is using a combination cleaner-candler that will handle 1,000 eggs per hour. This "dry" cleaner cleans the eggs as they pass underneath an abrasive belt. The eggs are returned to the operator after being run through the machine.

The cost of this 1-hp cleaner is approximately $800. It will clean a case of eggs while using about 1 kwh of electricity. The major service required to keep this cleaner operating is to adjust and replace cleaning belts. Some minor adjustments of parts may be needed occasionally. Also, the motor may require lubrication. See the operator's manual for instructions.

Washer-type cleaners are also available. The better ones wash the eggs in hot water and dry them in one operation.

A washer or dry-egg cleaner can be set up to operate in conjunction with an egg grader as well as with a candler.

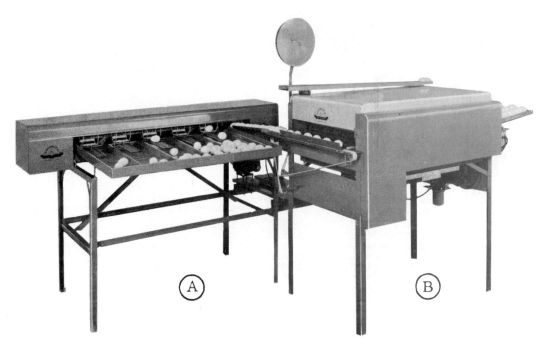

FIGURE 323. *Top:* This egg cleaner-candler-grader handles 1,800 eggs per hour.

FIGURE 324. *Bottom right:* Refrigerated egg storage can be constructed by the farmer. Note the compressor unit on top of the cabinet. A room air-conditioning unit can be used for cooling.

It is thus possible to clean and grade 1,000 eggs per hour with these two moderately-priced machines. Note that egg washers usually cost slightly more than dry-egg cleaners of the same capacity. Also, the washer requires a supply of hot water. You can make your own egg candler for a few dollars worth of materials.

Egg Graders. As mentioned in the preceding topic, an egg grader can be operated in conjunction with a cleaner. In the model shown in Figure 323, a $\frac{1}{20}$-hp motor operates the mechanism that separates eggs into five grades. The capacity is 1,800 eggs per hour.

The price of egg graders of the type illustrated ranges from $300 to $500. The

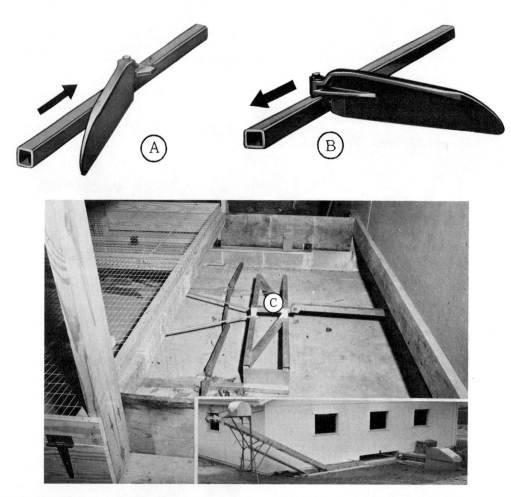

FIGURE 325. In *A* and *B*, the shuttle stroke of the paddle moves the litter about 24 inches with each stroke; *(C)* the drive shaft and header installation connect with the paddle unit for operating the paddle mechanism.

more expensive models will grade up to 5,600 eggs per hour. Ball bearings in this machine should give it long life. Only 1 kwh is required to grade 100 cases of eggs.

Egg Storage. Gone are the days of storing eggs for several days on the pantry shelf "until the next trip into town." Since eggs begin to deteriorate the minute they are laid, a refrigerated storage is necessary to meet the market demands for good quality eggs.

You can purchase a commercial egg cooler, or you can build your own. Homemade coolers, which use room air conditioners as the source of refrigeration, are satisfactory. The cabinet can be built with ordinary carpenter's tools.

The Oklahoma State University reported satisfactory results from a ¼-ton refrigeration unit installed in a well-insulated cabinet.* The size was sufficient to accommodate six egg cases and two half-case egg baskets (for cooling). This was deemed adequate for a 300- to 400-hen laying flock. For construction details order the leaflet listed in the footnote. By building this cooler at home you can save about $200 in comparison with the cost of a commercial cooler of equal capacity.

Care and Operation. Eggs should be stored in 50 to 60 degree temperature and in 80 to 85 percent humidity. Eggs should be gathered frequently and cooled in wire baskets (not in boxes or pails) where faster cooling will take place. The cooling unit should be placed so that the sun does not shine directly on it.

The price of egg coolers varies greatly depending upon type and size. Check with your local appliance dealer for quotations. A ¼-ton unit powered by a ½-hp motor, operating 8 hours in 24, would use about 4 kwh of electricity.

Litter-Cleaning Equipment for a Laying House

Figure 325 shows an installation of a shuttle-stroke litter cleaner that operates on the same principle as the dairy barn cleaner previously described. The droppings and litter can be pushed into the gutter where the push-pull action of the paddles carries it out to the end of the conveyor. There it is loaded directly onto a manure spreader.

A second- and third-floor arrangement can be fitted into this system by installing chutes to empty from those floors into the first-floor gutter. The motor size required depends on the size of the cleaning system; however a 1- to 2-hp, single-phase motor is usually large enough. The cost of electricity for cleaning a multiple-story laying house is only a few dollars a year. The equipment should last at least 10 to 12 years.

FIGURE 326. A poultryman on the second floor of a laying house pushing litter into a chute leading to the gutter cleaner on the first floor.

SUMMARY

Important points to consider in equipping the farm include (1) expand-

---

* *A 7-Case Egg Cooling Cabinet.* Oklahoma State University Leaflet L-22, Stillwater, Oklahoma (no date).

ing farm production, (2) dependable markets, (3) the farm labor situation, (4) trends in consumer demands, (5) improvement in the quality of farm products, and (6) reduction of risk in storage. Consider also the reliability of the dealer in maintaining vital equipment. Remember to include both fixed and operating costs in determining the total annual cost of an appliance.

Major electric appliances for a modern dairy include the following items: (1) milker—conventional one- or two-pail types, cow-to-can type, or combine type, depending upon the size of the herd and the labor situation; (2) milk cooler—wet type or bulk-tank cooler, depending upon the volume of milk and the market demands; (3) water heater—50- to 100-gallon pressure type preferred for a modern dairy (also consider the type equipped with house heating elements); (4) gutter cleaner—either the endless-chain type or the shuttle-stroke paddle type is satisfactory.

Automatic electric feed-handling systems now on the market can save up to 90 percent or more of the human labor required in feeding livestock or dairy cattle. Some of these systems do every operation involved in handling feedstuff from the time it leaves the field until it is dispensed into the feed trough.

The basic units that go into an automatic feed-handling system include self-unloading wagons, cup-type or screw-type elevators, overhead holding bins, metering devices, grinders, mixers, and dispensers.

A complete feed-handling unit consists of a combination of several of these individual units geared to operate with little or no human labor. Many livestock and dairy farmers build some parts of their feed-handling equipment themselves.

Other electric feed-handling equipment includes elevators of the screw, endless-chain, and cup types. One of these will handle almost every type of feedstuff on the farm. Either vertical or horizontal elevators-conveyors can be built in the home farm shop or in the school shop and have proven to be one of the most useful projects in farm electrification.

A silo unloader, in conjunction with distributing machinery, can relieve the farmer of much hard labor. And, if the herd is sufficiently large, the equipment will be economical. The addition of a time switch will allow you to go about other work while the silage feeding chore is being done. Several well-known makes of silo unloaders are available. These include both surface and bottom-type unloaders.

In all poultry producing areas, the poultry industry is being electrified on a rapid and widespread scale. This development has resulted in larger flocks. Electric equipment needed to handle these large operations includes (1) brooders of the hover type, infrared heat lamps, and underheat brooding; (2) automatic feeding systems for all types of poultry production; (3) automatic watering system; (4) egg cleaner-candler machines; (5) egg graders; (6) combination egg cleaner-candler-grader machines; (7) egg coolers; and (8) litter-cleaning equipment. Caged-type houses are increasing.

The proper combination of electric equipment will enable you to handle 36,000 or more head of laying hens or up to 96,000 broilers (per turn) without outside help.

QUESTIONS

1 Why should trends in market demands be considered in selecting new electric equipment?
2 What should you know about a dealer before buying a major electric appliance from him?
3 What are the advantages of a combine milker? A bulk-tank cooler?
4 What are the basic parts in an automatic feed-handling system?
5 What size herd of dairy or beef cattle is needed to justify an electric silage-feeding system?
6 What are the advantages of underheat brooding? Infrared brooding?
7 What major pieces of electric equipment are necessary to establish a one-man, 36,000-laying-hen operation?

ADDITIONAL READINGS

Barn Cleaner, Cattle Feeder and Silo Unloader Association, *Modern Electric Choretime Equipment.* Bulletin No. 2, Chicago, Ill. (no date).
Edison Electric Institute, *Farm Electrification.* New York, N.Y. (no date).
Edison Electric Institute, *Farm Electrification Manual.* New York, N.Y., 1953.
"Electricity on the Farm" (all issues). Reuben H. Donnelly Publications, New York, N.Y.
Great Lakes Steel Corporation, *Loose-Housing Dairy Barns.* Detroit, Mich., 1953.
Leonard, Harry and Paul Johnson, *Aids to Using Electricity on Indiana Farms.* Agricultural Engineering Department, Purdue University, Lafayette, Ind., 1962.
McPartland, J.F., *Modern Electric Controls.* Batavia, Ill., 1961.
McPartland, J.F. and W. Novak, *Electrical Equipment Manual,* Third Edition. McGraw-Hill Book Co., New York, N.Y., 1965.
Pennsylvania Electric Co., *Farm Electrification Reference.* Johnstown, Pa. (no date).
U.S. Department of Agriculture, Rural Electrification Administration, *How Electric Farming Can Help Your Business.* R E A Bulletin No. 140-1. Washington, D.C. (no date).
Westinghouse Electric Corporation, *Electrical Equipment You Can Build.* Pittsburgh, Pa. (no date).

14 HOW TO SELECT AND CARE FOR ELECTRIC EQUIPMENT FOR CROP DRYING AND FOR THE HOME AND FARM SHOP

The savings from the use of a crop dryer can run as high as 100 percent, when, for example, weather conditions would result in a total loss of the crop in the field. Of course, such high savings should not be expected every year. You may, however, expect to save from 5 to 15 percent more grain in harvesting and up to 25 percent more nutritional value of hay when crop dryers are used. These savings are the result of a better choice of harvesting period and of better storage. The net result is better prices for the product and a reduction of risk while the crop is in storage.

Since crop drying is a rather technical problem, you should seek the advice of a competent person in this field if you are planning a dryer system. Of course, you can obtain thorough technical information on this subject by writing to the nearest land-grant college or to the USDA. The information presented in this chapter is of a practical nature, dealing with basic principles rather than going very deeply into the technical phases of crop drying. Emphasis is placed on the selection and care of crop-drying equipment that is electrically operated.

The last section in this chapter presents information needed in selecting basic power tools for the farm shop. The upkeep of the equipment on a modern farm, together with the need for constructing other farm equipment, demands a few basic power tools. In fact, it is uneconomical for most farms to be without these few basic power units.

WHAT ELECTRIC EQUIPMENT IS NEEDED FOR DRYING FARM CROPS?

Grain, hay, seed, and other crops can be dried on the farm by using either heated or unheated air. The procedures and equipment required for the two methods are quite different, however. Although both methods are discussed in later topics, it should be mentioned here that the use of unheated air is more common to the average farm where crops are dried.

Things Involved in a Crop-Drying Problem

The drying of any crop with unheated air involves several factors that vary from one situation to another. In getting ready for drying hay, grain, or seed, you must consider the following things: (1) the type, size, and cost of storage structure required; (2) the type, size, and cost of drying equipment required; (3) the wiring required; (4) the kind and amount of product to be dried, its moisture content at the beginning of storage and the final moisture content desired; (5) local weather conditions during the drying period; (6) drying procedure; and (7) cost of electric energy.

How Resistance to Air Flow Affects Size of Fan Required

A fan that will move 30,000 cfm (cubic feet of air per minute) in open air will move only 20,000 cfm through a 5-foot layer of shelled corn. The reason for this is that the corn offers resistance to the movement of the air. For example, resistance would be almost 100 percent in a one-foot layer of sand. A five-foot layer of hay has much less resistance in it than does a five-foot layer of corn. The resistance to air flow is called *static pressure* and is sometimes expressed as

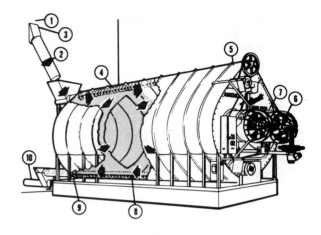

FIGURE 327. Modern Batch-Dryer. From wet grain storage bin (1) to storage spout (2) onto dryer inside hopper. Wet grain pressure switch (3) senses presence of grain; leveling augar (4) loads itself until grain releases pressure switch (5) stopping top auger (4) and turning on fans (6) and furnace units (7); drying begins. When grain is dried, furnace units (7) automatically stop; fans (6) continue to operate, cooling grain. When cool, unloading auger (8) transports dry grain to storage (10). Pressure switch (9) cuts off when dryer has been emptied. Capacity 9,400 bushels per 24 hours.

"inches of water." The average resistance in grain and seed is from ¾ to 2½ inches and in hay from ½ to 1½ inches.

Since resistance greatly influences the fan size required for a given drying situation, it must always be taken into account. Three things can increase resistance to air flow: (1) denser, closer-lying materials, (2) greater depth in storage, and (3) faster flow of the drying air. Several of the references listed at the end of this chapter include resistance or static-pressure tables for different crops, depths, and so on.

Equipment for Drying Hay With Unheated Air

Why must hay-drying fans have such a large capacity? The answer is that 1 ton of dry hay (15 percent moisture) weighs 2,700 pounds *before drying* (at 35 percent moisture). Therefore, in drying 60 tons of hay from 35 percent moisture to about 15 percent, around 40,000 pounds of water must be removed. Adequate ventilators for the escape of moist air must be provided.

In order to do this job in 1 to 3 weeks, a 25,000 cfm fan is required, and it must work against a resistance determined by the kind of hay and its depth in storage. A 7½- to 10-hp motor is needed to operate a fan this size. This example is based on two general rules of thumb: (1) provide 350 cfm per ton of "dry" hay, or (2) provide 15 to 20 cfm per square-foot of floor area in the haymow. (Rules of thumb vary on this.)

The temperature and humidity of the outside air, together with the moisture content of the crop to be dried, determines the amount of time required to dry hay, provided, of course, that the drying structure, equipment, and procedure are proper. A fan does little or no good when the outside temperature is below 50 degrees or the humidity is above 70 percent.

Hay-Drying Systems

Hay-drying systems are classified according to the method of air circulation. The three most popular systems are (1) a main (or central) duct, (2) a main duct with side laterals, and (3) a tiered duct arrangement.

Main Duct System. For haymows from 20 to 30 feet in width, the least expensive and simplest system is a central A-frame. Figure 329 shows an installation of this type filled with chopped hay. Many farmers make their own A-frames.

For mows from 30 to 36 feet in width, a rectangular or oval-shaped frame should be used. Figure 330 illustrates this type of installation.

Main Duct with Side Laterals. For mows wider than 36 feet, lateral ducts are used in addition to a main duct. This arrangement provides better distribution of air. The installation in Figure 331 has a main duct in the center of the mow with laterals to each side. Another arrangement is to have a main duct at one side of the mow with laterals leading across the building. Ducts should end about 10 feet from the ends and sides of the mow.

FIGURE 328. A crop-dryer fan must have adequate capacity and power for moving large quantities of air, since moisture is removed from the crops in this manner. The fan shown has a 25,000-cfm capacity and is operated by a 5-hp single-phase motor.

ELECTRIC EQUIPMENT FOR CROP DRYING AND SHOP

FIGURE 329. *Top:* Homemade A-frame, main duct, hay dryer installation. Note the slatted sides that allow air to circulate through the hay.

FIGURE 330. *Bottom left:* An oval main duct system is suitable for haymows 30 to 36 feet in width. Note also the metal tie rods that help the walls withstand the lateral pressure of the hay.

FIGURE 331. *Bottom right:* Oval main duct with laterals A and B. It is used in haymows wider than 36 feet.

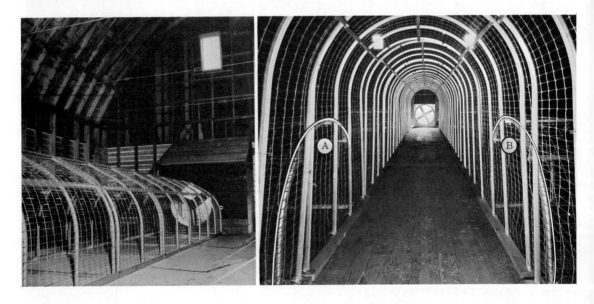

Tiered Duct System. A recent development in hay drying is to install portable ducts as the depth of hay is increased in the mow. This arrangement makes it possible to dry hay to any depth. Figure 332 shows an installation of this type. The portable ducts are made of expanded metal or strong wire on lumber framing.

Hay-Drying Fans and Motors

The six-blade, propeller-type fan in Figure 333 is rated at 25,000 cfm at 1 inch static pressure. This fan is adequate for a 36 x 55 foot mow (or equivalent) containing up to 75 tons of hay. Tables giving information on fan capacity can be found in USDA bulletins and commercial publications.

Figure 334 shows a centrifugal fan, which is also a popular type for crop drying. Again, the capacity of a centrifugal fan is rated in cubic feet per minute. Usually, the motor is designed for a fan of given type and size.

The single-phase motor in Figure 334 is rated at 7½ hp. A magnetic switch protects it from overloads and low voltage as well.

FIGURE 332. Tiered duct system of hay drying. Note the vertical arrangement of the fans and ducts in *A* and *B*. The workman is laying an expanded-metal type lateral *(D)* in proper position with main duct *(C)*.

ELECTRIC EQUIPMENT FOR CROP DRYING AND SHOP

A quick rule of thumb on motor size is to allow 1 hp for each 9 or 10 tons of dry hay. For example, a 75-ton mow requires a 7½-hp motor (75 ÷ 10 = 7½).

Equipment for Drying Hay with Heated Air

The oil-fired dryer shown in Figure 335 is powered by a 5-hp, single-phase motor, which provides air movement. The fan is a 36-inch, propeller type, rated at 25,000 cfm.

The advantages of this dryer over the unheated air types are wider choice of harvesting period despite unfavorable weather and quicker, more thorough drying.

Baled hay, at 35 percent moisture, is stacked on a slatted floor connected to the header duct at the left end. The dryer is connected to the header in such a way that air must circulate through the floor and through the bales of hay.

A tarpaulin protects the hay against rain while the drying is taking place. Notice that the sides of the stack are left open, thus providing an escape for moist air.

The batch shown in Figure 335 will be dry and ready to place in storage within 24 hours after the hot air begins to circulate. Another batch from the field will replace this one.

Cost of Drying Hay

The cost of an A-frame installation for a 75-ton mow, using a 25,000 cfm fan equipped with a 7½-hp, single-phase motor, ranges from $1,200 to $1,500. This includes lumber and the labor needed for construction of the A-frames.

FIGURE 333. A farmer discussing a fan and motor installation with a technician. This 25,000-cfm fan and 7½-hp single-phase motor were selected to take care of a 70-ton mow.

FIGURE 334. The centrifugal-type fan shown is rated at 25,000 cfm; the motor is a 7½-hp single-phase type. Note the magnetic starter-protector near the top of the fan housing.

FIGURE 335. This oil-fired dryer, equipped with a 25,000-cfm fan and a 5-hp motor, turns out a batch of dried bales every 24 hours.

The same size fan and motor with a main and lateral system for a 75-ton mow runs from $1,450 to $1,800.

The cost of electric energy for drying hay with unheated air averages about $1.50 per ton, at 2½ cents per kwh.

The total cost of drying hay with unheated air, including depreciation of equipment, ranges from approximately $3.75 to $4.50 per ton. The total investment for this type of drying ranges from about $18 to $22 a ton of mow capacity.

Data on the cost of drying crops with heated air can be obtained from the USDA. The per-ton cost of fuel or energy is somewhat greater with heated-air equipment.

How to Operate and Care for Hay Dryers

Once the dryer is started, keep it running until the top layer of hay is dry (feels dry to your cheek or crackles in your hand). The exception to this is when the temperature goes below 50 degrees F. and/or humidity gets above 70 percent.

Check for thin areas of hay or open holes around posts. Add hay to these spots and pack firmly until no excess air movement can be felt.

In adding green hay to the mow, limit the amount to 6 feet of chopped hay, 8 feet of long hay, or 3 tiers of bales. Allow each layer to dry before adding another. Limit the total depth over a duct to 15 feet of long hay, 12 feet of chopped hay, or 8 layers of bales.

For unheated-air drying, figure 1 to 3 weeks for drying a mow full of hay; for heated-air drying, figure 3 to 24 hours per mow or per batch of bales. The system should be shut off when the very top layer of hay is dry.

Keep the fan and motor lubricated in accordance with the operator's manual. Clean the motor once each season;

ELECTRIC EQUIPMENT FOR CROP DRYING AND SHOP

do not blow it out with an air hose since this forces dust into bearings and windings. Disassemble the motor and clean as directed by the manual. You can use the dryer motor for other farm operations during the fall, winter, and spring.

Grain Drying Structures and Equipment

There is a trend throughout the country toward the use of metal bins and quonset huts for drying and storing grain. Figure 336 shows a cutaway view of a grain storage bin with a perforated floor. Forced air circulates through the floor openings. The metal walls prevent the air from escaping at the sides, thus forcing it through the grain. For large farming operations, the use of quonset metal buildings and multiple fans is becoming popular.

In comparison with hay, the volume of grain is less for equal weight. Therefore, on the average farm, the amount of air required for drying grain is somewhat less than that required for drying hay. However, the higher resistance of grain offsets this insofar as horsepower is concerned.

Many farmers have converted existing buildings into grain-drying structures. The principal precaution to take is to build extra supports underneath

FIGURE 336. Illinois farmer and electric power company experts examine a 5,700 bushel on-the-farm type of grain storage bin. Inset shows cutaway section of storage bin floor which must be perforated in order that heated air can circulate uniformly through grain as fresh layers of "wet" grain are added. Heating unit is rated at 24,000 watts.

floors that might otherwise give way because of the excess weight of grain. Also, install metal tie rods to prevent the walls from bulging (see Figure 337). Line the inside of old buildings with one or two layers of 30-pound asphalt paper to make them airtight. Inverted V-shaped troughs with perforations may be installed on the floor for ventilators. Take care to provide openings for the escape of moist air. Your fan motor should be located outside the building where good ventilation is possible.

How to Determine Amount of Air and Fan Size Needed

The rule of thumb for amount of air for drying grain is 2½ to 5 cfm per bushel, depending upon the size of the grain or seed to be dried. For example, a storage containing 1,000 bushels of corn or beans would require 5,000 cfm of unheated air; 1,000 bushels of grain sorghum would require 2,500 cfm. The fans have to move the air against a certain resistance, however, so the fan size, in both examples, must have greater capacity than the figures indicated so as to offset this resistance. You can easily read and interpret static pressure tables, which can be found in grain-drying publications issued by the USDA and by land-grant colleges.

Heated Air Dryers. There is practically no difference between a heated-air dryer for grain and one for hay. Refer to page 335 for information on this type.

FIGURE 337. Quonset-type granary for large grain farms. Note the metal tie rods to support the side walls. Note also the expanded metal ducts on the floor. A fan and motor for each duct are installed outside the building.

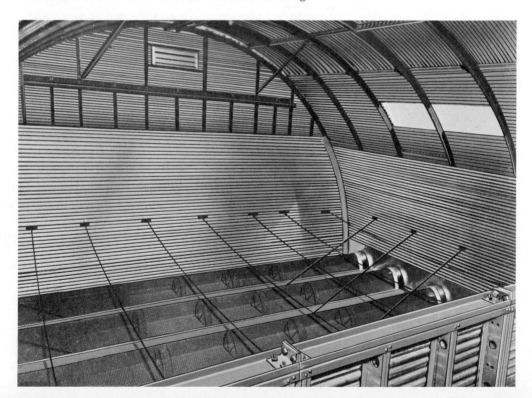

ELECTRIC EQUIPMENT FOR CROP DRYING AND SHOP

FIGURE 338. *Top:* Shelled corn being dried on wagons by a batch drier. The same dryer can be used for drying hay. At lower right, early settlers harvesting and threshing grain by hand.

FIGURE 339. *Bottom:* Electric moisture tester for grain.

How to Operate and Care for Grain Dryers

Once the dryer is started, operate it continuously until the top layer of grain is dry (use a moisture tester). The exception is when the temperature is below 50 degrees F. or when the humidity is above 70 percent. Check the air movement on top of the grain; fill thin spots to make the air flow evenly. Clean and lubricate fans and motors as directed in the operator's manual.

WHAT POWER TOOLS ARE NEEDED FOR THE HOME AND FARM SHOP?

There is almost no limit to the possibilities for farm improvement through a well-equipped shop. A good shop should be more than a repair center. It should be an improvement center as well.

FIGURE 340. *Left:* A bench-type tool grinder equipped with a 6-inch wheel and a ¼-hp motor. This machine is one of the most useful for the farm shop.

FIGURE 341. *Right:* The heavy-duty grinder shown here is valuable for doing heavy farm grinding. This type grinder should be mounted on a pedestal or oil drum filled with concrete. Note the adjustable eye shields.

The power tools described in the remaining section have been recommended by a well-known authority on farm shops. The descriptive material has been drawn largely from an article written by him.[*]

Grinders

A grinder is one of the most useful power tools for the shop and should be one of the first pieces of equipment to be purchased. The three types generally found in farm shops are bench, pedestal, and portable grinders.

Bench Grinder. A good farm grinder should meet the following specifications: motor (½-hp, 3,600 rpm, 115-230v, sealed ball bearings), grinder wheel (7 x 1 inch), ⅝-inch arbor, guard, tool rests, and lighted shield. (NOTE: A lighted shield provides both protection and good vision.)

Pedestal Grinder. Heavy veeing, grinding plow shares for hardfacing, and other difficult grinding jobs can be done best with a large grinder with plenty of power. A 1,750 rpm, 1- to 2-hp motor should be large enough for most grinding jobs on the farm. The grinding wheel should be at least 1¼ x 10 inches, and should have sealed ball bearings. This grinder should have an illuminated shield.

If a large grinder is available, a ⅓-hp, bench grinder should be available for tool grinding.

Portable Grinder. A portable grinder is valuable for doing in-place grinding.

---

[*] V. J. Morford, "Hired Hands with No Demands," *Proceedings of the Seminar "Power Farming: A Better Way of Life."* (Huntley, Illinois, Thor Research Center for Better Living, 1956), pp. 33–35.

ELECTRIC EQUIPMENT FOR CROP DRYING AND SHOP

The preparation of metal for welding will alone justify the cost of this tool on the average mechanized farm. And a portable grinder will buff, wire brush, dress welds, grind bolts, dress rivets, and do other useful jobs. Specifications for a good portable grinder are wheel size, 5 x ¾ inches with ½-inch arbor; free speed, 4,500 rpm; sealed bearings; adjustable guard. The weight should be about 15 pounds.

Drills

The three types of drills that are common to the farm shop are two sizes of portable types and a drill press.

One-half Inch Portable Drill. This is one of the most useful tools on the farm. The following are suggested specifications: ⅓-hp universal motor; 400 to 500 rpm at free speed; ½-inch, three-jaw key chuck; grease-sealed bearings. Both metal and wood can be worked

FIGURE 342. The Missouri farmer demonstrates that he believes a well-planned, well-equipped, comfortable shop is an important factor in the success of any large farm or ranch.

FIGURE 343. The ½-inch portable drill at left is easily mounted in the light-weight frame at right for use as a drill press.

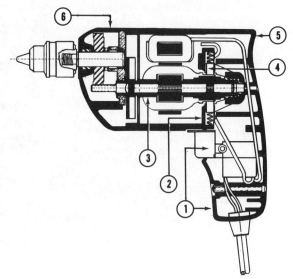

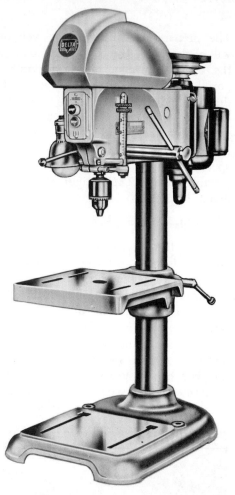

FIGURE 344. *Top:* Heavy-duty drill equipped with "shock-proof" insulation. Numbers indicate points where double insulation is used. Highly desirable where circuit is not 3-wire grounded type (green insulation indicates ground wire).

FIGURE 345. *Right:* A bench-type 14-inch drill press is needed in the farm shop for accurate heavy-duty drilling. Note the four-speed hookup, depth guage, adjustable drill table, lamp, slots for drill vise, vertical head lock, and other features.

with a portable ½-inch drill. Some well-known makes can be mounted in a bench stand and used as a drill press. Weight will average about 10 pounds. (Available in ¾-hp size.)

One-quarter Inch Portable Drill. For light work in metal or wood, you will need a ¼-inch portable drill with the following specifications: universal motor, 2,000 rpm; ¼-inch, three-jaw key chuck; capacity up to ¼-inch drill bits and up to 1-inch wood-working bits; grease-sealed bearings. (NOTE: Use special power wood-working bits in wood.)

Drill Press. For heavy and accurate drilling, you need a bench or floor-type drill press. The drill press should have a capacity from 12 to 17 inches with 14 inches suggested as a good choice (this measure refers to the largest diameter circle in which a hole can be drilled in the center). Most drill presses are equipped with ½-inch, three-jaw chucks. If heavier than ½-inch drilling is to be done, you can fit your drill with a No. 2 Morse taper-shank chuck which will handle up to 1-inch drill bits. The use of ⅝- to 1-inch drill bits requires a slow-

ELECTRIC EQUIPMENT FOR CROP DRYING AND SHOP

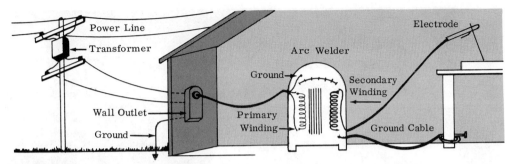

FIGURE 346. A farm arc welder takes 37.5 amperes at 230 volts from the transformer at the power line and steps it down to about 25–40 volts. The amperage is stepped up to a maximum of 180 amperes.

speed attachment that will operate at 200 rpm.

Other specifications for a good drill press are 4-inch spindle travel, with spring return and depth gauge; sealed ball bearings; six splines in spindle and sleeve; ½-hp capacitor motor. Floor models should have foot feed, and both floor and bench models should be equipped with a drill vise.

Arc Welder

The farm-type a-c welder is fast becoming one of the essential pieces of equipment for the mechanized farm. Figure 346 shows a diagram of how this welder converts 37.5 amperes at 230 volts into 180 amperes at 25 to 40 volts. The principle involved here is that a transformer in the welder "induces" the high-amperage, low-voltage current that will produce an arc hot enough to weld metal.

The farm-type welder can be purchased for approximately $200, including regular accessories and a carbon arc torch. Besides doing many types of welding, you can also use this carbon arc torch to braze, heat, and do other work that requires a hot flame. Figure 347 shows one of the popular makes of 180-ampere, limited-input welders.

Welders that operate on 115 volts have not proved very successful for farm use. Before purchasing a welder, you should check with your power supplier to make certain that you will be permitted to connect it to a power source. Be prepared to tell the power supplier what kind of welder you will have.

D-c welders are well adapted to the farm but the cost is two to four times greater than that of the a-c welder.

Refer to any good welding instruction manual for information on the use of arc welders. You may wish to enter one of the welding contests sponsored by various organizations. See your vocational teacher or your club agent for further information.

Electric Saws

You may wish to purchase a portable saw for general carpentry work, and a table saw for cabinet work and for ripping lumber.

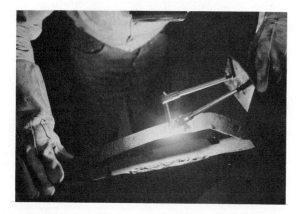

FIGURE 347. *Top:* The farm-type arc welder at left will do most of the welding jobs required on the farm and will operate on rural lines. The workman at right is using a farm welder to build up a worn axle shaft.

FIGURE 348. *Left:* A farmer using a carbon arc torch to heat steel for bending. This torch operates off the farm welder and provides an open flame for many heating jobs on the farm.

Portable Saw. The blade of a portable saw should cut through at least 1⅞ inches at 45 degrees and preferably 2¼ inches at 45 degrees. The motor should be of the universal type with 5,500-rpm free speed. An automatic, telescopic guard should be provided. The saw should weigh 10 to 12 pounds.

Table Saw. A table saw that is to serve any very useful purpose on the farm must have at least a 10-inch blade and preferably a 12-inch one; the motor should be 1 to 2 hp and should have sealed ball bearings. The saw arbor should also be equipped with sealed ball bearings and should tilt to 45 degrees; maximum depth, 3¼ inches at 90 degrees.

Air Compressor

Modern farming with its tractors, trucks, and other needs for compressed air requires an air compressor. Many farmers and farm boys have built their own compressors from old, discarded refrigeration compressors, water tanks, and other materials. If you undertake the job of building a homemade air compressor, make certain to observe all safety precautions in its construction. Improperly constructed compressors have been known to explode and cause serious injuries and death.

Your compressor should be portable (mounted on skids or wheels) so that paint spraying and other jobs can be done in place. Other specifications include a ⅓- to ½-hp capacitor-start or repulsion-induction motor; a single-stage compressor; a pressure gauge; a 20- to 30-gallon tank; a displacement of 2 feet per minute; a cutout that operates at 150 pounds per square-inch; a total weight of 150 to 175 pounds.

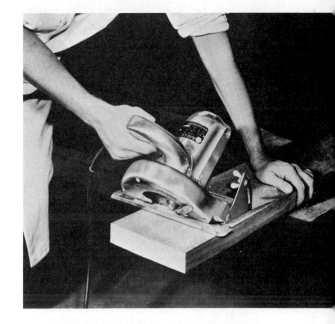

FIGURE 349. A portable electric saw operating at a 45° angle. This 6-inch saw will cut through 1 13/16-inch lumber at this angle.

FIGURE 350. This combination table saw and jointer are powered by the same 1-hp motor.

FIGURE 351. A farm boy operating a paint gun with a homemade air compressor.

SUMMARY

The risk of losing hay, grain, and seed in storage can be eliminated or greatly reduced by a drying system for the farm. In the field, savings of 5 to 15 percent on hay, grain, or seed are possible. Up to 25 percent more food value can be retained in hay by curing it in the barn.

Hay-curing structures include the central duct with laterals, the A-frame, and the tiered duct method. Motor and fan size for all types of structures are determined on the basis of the tonnage to be dried, the depth of the hay in storage, and the type of hay (chopped or long). Also, the amount of moisture to be removed affects the requirements for air volume. Chopped hay offers greater resistance to air flow than does long hay.

A fan large enough to provide 350 cfm per ton of hay (dry) or 15 to 20 cfm per square-foot of floor area is required. For a 75-ton mow, a 25,000-cfm fan, powered by a $7\frac{1}{2}$-hp motor, is required. In good haying weather, this fan will cure the 75 tons of hay in 3 weeks. About 6 feet of chopped hay at a time can be added to the dried hay in the mow. The cost of electric energy averages $1 to $1.50 per ton, while the total costs run from $4 to $5 per ton.

Stored grain offers greater resistance to air flow than does hay; therefore, fans for drying grain must be rated to operate at a higher "static pressure" or resistance.

Structures for drying grain must be practically airtight at the sides and bottom, except for the air inlets. Air entering from the bottom is forced upward through the grain. Moist air outlets must be provided.

From $2\frac{1}{2}$ to 5 cfm of unheated air is needed per bushel of grain or seed. A movement of 5,000-cfm is sufficient to take care of 1,000 bushels of shelled corn. Grain sorghum requires less air—2,500 cfm for 1,000 bushels.

Heated air dries grain and seed much faster than does unheated air. A batch dryer requires only a few hours to dry a wagon load of shelled corn or wheat in any kind of weather. Heated air is used also to dry hay, especially baled hay. Usually, heat is supplied by an oil or gas burner, while an electric fan provides the air movement.

A well-equipped farm shop pays off in keeping vital equipment in production. It also results in the improvement of the farm through the construction of labor-saving equipment.

The basic power machines for the farm shop include a bench grinder, a pedestal grinder, a portable drill, a drill press, a bench saw, a portable saw, an arc welder, and an air compressor. Additional power tools may be justified, depending on the size of the farm.

"Cheap," lightweight power tools should be avoided. A good standard machine will usually last a lifetime if given proper care. One of the best features to look for in a power tool is sealed ball bearings. A good grade of extension cord with an adapter for polarizing the plug-in with the circuit wires is another excellent feature to have for operating portable power tools.

QUESTIONS

1. Why does the depth of the grain or seed increase the requirements for volume of air flow needed in drying?
2. In what kind of weather is it best to cut off the drying fan?
3. Why are sealed ball bearings generally recommended for power tools in the farm shop?
4. What is meant by "total confinement feeding"?
5. What electrical shop equipment can you construct?

ADDITIONAL READINGS

Butler Manufacturing Co., *Grain Drying On-the-Farm With Butler Buildings and Force-Aire Drying Equipment.* Kansas City, Mo., (no date).

Edison Electric Institute, *Farm Electrification.* New York, N.Y. (no date).

"Electricity on the Farm" (all issues). Reuben H. Donnelly Publication, New York, N.Y.

Leonard, Harry and Paul Johnson, *Aids to Using Electricity on Indiana Farms.* Agricultural Engineering Department, Purdue University, Lafayette, Ind., 1962.

McPartland and Novak, *Electrical Equipment Manual.* McGraw-Hill Book Co., New York, N.Y. (no date).

Pennsylvania Electric Company, *Farm Electrification Reference.* Johnstown, Pa. (no date).

U.S. Department of Agriculture, Rural Electrification Administration, *How Electric Farming Can Help Your Business.* R E A Bulletin No. 140-1. Washington, D.C. (no date).

Westinghouse Electric Corporation, *Electrical Equipment You Can Build.* Pittsburgh, Pa. (no date).

SUGGESTED PROJECTS FOR PROBLEM-UNIT SIX

Many of the items discussed throughout Chapters 13 and 14 are suitable for project work, both in the school shop and on the farm. For example, the homemade compressor in Figure 351 was made as a club project by a farm boy in Alabama. He used old, discarded parts in putting this machine together. Also, the homemade elevator discussed in Chapter 1 is easy to build and makes a

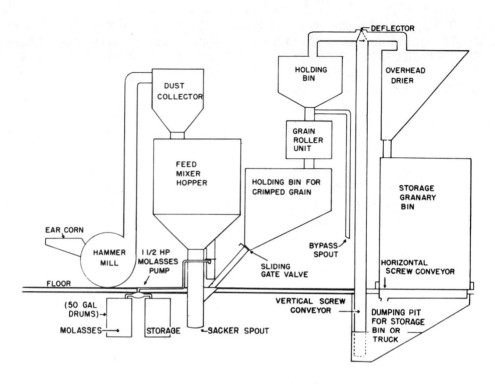

FIGURE 352. Diagram of a complete feed-handling system, including a molasses dispenser. In this system, the feed is not touched by human hands at any time.

valuable addition to the feed-handling equipment for the farm. Study the illustrations throughout Chapters 13 and 14 with a view to using some of them for project work in farm electrification.

Electric Feed Handling. There are hundreds of possible variations in feed-handling systems. One basic arrangement is shown in Figure 352. By studying this layout you may find several ideas that will help you to improve your own feeding system. As for individual projects in feed handling, refer back through Chapters 13 and 14.

Additional Projects for Electrifying the Farm (Hog System). Figure 353 shows an overview of confinement-type production of hogs. Bulk-feeding system, including mix-mill, is fully electrified and automated. Electric brooding, lighting, and watering make this a completely "electrified," confinement-production unit. You may want to electrify your swine production in a similar manner.

Total Confinement Beef System. Figure 354 shows how beef cattle can be produced in total confinement. Silage, other feed, and water are dispensed by electric, automatic equipment. This unit handles 200 steers at the time; ten minutes man-labor per day.

PROJECTS FOR PROBLEM-UNIT SIX

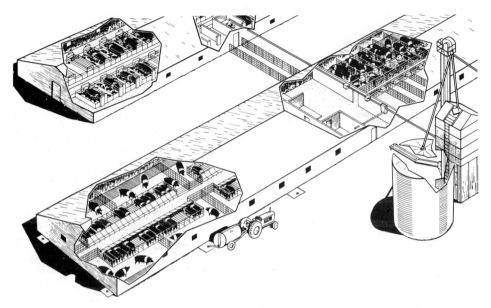

FIGURE 353. *Top:* Cutaway view of completely electrified, automated system of confinement production of swine. Bulk feed supply at right; liquid manure disposal tank is hitched to tractor.

FIGURE 354. *Bottom:* Cutaway view of electrified, automated beef cattle system. Electric-powered silage setup is shown at top right. Liquid manure disposal tank is hitched to tractor. Cattle never leave barn area.

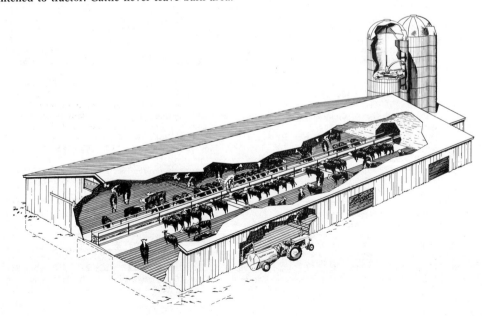

ILLUSTRATION CREDITS

The author wishes to thank the following for permission to reproduce illustrations (identified by figure numbers): Aerovent Fan & Equipment Inc., 329, 333, 338, 339. Alabama Power Co., 23, 351. Allis-Chalmers Manufacturing Company, 80, 93, 207, 212, 215. American Farm Equipment Co., 327. Anaconda Wire & Cable Company, 54, 55, 132, 133, 138. Charles Anderson, 1, 322. Mrs. Charles Anderson, 322. Paul M. Anderson, 332. Cletus Armacost, 343. Babson Bros. Co., 13, 105. Badger Northland Inc., 300, 302, 303, 313. Drawing by Zeno Bailey, 41, 43. Baldor Electric Company, 200. Barnes Manufacturing Co., 8. Belle City Engineering Company, 19, 307. The Black & Decker Mfg. Co., 46, 60, 340, 343, 349. Clifton Boyd, 18, 293. The Brown-Brockmeyer Company, 210, 214, 341. Browning Manufacturing Company, 217, 218, 223, 234. The Bryant Electric Company, 53, 113, 114, 115, 116, 142. David Bush, 92. Bussman Mfg. Co., 98. Butler Manufacturing Company, 2, 328. Carlon Products Corporation, 254. Century Electric Company, 208. Cherry-Burrell Corporation, 295. Chicago Flexible Shaft Co., 304. Clay Equipment Corp., 3, 17, 20, 313, 330, 331, 334, 353, 354. Commonwealth Edison Company, 11, 15, 85, 92. H. C. Davis Sons' Mill Machinery Co., 305, 306. DeLaval Separator Company, 12, 13, 289, 291, 292, 293, 296, 297. Delco-Remy, 42. Delta Power Tool Division, Rockwell Mfg. Co., 345, 350. The Deming Company, 245, 246, 248. The Detroit Edison Company, 85, 130, 282, 283. Eagle Electric Co., 116. East Texas State College Photo Laboratory, 45, 50. Thomas Alva Edison Foundation, Inc., 4, 5, 40, 258. Robert Eifert, 329. *Electricity on the Farm,* 37 69, 79, 204, 211, 255, 256, 266, 286, 317, 342, 344, 352. A. E. Evans and Robert Bradley, 199. Aubrey Evans, 18, 27, 312. Martin and Victor Eversgerd, 336. Fairbanks, Morse & Co., 45, 50, 240, 243, 247. Farm Electrification Bureau, 151, 152, 153, 165. Farm Equipment Institute, 2, 9, 12, 15, 338. Farm Fans, 231. *Farm Journal,* 67. *Farmstead Wiring Handbook,* 120, 121, 122, 123, 124, 125, 126, 127, 128, 129. The Five Mfg. Co., Inc. and Jessee B. Brooks, 308. The Gates Rubber Company Sales Division, Inc., 219, 221, 235. General Electric Company, 5, 35, 260, 267, 268, 269, 270, 272, 273, 274, 275, 276, 277, 278, 279, 280, 283, 309. General Mills, Inc., 14, 64, 239, 324. W. E. Gould 92. Waldemar Grewe, 14, 324. Mrs. J. R. Hamilton, Jr., 110. Mrs. Roland Hamilton, 21. James R. Hamilton, Jr., 230. Roxane Hamilton, 316. E. Grady Hendrix, 23. Henry Holt and Company, Inc., 47. Drawing by Annie B. Hudson, 29, 36. Illinois Electric Cooperatives, 201, 253, 259, 322. Illuminating Engineering Society, 271. Independent Protection Co., Inc., 26, 39. Industry Committee on Interior Wiring Design, 118. Ingersoll-Rand Company, 46. International Harvester Company, 44. James Mfg. Co., 3, 10, 11, 301, 304, 311, 315, 318, 319, 320, 321, 325. William Junker, 326. Kennecott Copper Corporation, 82. LGM Electronics, Inc., 286 Drawing by Oran Lewellyn, 141. The James F. Lincoln Arc Welding Foundation, 7, 58, 65, 346, 347, 348. Marathon Kleen-Stall Co., 304. Markel Electric Products, Inc., 47. John A. Martin & Co., 299. A. Y. McDonald Mfg. Co., 239. Dennis McGuire and M. Huested, 333. Mrs. John I. Mitchell, Jr., 111. John I. Mitchell, Jr. and John I. Mitchell, III, 22. John I. Mitchell, III and Jamie Mitchell, 112. Mississippi Federated Cooperatives, Food Processing Division, 4. Mississippi State University, Extension Service, 244. Montgomery Ward, 53, 72, 84, 95, 119, 134, 139, 144, 145, 148, 149, 154, 156, 157, 161, 163, 164, 167, 169, 172, 173, 175, 176, 178, 179, 180, 181, 184, 185, 187, 189, 190, 192. A. H. Myers, 2. F. E. Myers & Bro. Company, 249. National Association Domestic and Farm Pump Manufacturers, 241, 242, 243, 245. National Cooperatives, Inc., 290, 298. National Safety Council, 66, 68, 103, 316. New Holland Machine Company and Leroy Leisey, 335. Niagara Mohawk Power Corporation, 30, 62, 71, 83, 86, 90, 94, 97, 100, 101, 102, 106, 131, 150, 158, 193. Otto Niederer & Sons Inc., 323. Onan Corporation, 18, 27, 312. Robert Pendell, 201. The Pennsylvania State University, 15, 332. Portland Cement Association, 285. *Poultry Tribune* and Theron Campbell, 321. Progress Manufacturing Company, Inc., 55. Robert W. Pimpen, 336. Forrest Puellmann, 342. M. A. Roney, 27. Royal Electric Corporation, 53, 60, 83, 132, 134, 228. Donald Sanford, Jr., 23. Howard Schweighart, 1. Carroll Shelton, 18, 27. Sears, Roebuck and Co., 52, 100, 104, 107, 113, 116, 138, 140, 143, 162, 166, 168, 170, 171, 177, 183, 186, 191. John Shultz Family, 266. R. A. Smitherman, photo by George Laycock, 19. O. L. Snowden, 286. Southern Association of Agricultural Engineering and Vocational Agriculture, 88, 108, 216, 220, 222, 224, 227, 229, 233, 251. Square D. Company, 71, 96, 159, 160, 226. Steber Mfg. Co., 287. Stran-Steel Corporation, 9, 337. *Successful Farming,* 257, 304, 310, 326. Sylvania Electric Products Inc., 79, 261, 262, 281, 283, 284, 287, 288. Tennessee Valley Authority, 75, 99, 144, 146, 154, 155, 161, 182, 216. Andrew Thompson, 18. J. M. Thompson, 69. Robert Titus and Joe Bob Hinton, 28, 41, 43, 44, 75, 155, 194, 195, 197, 198. G. V. Troyer, 63. Drawing by Charles Turner, 88. Underwriters' Laboratories, Inc., 38, 59, 73, 74, 107, 131. U.S.D.A. Photograph, 63, 143, 144, 150, 188, 250, 252. C. D. Walters, 201. Wagner Electric Corporation, 34, 49, 56, 203, 206, 209, 236, 237. Sid West, 27. Westinghouse Electric Corporation, 8, 25, 30, 31, 33, 48, 51, 57, 61, 70, 76, 77, 78, 81, 91, 109, 116, 136, 142, 147, 174, 196, 202, 205, 213, 225, 227, 232, 235, 238, 263, 264, 266, 283. Weston Electrical Instrument Corporation, 32. M. O. Whitehead, 117. The Wiremold Company, 107, 135. University of Wisconsin, 16, 65. Joe Young, 352. Zero Sales Corporation, 18, 294. Paul and Laura Hamilton, 110.

INDEX

Directions for doing electrical work are indexed under "How to."

Accidents, electrical, 61–65
  causes, 61–62
  deaths and injuries, 61, 72
  safety hints, 63–65
Adapter plug-ins, 55
Air compressors, 345
Air-conditioners, wiring for, 122, 124
Air flow, resistance to, 331
Alternating current (a-c), 29
Alternators, 47–50, 59
Aluminum conductors, 56
American Wire Gauge, 93
Amperage, 32–33, 41
  factors affecting, 33
  to measure electric current, 32
Amperes (amps):
  definition, 77
  formulas, 34
Appliance circuits, 202
Appliances, electric, 18
  dairy farming, 302–311, 328
  faulty, 66, 68
  voltage, 30
  wattage ratings, 33
  wiring for, 114, 120, 124, 144
Arc welders, 59, 60, 343
  a-c welders, 343
  d-c welders, 343
  safety hints, 64
Armatures, 51
Atoms, 25, 27
Attraction, magnetic, 27, 46–47
Automatic farming, 10
AWG (*See* American Wire Gauge)

Bare ground wire, 147, 158
Barn cleaners, 309–311
  "shuttle stroke," 310–311
Barns, outlets, 133–134, 138, 142–143
Batteries, 28
  storage, 45–46, 59
Bearings:
  electric motors, 206
  lubricating, 236–238
  sleeve and ball, 216
Belts; *see* Pulley and belts

Bill of materials:
  preparing, 154–155, 158
  for twelve-circuit home, 155–157
Bills, electric, 35–40
  discounts for prompt payment, 40
  meter readings, 37–38
Bonding wires, 163
Branch circuits, 113–144
  accessories, 121–124
  appliances, 114, 120, 124, 144, 202
  electric motors, 118
  farm service buildings, 130–143
  fuses, 118, 120
  general purpose, 120, 124, 143, 202
  grounding, 131–132
  high-wattage equipment, 115, 121
  for houses, 114, 124–130
  individual equipment, 121, 124, 144, 202
  outlets, 114, 122, 130
  planning, 114–118
  problems and faults, 114
  protection devices, 115, 118
  purpose of, 113
  safety precautions, 131, 132
  three-wire (230-volt), 54, 114–115, 118, 202
  two-wire circuits, 52–54, 114, 118, 202
  types of, 118–129
  voltage ratings, 118–120
Breaker panels, 109
Breakers, circuit, 203
Brooders, 15–16, 116, 319–321
  hover brooding, 319
  infrared lamp, 319–321
  underheat brooding, 320–321
  wiring for, 136–137
Brushes, electric motors, 247
Bulk tank coolers, 302–307
  direct-expansion, 305–306
  ice-bank, 305–306
  vacuum tank, 305
Burglar alarms, 16
BX cable; *see* Cable, flexible armored

Cables, 147
  flexible armored, 152–153
    fastening to boxes, 180–183
    installation of, 180–183
  non-metallic sheathed, 151–152, 177–178
    connecting to boxes, 178
    fastening to supports, 178

  plastic, 150
  power, 149
  service entrance (SE), 172, 173–174
  soil-heating, 320
  strippers, 154
  Type-W cable, 150
  underground, 150, 171
Capacitors, electric motors, 206
Charges:
  laws of attraction and repelling, 44
  negative, 27, 43–44
  positive, 27, 43–44
Circuits, 25, 28, 29–34, 77
  amperage, 32–33
  branch, 113–144, (*see also* Branch circuits)
  breakers, 203
  closed, 202
  constructing simple circuit, 73
  feeder, 166–171; (*see also* Feeders
  open, 203
  overloading, 80
  short, 203
  three-wire, 54, 114–115, 118, 202
  wiring, 193–195
  two-wire, 52–54, 202
  voltage, 30–31
Clothes dryers, wiring for, 195
Codes, electrical, 66–69, 70–71, 145–146, 158
  local, 145–146, 159
  National Electrical Code, 145–146, 158
Colors, light reflected by, 279–280
Commutators, electric motors, 247
Condensers, electric motors, 206, 247
Conductors, 25, 28, 50, 56, 77
  feeders, 149
  materials for, 56
Conduits:
  service entrance, 174–175
  thin-wall, 152
    installation of, 178–180
  wiring systems, 95–96, 102
Connecting wires to terminals, 161–164
Connectors, wire, 164
Converters, phase, 210, 216, 226, 247
Conveyors, for feed handling, 314, 317
Coolers:
  bulk tank, 302, 304–307
  wet-type, 307

351

352 INDEX

Copper conductors, 56
Copper wire, 54, 56
Cords:
  asbestos heat-resisting, 150
  extension, 74, 201
  heater, 149
  kitchen-unit, 149
  thermo, 150
  TV cord, 150
Cornice lights, 285
Costs:
  of electricity, 6, 7
    computing, 35–41
  electric bills, 35–40
  of electric equipment, 7, 301
  fixed costs, 301
  of hand labor, 6
  operating, 10, 301
  rates, 36–40
  repair and maintenance, 301
Crimper mills, 312
Current, electric, 25, 40
  alternating (a-c), 29, 46, 49–50
  amperage, 32–33
  cycles, 77
  direct (d-c), 50
  flow of, 25, 38, 40
  induced, 45, 47
  single-phase, 49, 53, 59
  60-cycle, 49
  three-phase, 53, 58–59, 77
  two-phase, 77
Cutting wires, 161–164
  splicing wire, 161–163
Cycles, definition, 77

Dairy farming, 11–13, 302–311, 328
  barn cleaners, 309–311
    "shuttle stroke," 310–311
  bulk tank coolers, 302, 304–305
    direct-expansion, 305–306
    ice-bank, 305–306
    vacuum tanks, 305
  effect of labor problem on electrification, 17
  electric appliances, 12–13, 302–311, 328
  feed-handling equipment, 311–318
  gutter cleaners, 309
  milkers, 302–304
    combined with cleaning system, 302–304
  miscellaneous equipment, 311
  outlets in barns, 133–134
  water heaters, 307–309
    combined with house heater, 309
    double-element, 308
    non-pressure type, 307
    pressure type, 307
    single-element, 308
    wet-type coolers, 307
Davenport, Thomas, 206
Demonstration boards, wiring, 201
Depreciation, useful life basis, 10
Diffusers, 282, 287
Diffusion of light, 281
Disconnecting means, 82, 94, 101, 109
Disconnects, individual, 84
Dishwashers, wiring for, 144
Distribution equipment, 164–175
  installing feeder circuits, 166–167
  installing grounding systems, 175
  installing service entrance, 171
  metering at building, 165–166
  wiring service switch, 175
Drills, 341–343
  drill press, 342–343
  one-half inch, 341–342
  one-quarter inch, 342
Drives for electric motors, 207, 219–225
  pulleys and belts, 207, 219–225
Drying farm crops, 15, 330–339
  grain drying, 337–339, 346
    fans and motors, 338
    heated dryers, 338
    structures and equipment, 337–338
  hay dryers, 332–334, 346
    chopped hay, 346
    cost of, 335–336
    fans and motors, 334–335, 346
    with heated air, 335
    main duct system, 332
    main duct with side laterals, 332
    operation and care of, 336–337
    tiered duct system, 334
    with unheated air, 332
  heated air, 331
  resistance to air flow, 331
  size of fan required, 331, 338
  static pressure, 331
  unheated air, 331
Duplex receptacles, 122

Edison, Thomas A., 277
Egg-handling equipment, 324–325
  cleaner-candlers, 324–325
  coolers, 16, 326–327
  graders, 325–326
  storage, 326–327
Egg storage and handling room, electric outlets, 137
Electric motors (*See* Motors, electric)
Electric shock, treating, 71–72
Electricity:
  characteristics, 25
  converted into heat energy, 50–51
  converted into light energy, 51
  converted into motive power, 51–52
  definition, 25, 28
  electric current, 25, 40
  electron theory, 27–29, 40–41
  energy produced, 25, 28
  to improve farms and farming, 1–23
  laws of attraction and repelling, 44
  principles of, 24–42
Electrification:
  amount of electrified farming, 5, 11–13
  cost and returns, 13, 20
  deciding what projects to undertake, 8
  effect of labor problem, 17–18
  effect on agricultural progress, 1–23
  growth of, 79
  home and farmstead improvement, 23
  insurance against risk, 15–17
  investment in electric equipment, 12
  market considerations, 14–15
  national awards programs, 21–22
  opportunities in, 1–23
  size of farm operation, 10–11
  survey of present and future needs, 6, 9
Electrocution, danger of, 55
Electromagnetics, 52
Electron theory of electricity, 27–29, 40–41
Electrons, 25, 27
  "free," 27–28
  rate of flow, 28
Elevators, feed-handling, 314
  construction of, 10–11
Emergency lights, portable, 299
Energy, electrical, 25
  converted to heat energy, 50–51
  converted to light energy, 51

INDEX 353

Equipment and machines:
  costs of, 7, 301, 328
  for dairy farming, 302–311, 328
  depreciation, 7, 10
  dryers for farm crops, 330–339
  economic decision, 300, 328
  effect of market trends, 301
  feed-handling, 311–318
  investment in, 12
  labor requirements, 301, 302, 329
  management problems, 200–201, 328–329
  poultry farming, 319–327
  power tools, 339–346
  purchasing, 301
  repair and maintenance, 301
    following operator's manuals, 307, 311
  storage facilities, 301
Extension cords:
  heavy-duty, 201
  repairing, 74
Exterior wiring:
  distribution equipment, 164–175
  installation of, 164–175
  wires used for, 148–149

Fans:
  for drying crops, 331, 334–335, 346
  for grain dryers, 338
Faraday, Michael, 47
Farms:
  number and percent of farms electrified, 5
  total farm employment and horsepower per farm worker (1870–1970), 6
Farmstead Wiring Handbook, 79, 87, 97, 105, 109
Farmsteads:
  lighting system, 294–296
  wire systems, 12, 97–104
Feed-handling equipment, 17, 311–318, 329
  combination units, 311
  conveyors, 314, 317
  crimper millts, 312
  electrical appliances, 311–318, 329
  elevators, 314
  feed bunkers, 317
  grain-handling, 314–315
  hammer mills and dial controls, 313
  hay-handling equipment, 314
  for poultry, 321–324
    automatic feeders, 322
  feed troughs, 321–322
  silo unloaders, 315–318, 328
  use of elevators, 10–11
  watering systems, 323–324
Feed handling room, 141–142
Feeder conductors, 149
Feeders, 82–84, 85, 102
  attachment to buildings, 169
  attachment to power source, 169
  connecting wires to service entrance, 175
  definition, 202
  installation of, 166–171
  two-wire feeder circuit, 171
  underground, 85, 96, 171
  wires for, 169
Fires:
  due to faulty wiring, 63, 65–69, 70–71
  water wells for protection from, 253
First aid, 71–72
"Fish tape," 154
Floor plans, 154
  diagram of electrical system, 154–155
Fluorescent tubes 281–286, 291, 292, 295, 296
Food freezers, 114, 120, 124
Foot-candles, 278, 296
Formulas:
  for amperes, 34
  for volts, 34
  for water horsepower, 266–267
  for watts, 34
4-H Clubs, national awards programs, 8, 22
Franklin, Benjamin, 44
Friction loss:
  motors, 34
  water pipes, 255–257
  water pumps, 265–267, 275
Friction tape, 163
Fuses and fusing, 84, 94, 103–104, 108
  changing blown, 76–77, 80
  definition, 203
  fuse-blowing trouble, 80
  Fusetron, 195–196, 230
  Fustats, 195–196, 230
  for motor protection, 195–196, 230, 244–245
  panel, 108–109
  safety hints, 67–68
  sizes, 94, 244–245
  time-delay, 120, 144, 230, 247
Future Farmers of America, national electrical awards programs, 21

Garages, 130
  electric outlets, 141
Generators, 28, 41
  a-c electricity, 47–50
  magnetic induction, 47–50, 51
  operation of, 46–50
  to recharge storage batteries, 45, 59
Grain drying equipment, 15, 337–339, 346
  fans and motors, 338
Grain-handling equipment, 314–315
Grinders, 340–341
  bench, 340
  pedestal, 340
  portable, 340–341
Ground wires, 53, 60
Grounds and grounding, 53, 54–56, 60
  branch circuits and outlets, 131–132
  Code requirements, 132
  definition, 203
  service entrance, 175
  wiring-systems, 84–85, 95, 102, 103, 110
Gutter cleaners, 309

Hammer mills, 313
Hay dryers, 16–17, 332–334, 346
  chopped hay, 346
  cost of, 335–336
  fans and motors, 334–335, 346
  with heated air, 335
  main duct system, 332
  main duct with side laterals, 332
  operation and care for, 336–337
  tiered duct system, 334
  with unheated air, 332
Hay-handling equipment, 314
Heat, 25
  electric energy converted to, 50–51
Heaters and heating equipment, 16, 124, 132–133
  heat lamps, 283, 295
  portable, 299
  water heaters, 307–309
    non-pressure type, 307
    pressure type, 307
  wattage ratings, 33, 51
  wiring for, 194–195
High lines, 28, 81
  two-wire, 53–54
Hog and farrowing houses, 139

INDEX

Horsepower:
  electric motors, 33, 34, 207, 208–210
  total farm employment and horsepower per farm worker (1870–1970), 6
  water, 266–267
Hot wires, 54–55, 60
Houses:
  branch circuits, 114, 124–130
  interior wiring, 175–197
  outlets, 125–130
How to:
  arrange feed-handling system, 312–313, 348
  bond electric wires, 163
  change blown fuses, 76–77, 80
  compute maximum demand, 101–104
  compute water horsepower, 266–267
  connect wire to terminals, 161
  construct a lighting project, 22
  construct an elevator, 10, 347
  construct emergency light, 299
  construct portable heat lamp, 299
  construct simple circuits, 73
  construct test lamp, 196–197
  construct trouble lamp, 199
  construct watering tank, 298
  construct wiring demonstration board, 201
  cut wires, 161
  demonstrate voltage drop, 74
  determine capacity of pumps, 265–266
  determine cost of electricity, 34–40
  determine if circuit is overloaded, 67, 68
  determine location and type of wells, 249–251
  determine pulley size, 219–221
  determine size of water pipe, 262–265
  determine type of V-belt, 222–224
  determine type of water pump, 251–262, 264
  draw wiring diagrams, 154–155
  figure kilowatthours, 33, 35
  figure monthly electric bills, 34–40
  fish wire into a conduit, 154
  ground electrical system, 54–56, 60, 95, 102
  install convenience outlets, 183–184, 188–190
  install electric motors, 231–234
  install exterior wiring, 97–100, 164–175
  install feeder circuits, 166–171
  install flexible armored cable (BX), 180–181
  install interior wiring, 175–197
  install light fixtures, 190–193
  install metering service, 165–166
  install non-metallic sheathed cable, 177–178
  install power poles, 102
  install service drop, 94–95, 102
  install service entrance, 171–175
  install service switches, 102
  install thin-wall conduit, 178–180
  install tractor lights, 76
  install wall switches, 183–188
  install yard light, 76, 201
  lubricate electric motors, 236–238
  make test lamps, 154
  measure electric power, 33
  operate hay dryers, 336–337
  place lamp for proper light, 280
  polarize wiring, 54–56
  prepare bills of materials, 155–157
  protect electric motor circuit, 195–196
  protect water pipes from freezing, 298
  read electric meters, 37–38
  repair cord ends, 74
  repair extension cords, 74, 201
  replace plugs, 74, 210
  select electric motors, 207–210
  splice wire, 161–163
  survey present and future needs, 6, 9
  tap into a circuit, 183–184, 188–190
  wire clothes dryer, 195
  wire pump motors, 268
  wire range circuit, 193–195
  wire three-way switch control, 193
  wire water heater, 194–195
  wire 230-volt circuits, 193–195

Incandescent light bulbs, 51, 277, 278, 281, 283
Induced current, 45, 47
Inductance coils, 50
Induction, magnetic, 51–52
Infrared lamps for brooders, 319–321
Installation (*See under* How to)
Insulation materials, 28
  use of colors, 54–55
Insulators, 28, 54, 60, 77, 85, 95
Interior wiring:
  installation of, 175–197
  wires used for, 150–153
Irons, electric:
  amperage, 33
  flow of electricity, 29
Irrigation systems, 270–274, 275
  cost of, 274, 276
  crops and soil types, 270–271
  electric motors for, 208–209
  labor available, 271–272
  motor size, 273–274
  power required, 272
  water supply, 270

Junction boxes, 55, 121

Kilowatthours, 33
  definition, 77
Kilowatts, 33, 35, 41
  definition, 77
Kitchen, light fixtures for, 291

Labor requirements:
  effect of electrification, 7, 17–18
  electrical equipment and, 301, 302, 329
  total farm employment and horsepower per farm worker (1870–1970), 6
Lamps and fixtures, 282, 285–286, 288–289
  cords, 147–148
  placement of, 280, 285–286
Laundry, light fixtures for, 292–293
Lead-in wires, 81
Light, 25
  electric energy converted to, 51
Light bulbs, incandescent, 51, 277, 278, 281, 283
Light fixtures, installation and wiring, 190–192
Light meters, 278
Lighting systems, 277–299
  amount of light needed, 278–280
  brightness of light, 278
  bulb protectors, 295
  Circlines, 281, 283–284, 291–292
  for close work, 286–293
  color schemes for interiors, 280
  cornice lights, 285
  diffusers, 282, 287
  diffusion of light, 281, 282, 287

INDEX

exterior, 286, 293, 296–297
farm service buildings, 294–296
fixtures and lamps, 282, 285–286
    placement of, 280
fluorescent tubes, 281–285, 286, 291–292, 295, 296
foot-candles, 278–296
heat lamps, 283, 295
for homes, 284–293, 296
incandescent, 51, 277, 278, 281–283, 285
indirect lighting, 282, 296
interior, 284–293, 296
for the kitchen, 291
lamp shades, 288–289
    close to ceiling, 285
    type and placement of, 285–286
for the laundry, 292–293
light meters, 278
porch and step ilghts, 286
quality of light, 282
for reading, 289
reflectors or reflector bulbs, 280, 294–295
semi-indirect, 282
for sewing, 290
size of bulbs, tubes and Circlines, 281
sources of, 282–284
for studying and writing, 290
three-lite arrangement, 287–288
wall lighting, 286
yard lights, 281, 293, 296–297
Lightning, 43–44, 59
Lightning rods, 44
Litter-cleaning equipment, 327–328
Livestock farms, electric equipment, 12, 17, 311–318, 329
Lott Farm, amount and cost of electricity, 39
Lubrication, electric motors, 236–238, 245

Machine shops, 140
Magnetic attraction, 27
Magnetism, 46–47
    induction, 51–52
    positive and negative poles, 47, 52
    relation between electricity and, 47–50
Manuals, operators, 307, 311
Matter, atoms and molecules, 25, 27
Metallic raceway, 151
Meter loops, 81
Meters, electric, 31
    computations, 33–34, 41
    installation of, 165–166
    location of, 105–107
    pole metering, 86, 109
    wiring yardpoles, 166
Milkers, 302–304
    care of, 304
    combined with cleaning systems, 302–303, 304
    cow-to-can, 303–304
Molecules, 27
Motive power, 25
    electric energy converted to, 51–52
Motors, electric, 204–247
    accessories, 216–219
    bearings, 206, 207, 216
        lubricating, 236–238
    brushes, 247
    capacitors (condensers), 206, 247
    care of, 228–247
        protection devices, 228–231
        routine maintenance jobs, 235
        trouble-shooting chart, 238–243
    commutators, 247
    condensers, 206, 247
    construction, 51–52
    converting electric energy, 51–52
    cost of operating, 10, 204, 207
    drives for, 207, 219–225
        flexible hose, 224
        flexible shaft, 225
        positive and direct, 224–225
        pulleys and belts, 207, 219–225
        rigid flange, 224–225
    efficiency of, 204, 225
    electric service and, 215–216
    friction loss, 34
    full-load currents in amperes, 111
    for grain dryers, 338
    grounding, 54
    for hay dryers, 334–335, 346
    housing or covers, 207, 219
    installation of, 231–234, 243
        mounting, 231
        portable, 233–234
        position of, 231
        rigid shaft, 232
    lubrication system, 206, 236–238, 245
    magnetic induction, 51–52
    operation of, 51–52
    overheating, 210
    parts, 206–207, 247
    phase converters, 210, 216, 226, 247
    poles, 247
    protection devices, 195–196, 228–231
        automatic reset protectors, 229, 243
        built-in protectors, 229, 243
        magnetic switch, 230
        manual switch, 229, 243
        manual-start switch, 230
        resistor starters, 231
        special starter, 230–231
        time-delay fuses, 230
    pulleys and belts, 207, 219–225
        care of, 234–235
        determining size of, 219–220
        four-step cone, 220–222
        length of belt, 223–224
        types, 220–222
        V-belts, 222–223
        V-pulleys, 220
    for pumps, 253–254, 265–267, 275
    reversing direction of split-phase, 246
    rotors, 51–52, 206, 247
    selection of, 206–227
    "shorted," 54, 56
    size and wattage, 35
    size in horsepower, 207, 208–210
    starting torque, 206, 207, 210, 226
    stators, 51, 206, 247
    time-delay fuses, 247
    torque, 247
    types, 207, 210–215, 217
        capacitor-start, 210, 212–213, 217, 267
        capacitor-start capacitor-run, 210, 214, 217
        comparison of, 217
        for heavy loads, 214–215
        for light starting loads, 210–211
        for moderate loads, 211–214
        repulsion-capacitor, 210, 215, 217
        repulsion-induction, 210, 214–215, 217, 267
        repulsion-start, 210, 214, 217
        single-phase, 210
        split-phase, 210, 211, 217
        three-phase, 210, 213, 216, 217, 226
        universal, 210, 215, 217
    voltage, 30
    wattage, 33, 34, 35
    wire and fuse size for, 244–245

356 INDEX

Motors (cont.)
  wiring box, 206
  wiring for, 118, 144

National Board of Fire Underwriters, 69
National Electrical Awards Program, 8, 21–22
National Electrical Core, 55, 66–69, 70–71, 72, 105, 109, 145–146, 158
  wire size, 87, 105
National Safety Council, 61
Negative charges, 27, 43–44
Negative poles, 47
Neon test lamps, 154
Neutral wires, 53, 54, 60
Non-conductors, 28
Nucleus, 27–28

Oersted, Hans Christian, 47
Ohms, definition, 34
Outlets, 113–144, 122, 131
  boxes, 121
  convenience, 183–184, 188–190, 202
  grounded receptacles, 122, 131
  installation of, 183, 188–190
  installing boxes, 184
  lighting, 202
  marking location, 183
  outdoor, 201
  for residences, 125–130
  special-purpose, 202
  three-prong plugs, 122
  wiring, 188–190
Overloading circuits, 66–67, 80

Phases, 77
  converters, 210, 216, 226, 247
  three-phase current, 77
  two-phase current, 77
Pipes:
  anti-freeze heating tape for, 298
  friction loss, 255–257
  long runs, 262
  size required, 262–265
Planning wiring jobs, 145–159
Plastic tape, 163
Plug-ins, 55–56
  "big tail," 194
  duplex, 122
  polarizing, 122
  replacing plugs, 74
  three-prong, 56, 122, 131
Polarizing electrical system, 54–56, 60

Poles:
  electric motors, 247
  metering at, 86, 109, 166
  positive and negative, 47
  selection of, 166
  wiring systems, 85, 95, 102, 103
Porch and step lights, 286
Positive charges, 27, 43–44, 47
Post lanterns, 201
Poultry farming equipment, 138, 319–327, 329
  brooders, 319–321
    hover brooding, 319
  infrared lamps, 319–321
    underheat brooding, 320–321
  caged-type house, 324, 328
  egg cleaner-candler, 324–325
  egg coolers, 326–327
  egg graders, 325–326
  egg-handling equipment, 324–325
  egg storage, 326–327
  electric appliances for, 319–327, 328
  feed-handling equipment, 311–318, 321–324
    automatic feeders, 322
    feed troughs, 321–322
    mechanical feeders, 321–322
    operation and care, 323–324
  investments in, 12, 13
  litter-cleaning equipment, 327–328
  outlets in poultry houses, 135–136, 138
  ventilation, 324
  watering systems, 323–324
    automatic, 323
    operation and care, 323–324
Power plants, generators for, 47–50
Power tools, 339–347
  air compressors, 345
  arc welders, 343
  drills, 341–343
  grinders, 340–341
Pressure, electrical, 28
  static, 331
Projects, 298–299, 347–348
  anti-freeze heating tape for water pipes, 298
  changing blown fuses, 76
  concrete watering tanks, 298
  constructing simple circuit, 73
  construction of air compressors, 347
  construction of elevators, 10, 347
  demonstrating voltage drop, 74
  electric feed handling equipment, 348

  electric motors, 246
  electric saws, 343–345
  extension cords, 201
  Fusetron protection, 246
  home-made saw, 22
  installing tractor lights, 76
  installing yard light, 76
  lighting projects, 22
  outdoor plug-ins and post lanterns, 201
  permanent mount for sanders, 246
  portable electric motors, 246
  portable emergency light, 299
  portable heat lamps, 299
  repairing extension cord, 74
  replacing plugs, 74
  reversing direction of split-phase motors, 246
  total confinement beef system, 348
  trouble lamp, 199
  wiring demonstration board, 201
Property damage:
  electrical codes, 66–69
  from electrical fires, 61, 63, 65–69, 72
  safety measures, 65–69
Protection devices:
  branch circuits, 115, 118
  electric motors, 195–196, 228–231
Protons, 25, 27
Pulleys and belts, 207, 219–225
  care of, 234–235
  determining size of, 219–220
  four-step cone, 220–222
  length of belt, 223–224
  types, 220–222
  V-belts, 222–223
  V-pulleys, 220
Pumps and pumping, 132, 251–254
  peak-use periods, 253, 254
  capacity, 251–252, 265–266
  care of, 143, 268–269
  centrifugal, 258–259, 275
  centrifugal jet, 260, 264, 275
  deep-well, 254–258, 260–262, 264, 275
  discharge head, 265–266
  friction loss, 254–261, 275
  houses for, 143, 267–268, 275
  irrigation system, 270–274
  motors for, 253–254, 265–267, 275
    formula for computing horsepower, 266–267
  number of faucets, 253, 254, 275
  piston, 259–260, 264, 275

plunger-type, 260–261, 264, 275
power supply, 270
rotary, 262
selecting, 254–262
shallow-well, 258–260, 264, 275
submersible, 261–262, 264, 275
suction head, 265
suction lift at different altitudes, 258
trouble-shooting charts, 269, 275
turbine, 262
types of, 254–262, 264
characteristics, 274
well capacity and, 253–254
wiring, 268
working head, 275

Ranges:
branch circuits, 114–115, 118, 120, 122, 124, 144
wiring 230-volt circuits, 193–195
Rates, electric, 36–40
demand charges, 40, 41
minimum charges, 36
off-peak, 40
reading meter, 37–38
sliding scale, 36, 41
special, 40
variations in, 36–40
Receptacles, three slots, 122, 131
Reflectors or reflector bulbs, 280, 294–295
Resistance, 30
to air flow, 331
heat produced by, 50–51
natural, 31
varies with wire size, 34
voltage drop, 31
Rotors:
in alternators, 50
electric motors, 51–52, 206, 247
Rubber insulation, 56

Safety measures, 61–77
branch circuits, 131
deaths and injuries, 61, 72
electrical accidents, 61–62
fires, 61, 63, 65–69, 72
first aid and treatment, 71–72
grounding, 53, 54–56, 60
insulators, 54, 60
national protection agencies, 68–71
safety hints, 63–65
treating electric shock, 71–72
Sanders, permanent mount for, 232, 246
Saws, electric, 343–345

portable, 345
table, 345
Service drops, 31, 81, 84, 94–95, 102, 105
Service entrances:
cable, 148, 173–174
conductors, 93–94, 101–102
conduit, 174–175
connecting wires to feeders, 175
definition, 202
flush-mounted switch, 171–172
grounding system, 54, 175
inspection of, 175
installation of, 171–175
installing meter, 165–166
preparing service wires, 172–173
service cable, 173–174
service switch, 84, 102, 171, 175, 203
sill plate or conduit ell, 172
wiring, 81–82, 84, 109
Service switches, 84, 102, 203
Sheep barns and lambing sheds, 139
Shops and shop equipment, 124, 140
"Shorts," 54, 56
avoiding, 65
Sill plates, 172
Silo unloaders, 315–318
surface or bottom unloading, 315–317
Single-conductor wire, 147, 158
Single-phase current, 53, 59
Soldering wires, 163–164
Splicing wire, 161–163
Stables, 139
Stairs and passageways, 141
Static electricity, 43–44, 50
Static pressure, 331
Stators:
in alternators, 50, 59
electric motors, 51, 206, 247
Storage batteries, 45–46, 59
Storage facilities, 301
Storage rooms, electric outlets, 142, 143
Surveys of present and future needs, 9
conditions of electrical system, 6, 9
cost of electric equipment, 7
records of man labor, 7
Switches, 29, 30
boxes, 66
definition, 203
for electric motors, 229–230
flush-mounted, 171–172
four-way hookup, 188
installation, 55, 183–188

pull-chain, 115, 122
service, 84, 102, 171, 175, 203
three-way hookup, 115, 121, 130, 131, 186–187, 193
two-way, 115, 121
wall, 115, 121, 131
installation of, 183–188
installing boxes, 183–184
marking location, 183
wiring, 184–199
wiring, 55, 184–199
Symbols, for type of wire, 203

Tapes:
friction, 163
plastic, 163
Tesla, Nikola, 206
Test lamps, construction, 196–197, 199
Three-phase current, 53, 58–59
Three-wire (230-volt) circuits, 114–115, 118, 220
grounding, 54
plug-ins, 56
voltages, 202
wiring, 193–195
Tools, electric, 55
for installing wiring, 153–154
power, 339–346 (See also Power tools)
Torque:
definition, 247
electric motors, 206, 207, 210, 226
Tractor lights, installing, 76
Transformers, 30, 31, 57–59, 60
high-line, 57
induced current, 58
"iron pot," 57–58
operation of, 57–59, 60
primary and secondary windings, 58, 60
size of, 59
Transmission lines, 30
"high lines," 28, 81
resistance, 30
Transmission of electricity, 43–60
materials and equipment, 56–57
from power plant to farm, 52–54
230-volt (three-wire) circuits, 114–115, 118, 202
grounding, 54
plug-ins, 56
voltages, 202
wiring, 193–195

INDEX

Two-wire electric circuits, 52–54, 171
  voltages, 202

Underwriters' Laboratories (UL) label, 68–70, 72

Ventilation, poultry houses, 324
Voltage, 28, 30–31, 41
  definition, 30
  drop, 31–32
    demonstrating, 74
  positive and negative, 47–49
  three-wire circuits, 202
  two-wire circuits, 202
Volts:
  definition, 77
  formulas, 34

Washer-dryers, 124
Water heaters, 124, 307–309
  wiring for, 194–195
Water horsepower, formula for, 266–267
Water pumps (See Pumps and pumping)
Water systems, 248, 249–276
  irrigation systems, 270–274, 275
  pipes, 262–265
  for poultry, 323–324
  pumps and pumping, 251–254; (see also Pumps and pumping)
  special pump service, 84
  wells, 249–251
    location of, 249–251, 274
      protecting from pollution, 251, 274
    types of, 250–251, 274
Watering tanks, construction of concrete, 298
Watthours, 33, 41
Watts and wattage, 33–34, 41
  converting to horsepower, 33, 41
  definition, 77
  formulas, 3, 34, 35, 41
  motors, 33, 34, 35
  rating on heating elements, 33
Wells, 249–251
  location of site, 249–251, 274
  protecting from pollution, 251, 274
  types of, 250–251, 274
Westinghouse Educational Foundation:
  national awards programs, 8, 22

Wires and wiring, 78–112, 146–153
  adequacy of, 79
  aluminum wire, 56
  asbestos heat-resisting cord, 150
  bill of materials, 104, 154–155, 158
  black (hot) wires, 29, 54–55, 184–185
  bonding wires, 163
  branch circuits and outlets, 113–144
  breaker panels, 109
  cables (See Cables)
  colors of insulation, 54–55
  conduits, 95–96, 102
  copper wire, 54, 56
  cutting and connecting wire to terminals, 161–164
  demand loads, 98–100, 105, 109
  development of new materials, 104–105
  diagram for, 154–155
  disconnecting means, 82, 94, 101, 109
  electrical codes, 70–71
  equipment and materials, 86–105
  estimating one-connected load, 93
  expandability of, 79
  exterior, 79–112, 148–149
    installation of, 164–175
  farmstead, 97–104, 160–203
  feeder conductors, 149
  feeders, 82–84, 85, 102, 166–171
  "fish tape," 154
  flexible armored cable, 152–153
  fuses and fusing, 80, 84, 94, 103–104, 108
  good "electrical insurance," 79–81
  green wires, 55
  grounding or polarizing, 54–56, 84–85, 95, 102, 103, 110, 147
  "high" lines, 28, 53–54, 56, 81
  high-pressure, 28
  hot, 28, 29, 53, 54–55, 184–185
  individual disconnects, 84
  installation, 87–92
  insulation materials, 56–57
  insulators, 85, 95
  interior wires, 150–153, 158
    installation of, 175–197
  lead-in wires, 81
  length of run, 92–93
  light fixtures, 190–192

"load" figures, 98–100, 105, 109
location of meters, 105–107
low-pressure, 28
metallic raceway, 151
neutral, 28, 29, 53–55, 184–185
non-metallic sheathed cable, 151–152
outlets and switches, 183–188
overhead wires, 148
overloading circuits, 66–67, 80
parallel lamp cord, 147
planning wiring jobs, 145–159
plastic cable, 150
pole metering, 86, 109
poles, 85, 95, 102, 103
power cables, 149
present and future needs, 96–104
"pre-wired" wire, 150–151
regulated by electrical codes, 55, 145–146
safety, 79
service drops, 81, 84, 94–95, 102, 105
service entrance, 81–82, 84, 109, 171–175
service switches, 54–55, 84, 102
single-conductor wire, 147, 158
soldering spliced wires, 163–164
special pump service, 84
special purpose wires, 149–150, 158
splicing wires, 55, 65, 161–163
thermo cord, 150
thin-wall conduit, 152
third wire, 55–56
three-wire (230-volt), 193–195
tools, 153–154
TV cord, 150
type-W cable, 150
weather-proof wire, 56
white (neutral) wires, 29, 54–55, 184–185
wire size:
  effect on resistance, 34
  effect on voltage drop, 32, 40
  for electric motors, 244–245
  selecting, 32
  tables, 87–93
wires, 146–153
  styles of, 147–148, 158
  types of, 92, 131
  symbols for, 203

Yard lights, 281, 293, 296–297
  installing, 76
  switches for, 122, 130
  wiring, 193
Yardpoles, metering at, 166